INDUSTRIAL NOISE CONTROL

MECHANICAL ENGINEERING

A Series of Textbooks and Reference Books

Editor

L. L. Faulkner

*Columbus Division, Battelle Memorial Institute
and Department of Mechanical Engineering
The Ohio State University
Columbus, Ohio*

Additional Volumes in Preparation

INDUSTRIAL NOISE CONTROL

FUNDAMENTALS AND APPLICATIONS

SECOND EDITION, REVISED AND EXPANDED

LEWIS H. BELL

Acoustics and Noise Control
Huntington, Connecticut

DOUGLAS H. BELL

Cavanaugh Tocci Associates, Inc.
Sudbury, Massachusetts

Marcel Dekker, Inc.   New York • Basel • Hong Kong

Library of Congress Cataloging-in-Publication Data

Bell, Lewis H.
  Industrial noise control : fundamentals and applications / Lewis
H. Bell, Douglas H. Bell. -- 2nd ed.
     p. cm. -- (Mechanical engineering ; 88)
  Includes bibliographical references and index.
  ISBN 0-8247-9028-6
  1. Industrial noise. 2. Noise control. 3. Vibration. I. Bell,
Douglas H. II. Title. III. Series: Mechanical
engineering (Marcel Dekker, Inc.) ; 88.
  TD892.B39 1993
  620.2'3--dc20   93-20902
                                                             CIP

The publisher offers discounts on this book when ordered in bulk quantities. For more information, write to Special Sales/Professional Marketing at the address below.

This book is printed on acid-free paper.

MARCEL DEKKER, INC.
270 Madison Avenue, New York, New York 10016

Current printing (last digit):
10 9 8 7 6 5 4 3 2 1

PRINTED IN THE UNITED STATES OF AMERICA

To Carolyn and Jeanne

PREFACE

In the 12 years since the publication of the first edition of this book, strides of progress have been made in digital computers and the miniaturization of electronic components. As such, computational methods have been extended and measurement methods expanded. Further, extensive developments in acoustical materials and exciting noise reduction techniques using sound to "cancel" sound have been brought to practicality.

To reflect this progress and these developments, a second edition has been compiled. The objectives remain the same: (1) to present, simply and concisely, the fundamental principles of noise and vibration control and (2) to illustrate these principles by direct application to actual and realistic problems found in industrial, building, and community environments.

As such, the book lends itself as

1. A textbook for a one- or two-semester introductory undergraduate course in noise and vibration control, and
2. A reference and self-study guide for the practicing engineer who has the task of meeting compliance with stringent federal, state, and local noise regulations.

In short, the book is intended for both students and professionals who have little or no knowledge of acoustics but who want to develop a practical and systematic approach to controlling noise and vibration.

The book is divided into four parts. In Part I we deal with the basic physics of sound and its measurement. In Part II basic noise control methods and materials are described along with a systematic design approach. In Part III we identify the major sources of industrial noise, their character, and their treatment. Part IV completes the coverage by presenting the basic fundamentals of noise control in buildings. In addition, because of rising environmental pressures, the elements of community noise and its description and measurement are discussed. Chapter 17 has been added to present an overview of regulations and standards relating to community noise in the United States.

Finally, despite the progress in noise control, many noisy environments remain where it is neither technically nor economically feasible to lower levels sufficiently to meet compliance. In these environments, personal hearing protection is the only measure available to reduce daily noise exposure. A new chapter (Chapter 18) describes the various types of defenders, their acoustical performance, and methods for their selection.

The appendices contain, for ready reference, conversion factors, tables of physical constants, international symbols and units, exerpts of regulations, and lists of applicable standards. In addition, many time-saving statistical methods for measuring daily noise exposure have been added (Appendix E), along with an extensive listing of commercially available machine enclosures (Appendix I).

It should be noted that most of the equations involving fundamental concepts of sound and sound propagation are derived from basic physics or engineering principles. However, many of the equations, especially in Parts II and III, are empirical in nature, and hence only the sources can be cited or referenced. In this way, the interested student can—and is urged to—pursue the technical background and recognize the inherent qualifying restrictions.

Most of the information and data included in this book were obtained from general noise control literature, commercial brochures, and the authors' years of experience as practicing acoustical consultants. We have attempted to organize and simplify some rather involved technical concepts and to develop systematic design and problem-solving methods.

We would also like to say that every effort was made to ensure that the information and date presented are accurate and authoritative. It should be emphasized that, in many cases, the performance of similar commercially available materials or devices differs. Rather than select one over the other, a range of typical performances has, in most cases, been presented. All sections of the text have been reviewed by professionals eminently qualified in their fields, and their suggestions and criticisms have been included.

Although this manuscript has been carefully read for errors, it is always possible that some may have been overlooked. Therefore, to correct subsequent editions, we encourage and thank in advance all those who take the time to call such errors to our attention.

Lewis H. Bell
Douglas H. Bell

ACKNOWLEDGMENTS

It is not possible to acknowledge or properly thank all those who have contributed or worked with us in preparing this book. We do wish to extend our sincere thanks specifically to Gregory C. Tocci, William J. Cavanaugh, Apryl Walker, Carolyn Bell, and Henry Boehm. We would also like to thank the following firms for their generous permission for us to use artwork and catalog material: Aeronautics Inc., American Society for Testing and Materials, Barry Wright Corp., Bilsom International Inc., Brüel & Kjær Instruments Inc., Eckel Industries, Industrial Acoustics Co., Larson Davis Laboratories, Mason Industries, Neiss Inc., Proudfoot Co., Scantek Inc., Soundcoat Co., and Universal Silencer Inc. We wish also to apologize to those people whose work we may have used but who unintentionally were not given proper credit.

CONTENTS

I

THE NATURE AND MEASUREMENT OF SOUND

It is absolutely essential to have an understanding of the basic physics of sound in order to develop a systematic approach to noise control. That is, the student must first have a good physical picture
of the nature of sound, sound propagation, and sound measurement
before proceeding to solving noise control problems.

In Part I (Chaps. 1 to 5) these fundamental physical concepts
are presented along with analytical methods and scales used to describe sound sources and fields. In addition, the reader is introduced to the basic instrument systems and methods of sound measurement.

With this background, the student will find the systematic steps
to noise and vibration control presented in Parts II, III, and IV easy
to understand and follow. It may seem to the reader, in Part I, that
attention was focused primarily on sound. However, many of the concepts and measurement methods developed for sound are common to
vibration control and can be applied directly.

1
PHYSICAL ACOUSTICS

1.1 SOUND WAVES

There are many approaches one can take to describe the wave motion
or vibratory characteristics of sound. For example, it is easy to
relate a plane sound field in appearance to expanding water waves
resulting from disturbing a quiet pond. Another conceptual picture
which is both popular and useful is to draw *rays* or arrows to illustrate the direction of propagating wave fronts. This ray concept is
especially useful, as we shall see later when describing the reflection
and refraction of sound waves. For industrial noise control, however,
describing sound as the vibratory motion of displaced molecules in an
elastic medium provides an easy and simplifying initial approach. As
a starting point, consider the motion of a piston in a cylindrical tube
containing air, as illustrated in Fig. 1.1a. As the piston moves quickly to the right, the molecules are compressed, with an excess of velocity and momentum being passed on from molecule to molecule. With
the compression of the molecules, there is also a sharp increase in
pressure locally at the face of the piston. As the piston moves to the
left (Figure 1.1b), a void or rarefaction occurs, with a corresponding
decrease in pressure. Since the air is an elastic medium (as are many
other common materials, i.e., steel, water, glass, etc.), the displaced
particles tend to "spring" back to their original position, and a propagating or traveling pressure wave is developed. In this case, the
propagating sound wave is plane in shape and travels down the tube
from left to right at a speed dependent to the first order on the particular medium. If one placed an ear at the end of the tube, a single
sound pulse would be perceived as the wave emerges. Waves of this
type are called longitudinal plane waves in that the motion is in the
same direction as the propagation of the wave. When particles of the
medium move perpendicular to the direction of wave propagation, the
waves are called transverse. As a matter of interest, the classic ex-

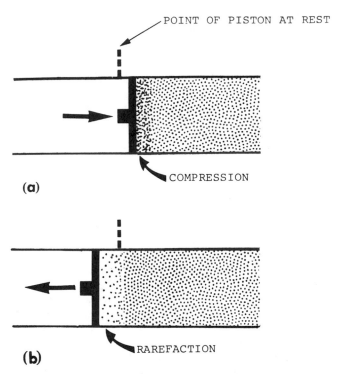

Figure 1.1 Compression and rarefaction of molecules due to the motion of the piston in the tube.

ample of a transverse wave is the circular diverging surface waves created when a stone is tossed into a calm pond. The transverse action is easily seen here by noting that the motion of a *bobbin* or leaf is perpendicular to the passing waves. We shall therefore deal exclusively with longitudinal sound waves in this text.

To see the vibratory character of sound, consider a rapid oscillation of the piston in the tube, as illustrated in Fig. 1.2. Let the time history displacement of the piston be simple harmonic motion, i.e., purely sinusoidal. Under a wide range of conditions, the dynamic pressure oscillations created will also be sinusoidal as the waves pass any stationary point in the tube. The resultant oscillating excess pressure or sound pressure at any point can then be described simply as

$$p(t) = P_0 \sin(\omega t - \phi) \qquad [\text{N/m}^2; \text{Pa*}] \qquad (1.1)$$

---

*Internationally accepted unit of sound pressure named in honor of the French physicist Pascal.

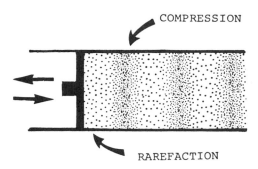

COMPRESSION

RAREFACTION

Figure 1.2 Sinusoidal wave motion traveling to the right due to the rapid oscillation of the piston in the tube.

where

P_0 = amplitude of sound pressure (N/m^2)

t = time (s)

$\omega = 2\pi f$, angular frequency (rad/s)

f = frequency of oscillation (c/s or Hz*)

ϕ = phase angle (dependent on initial conditions) (rad)

Throughout this text we shall preferably use the widely accepted International System of Units (Système International d'Unités, SI), as described in App. A. However, in some instances, especially where engineering usage is common, British units will also be used.

Equation (1.1) can also be illustrated graphically, as shown in Fig. 1.3. It is important to note also that the reciprocal of the frequency is called the period T and is given as

$$T = \frac{1}{f} \quad [s] \tag{1.2}$$

The period T of the wave motion is just the time required to complete one cycle of pressure oscillation. For example, if the time for one complete oscillation of the piston is 0.1 s, then the frequency or time rate change of pressure fluctuation is 10 Hz.

Intuitively, we can consider the progression of sound waves in the tube as a series of traveling waves and express more generally the sound field mathematically as

---

*Internationally accepted unit of frequency named in honor of the famous German mathematician Hertz.

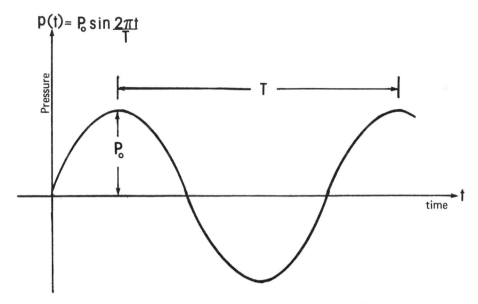

Figure 1.3 Graphic illustration of sinusoidal pressure fluctuation.

$$p(x,t) = P_0 \sin(\omega t - kx) \qquad (1.3)$$

where

P_0 = amplitude of pressure oscillation (Pa)

x = distance along horizontal axis of the tube (m)

$k = \omega/c$ = wave number (rad/m)

c = speed of traveling wave or speed of sound (m/s)

The travelinglike character of the waves can now be seen more easily by examining Eq. (1.3) more closely. Consider any single point x_0 in the tube; then Eq. (1.3) reduces simply to Eq. (1.1), since kx_0 can be considered a phase angle and we see a simple sine wave pressure oscillation with time throughout the tube. If, however, one selects a specific time t_0, a *snapshot* of the oscillating sound field is obtained.

The snapshot in a sense "freezes" the sound pressure field, which is again sinusoidal in amplitude along the horizontal axis of the tube with the following interesting relationship. The linear distance between peak values of pressure P_0 is called the wavelength λ. The wavelength is mathematically related to the frequency and the wave propagation speed as follows:

$$c = \lambda f \qquad [m/s] \qquad (1.4)$$

Equation (1.4) is one of the most fundamental relationships in all of acoustics, and we shall use and refer to it continuously throughout the text.

To illustrate this sinusoidal wavelike characteristic, consider the following example.

Example

The standard tone A produced by the vibration of the tynes of a tuning fork is a nearly pure tone whose frequency is 440 Hz. The speed of sound in air at room temperature is approximately 330 m/s or, in British units, 1100 ft/s. What is the corresponding wavelength of the sound produced?

Solution

From Eq. (1.4),

$$\lambda = \frac{c}{f} = \frac{330}{440} = 0.75 \text{ m}$$

or, in British units,

$$\lambda = \frac{c}{f} = \frac{1100}{440} = 2.5 \text{ ft}$$

Listed for easy reference in Table 1.1 is the speed of sound for some common materials or mediums.

In a later section of this chapter we shall deal with the speed of sound in more detail, especially with respect to the effect of medium temperature.

1.1.1 Complex Waves

As a starting point and for simplicity we have considered only simple sinusoidal pressure waves, and it might appear that this selection is

Table 1.1 Approximate Speed of Sound for Common Materials

Medium	Sound Velocity (ft/s)	m/s
Air (0°C and 0.76 m)	1,100	330
Wood (soft)	11,100	3400
Water (15°C)	4,700	1400
Concrete	10,200	3100
Steel	16,000	5000
Lead	3,700	1200
Glass	18,500	5500
Hydrogen (0°C and 0.76 m)	4,100	1260

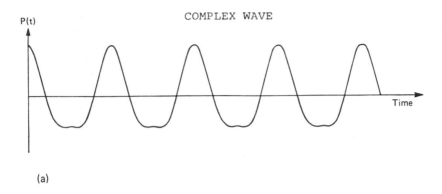

(a)

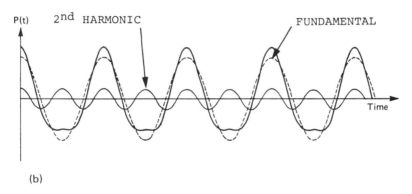

(b)

Figure 1.4 How a complex waveform can be "broken up" into a sum of harmonically related sine waves.

too specific to be of practical value in noise control. It can, however, be shown mathematically that complex periodic waveforms can be broken into two or more simple harmonically related sinusoidal waveforms, as illustrated in Fig. 1.4. In Fig. 1.4a we have a complex periodic waveform, and in Fig. 1.4b the complex wave is synthesized by adding the two simple sine waves. The harmonic relationship shown here is on a one-to-two basis; i.e., the frequency of one sinusoidal wave is just twice the other. Conventionally, the lowest-frequency sine term is commonly called the fundamental, and the next highest frequency the second harmonic, and the next the third harmonic, etc. We shall use this notation throughout the text. As a practical example of this synthesis, the sound pressure waves radiated from gears meshing and by hydraulic pumps are generally complex and periodic with easily discernible discrete tones or pure tones. When these complex waves are broken down or decomposed into simple sine waves, there are often 8 to 10 harmonics present with frequencies

related by integer-ordered multiples of the fundamental frequency.
In short, sound waves originating from rotating machinery are char-
acteristically complex and periodic and as such can be broken down
into simple sinusoidal terms.

To take advantage of this property, a more general equation can
be written to describe the sound pressure associated with a wide
variety of industrial noise sources:

$$p(t) = P_1 \sin(\omega t + \phi_1) + P_2 \sin(2\omega t + \phi_2) + P_3 \sin(3\omega t + \phi_3)$$

$$+ \cdots + P_n \sin(n\omega t + \phi_n) \tag{1.5}$$

where

P_n = amplitude of nth harmonic

ϕ_n = phase angle of each harmonic

Note that each term is a special case of Eq. (1.1) and that the harmonic relationships are integer multiples of the fundamental angular
frequency, i.e., ω, 2ω, 3ω, . . . , $n\omega$.

This series of sinusoids is just one form of the well-known Fourier
series named after the French physicist who used these series to describe complex functions more than 150 years ago. It must be emphasized that the concept of complex wave synthesis is one of the most
powerful analytical and diagnostic tools available to the engineer in
acoustics and noise control.

Unfortunately, it is beyond the scope of this text to discuss in
detail the mathematical methods required to synthesize or decompose
complex periodic waveforms. However, these topics are included in
any text in engineering analysis, and the student is strongly encouraged to pursue these basic analytical methods on his or her own to
assure a complete understanding. Fortunately, with recent technical
development in the area of digital electronic analyzers, the decom-
position or spectral analysis of complex waveforms from microphone
signals is rather routine. These electronic analyzers eliminate cumbersome mathematical and graphic methods used in the past and yield
the amplitude and phase of the harmonic terms in real time and in an
easy-to-use format. These techniques will be discussed in more detail in a later chapter.

Before leaving the subject of sound waves, it should be noted
that aperiodic sound such as the hiss from an air jet (broadband)
and impulsive sound, is common in industry. However, integer
harmonic relationships associated with aperiodic sound waves do not
in general exist, and the composition is more involved than a simple
series. Fortunately, most of the important and basic relationships
involving frequency, wavelength, phase, etc., are still preserved.

In summary, most of the sound sources found in industry can be characterized as periodic, impulsive, or broadband and combinations thereof. As such, the basic sinusoidal model selected herein is perfectly general and affords an easy understanding of the physics of sound.

1.1.2 Effective Sound Pressure

Because of the electronic characteristics of most sound measurement instruments, the term sound pressure is conventionally understood to be *effective pressure* or root-mean-square (rms) pressure. Now the root-mean-square value of a function of time is defined as

$$x_{rms} = \sqrt{\underset{T \to \infty}{\text{limit}} \; \frac{1}{T} \int_0^T x(t)^2 \, dt} \tag{1.6}$$

where T is the time duration over which the mean or average is taken. Further, the effective pressure or rms pressure can also be considered the root of the time-averaged sound pressure squared, or symbolically

$$X_{rms} = \sqrt{<X^2>} \tag{1.6a}$$

where the symbols $<\,;\,>$ denote time average. By comparing Eqs. (1.6) and (1.6a), the physical meaning of the rms pressure becomes easier to understand.

From Eq. (1.6a) then, the effective pressure of a simple sinusoidal sound wave is

$$p_{rms} = \sqrt{<P_0^2 \sin^2(\omega t + \phi)>}$$

where P_0 is the amplitude of the oscillating pressure. It can be shown (see Exercise 1.3 at the end of this chapter) that the time average for the square of a sine wave is just equal to 1/2. Therefore, the value of the effective or rms pressure is just

$$p(t)_{rms} = \frac{P_0}{\sqrt{2}} = 0.707 P_0 \qquad [N/m^2] \tag{1.7}$$

or 0.707 times the oscillating pressure amplitude.

These results can be extended to complex waves or more generally to linear combinations of sinusoidal waves, where the effective pressure or rms value of the complex wave is just the root of the sum of the squares of the rms values of each component wave:

$$p_{rms} = \sqrt{p_1^2 + p_2^2 + p_3^2 + \cdots + p_n^2} \tag{1.8}$$

Note that for periodic waves p_n is the rms value of the nth harmonic.

These results are fundamental to understanding the measurement of sound pressure which will be discussed in a later chapter. Hereafter, then, the term sound pressure will refer to the easy-to-measure effective pressure or root-mean-square pressure.

1.1.3 Superposition of Waves

Rarely in an industrial noise environment is there a single source of noise present, and as such the acoustical engineer must deal with the superposition of sound waves on a continuing basis.

As we have seen in the preceding section, when two or more sound waves are superimposed, they combine in a scaler fashion; i.e., they add algebraically at any point in space and time. Generally, the superposition yields a complex wave which can be broken down into simple sinusoidal spectral components. With the amplitude and frequency of each wave component, the major noise sources present can usually be identified and rank-ordered for subsequent engineering treatment.

There are two special cases of superposition that occur frequently that must be mentioned: beat frequency and standing waves. When recognized, they can be useful as a diagnostic tool or in some cases avoided as a possible source of error in measurement or analysis.

Beat Frequency

For the first case, we shall deal with the superposition of two sound waves which are nearly equal in amplitude but differ slightly in frequency. For example, consider the superposition of the following two sound waves:

$$p_1 = P_0 \sin \omega_1 t$$

$$p_2 = P_0 \sin \omega_2 t$$

where P_0, the pressure amplitude, is the same magnitude for each wave, but $\omega_1 \neq \omega_2$.

Upon superposition, we get

$$p(t) = P_0 \sin \omega_1 t + P_0 \sin \omega_2 t$$

$$= P_0(\sin \omega_1 t + \sin \omega_2 t) \qquad (1.9a)$$

Utilizing the well-known trigonometric identity

$$\sin A + \sin B = 2 \cos \frac{(A - B)}{2} \sin \frac{(A + B)}{2}$$

$$p(t) = 2P_0 \cos \frac{(\omega_1 - \omega_2)t}{2} \sin \frac{(\omega_1 + \omega_2)t}{2} \qquad (1.9b)$$

Since $\omega_1 = 2\pi f_1$ and $\omega_2 = 2\pi f_2$, upon substitution we get the final result in terms of the frequencies of each wave

$$p(t) = 2P_0 \cos 2\pi \frac{(f_1 - f_2)t}{2} \sin 2\pi \frac{(f_1 + f_2)t}{2} \qquad [N/m^2] \qquad (1.10)$$

In this form, the resultant wave can be considered as a complex sound wave whose frequency (from the sine factor) is the average $(f_1 + f_2)/2$ of the two superimposed waves and whose amplitude is

$$p'(t) = 2P_0 \cos 2\pi \frac{(f_1 - f_2)t}{2} \qquad (1.11a)$$

By examining the amplitude factor, it is easy to see that when the argument of the cosine takes integer values of π,

$$2\pi \frac{(f_1 - f_2)t}{2} = n\pi \qquad (n = 0, 1, 2, 3, \ldots)$$

the amplitude of the complex wave is a maximum and equal to $2P_0$:

$$p'(t)_{max} = 2P_0 \qquad (1.11b)$$

Following the same reasoning, the amplitude of the complex wave vanishes when the argument of the cosine takes on integer odd values of $\pi/2$:

$$2\pi \frac{(f_1 - f_2)t}{2} = \frac{(2n - 1)\pi}{2} \qquad (n = 1, 2, 3, \ldots)$$

This time-varying amplitude modulation is shown graphically in Fig. 1.5. It is important to note that the modulation or beat frequency is just the difference in frequency of the two superimposed waves, i.e., $f_1 - f_2$. To prove this, we solve the preceding equation for t or, more generally, for those times t_n when the amplitude of the sound pressure is zero:

$$t_n = \frac{2n - 1}{2(f_1 - f_2)} \qquad (n = 1, 2, 3, \ldots)$$

Now without loss of generality, if we take the difference in time between consecutive beats, i.e., the nth and nth + 1, we obtain

$$t_{n+1} - t_n = \frac{2(n + 1) - 1}{2(f_1 - f_2)} - \frac{2n - 1}{2(f_1 - f_2)}$$

$$= \frac{1}{f_1 - f_2} \qquad [s]$$

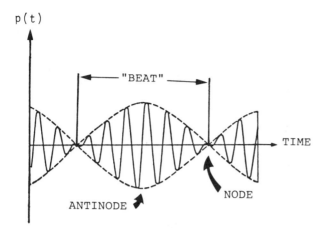

p(t)

"BEAT"

TIME

ANTINODE

NODE

Figure 1.5 Graphic representation of the superposition of two waves of slightly different frequencies but equal amplitudes. The beat is the period associated with the beat frequency.

which is the time duration between beats. But the time duration between beats is by definition the period T of the beat frequency, and taking the reciprocal of the period yields the beat frequency f_b directly:

$$f_b = f_1 - f_2 \quad [Hz] \tag{1.12}$$

This result is not just of interest but is an important diagnostic tool in many areas of noise control, as we shall see in later chapters.

In the more general case, when the amplitude of the superimposed waves are not equal, the amplitude of the resultant complex wave varies between the sum and difference of the component waves, as illustrated in Figure 1.6. This periodic variation in amplitude results in a rhythmic *pulsing*like sound, and when the frequency difference does not exceed 4 or 5 Hz, the beat is clearly discernible to the human ear. A common example of this beating phenomenon is the rhythmic sound variation when the engines of twin-engine aircraft are at slightly different rotational speeds or "out of sync." Other examples include the sound from diesel engines, gearboxes, compressors, pumps, turbines, etc., when they are operated (in pairs) at slightly different rotational speeds. An example from industry will best illustrate the beat characteristic from the superposition of two waves.

Example

The sound pressure radiated from two adjacent transformers operating singularly was determined to be, respectively,

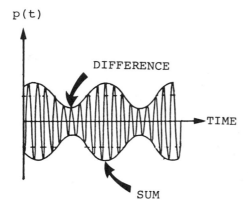

Figure 1.6 Graphic representation of the superposition of two waves of slightly different frequencies but unequal amplitudes. The arrow shows the range of variation of the beat amplitude.

$$p_1(t) = P_1 \sin 2\pi(120)t$$

and

$$p_2(t) = P_2 \sin 2\pi(116)t$$

What is the beat frequency of the resultant sound when they are operated simultaneously? What is the range of amplitude fluctuation for the beat frequency?

Solution

The beat frequency from equation (1.12) is just the difference in frequencies of the two waveforms:

$$f_1 - f_2 = 120 - 116$$
$$= 4 \text{ Hz}$$

The range of amplitude variations is the sum and difference of the amplitude of the waves, respectively:

$$P_1 + P_2 \qquad \text{and} \qquad P_1 - P_2$$

This is a common situation in power plants when a new generator is brought on-line and is not quite in synchronization.

Standing Waves

When two sound waves of the *same* frequency traveling in different directions are superimposed, a standing wave-type sound field is produced. To illustrate this phenomenon, consider the superposition of two sound waves traveling in opposite directions as given by

$$p_1(t) + p_2(t) = P_1 \sin(2\pi ft - kx) + P_2 \sin(2\pi ft + kx) \qquad (1.13)$$

Here the first sine term in the sum is a sound wave traveling to the right with amplitude P_1 and frequency f. The second term is a sound wave traveling to the left with amplitude P_2 and identical frequency f.

Using the well-known trigonometric identities for the sum and difference of angles, we get, upon substitution,

$$p_1(t) + p_2(t) = P_1 \sin 2\pi ft \cos kx - P_1 \cos 2\pi ft \sin kx$$

$$+ P_2 \sin 2\pi ft \cos kx + P_2 \cos 2\pi ft \sin kx$$

If the waves have equal amplitudes, i.e., $P_1 = P_2$, we get the following simplification:

$$p_1(t) + p_2(t) = 2P_1 \cos kx \sin 2\pi ft \qquad (1.14)$$

Now Eq. (1.14) can be considered just a simple sinusoidal function of time whose amplitude depends on space or the position x of the observer. In particular, when the argument of the cosine takes integer odd values of $\pi/2$, i.e.,

$$kx = \frac{\pi}{2}, \frac{3\pi}{2}, \frac{5\pi}{2}, \ldots \frac{(2n-1)\pi}{2} \qquad (n = 1, 2, 3, \ldots)$$

the sound pressure vanishes, and there are points in space of no sound or nodes. Now if we solve the preceding expression for x or rather x_n, we obtain the spatial locations of the nodes,

$$x_n = \frac{(2n-1)\pi}{2k} \qquad (n = 1, 2, 3, \ldots) \qquad [m]$$

and since $k = \omega/c = 2\pi/\lambda$, we have

$$x_n = \frac{(2n-1)\lambda}{4} \qquad (n = 1, 2, 3, \ldots) \qquad [m]$$

As should be expected, the location of the nodes is simply related to the wavelength of the superimposed waves.

More important, the nodes occur every half wavelength, which can be seen by taking the difference between successive nodal locations:

$$x_{n+1} - x_n = \frac{2[(n+1)-1]\lambda}{4} - \frac{(2n-1)\lambda}{4}$$

$$= \frac{\lambda}{2} \qquad [m] \qquad (1.15)$$

Returning again to Eq. (1.14), the antinodes or points of maximum sound pressure in the standing wave occur when the argument of the cosine takes on integer values of π:

kx = nπ (n - 1, 2, 3, . . .)

Further, the amplitude of the antinodes is just $2P_1$:

$$[p_1(t) + p_2(t)]_{max} = 2P_1 \quad [N/m^2] \tag{1.16}$$

These points of maximum sound pressure are stationary, located symmetrically between the nodes and correspondingly spaced one-half
wavelength apart.

A practical example will illustrate more clearly this standing wave
phenomenon.

Example

The noise from a hydraulic pump is periodic with a fundamental
frequency of 550 Hz. The pump is located several feet from a
hard masonary wall such that the radiated sound is reflected back
on itself with nearly equal amplitude. Intuitively, the amplitude
of the reflected wave is nearly the same as the original, especially
within a few feet of a hard wall. What is the spacing of the nodes
or position of minimum sound for the fundamental tone? What is
the amplitude of the resultant sound field for the fundamental
tone at the antinode locations? What is the spacing for the second
harmonic (1100 Hz)?

Solution

We first calculate the wavelength of the fundamental tone from
Eq. (1.4):

$$\lambda = \frac{c}{f} = \frac{1100}{550} = 2.0 \text{ ft}$$

Here the speed of sound c was taken at room temperature to be
1100 ft/s, a close approximation. From Eq. (1.15), the spacing
between the nodes occurs every one-half wavelength, or at distance intervals of 1.0 ft.

Correspondingly, from Eq. (1.16), the amplitude of the antinodes will be approximately twice, or, more accurately, the sum
of, the amplitudes of the original and reflected waves.

For the second harmonic at 1100 Hz, the spacing of nodes is
1/2 ft or 6 in., and the resultant amplitudes of the antinodes are
again approximately twice the amplitude of the original harmonic
wave.

This example provides important insight into the problems standing waves can create in measuring industrial noise. As seen from the
example, the amplitude of the standing waves associated with the
fundamental frequency varies from nil (nearly 0) to twice the amplitude
of the original wave within a distance of 6 in. in space. It is now
easy to see that severe measurement errors can and often do occur
in industrial environments where reflective surfaces are generally
present. In many cases, reflective surfaces are not in close proximity,

and the amplitude of the reflected wave is relatively small compared to the original. Naturally, if the reflected wave has a relatively small amplitude, the variation in amplitude of the standing wave is correspondingly small and can often be neglected. However, it should now be clear that cognizance and a reasonable understanding of the standing wave phenomena are essential.

Finally, to complete the discussion of beats and standing waves, it is of interest to note that one hears the rhythmic beat frequency throughout the entire radiated space, but one has to move the head or ear to hear the amplitude change associated with a standing wave.

1.1.4 Reflection

Another basic property of sound which plays an important role in industrial noise control is the ability of sound to *bounce* or reflect from surfaces. The classic example or reflection is, of course, the *echo*, which has intrigued and mystified mankind over the ages. In this treatment we shall deal only with reflections from plane, acoustically hard surfaces, since reflection from the more complex shapes or absorbing surfaces is highly analytical and beyond the scope of the text. Further, in most practical field situations, reasonably selected approximations to plane surfaces, i.e., surfaces with little curvature, will usually yield close first-order results.

A basic understanding of reflection follows from Huygens' principle which in essence says that advancing wave fronts can be considered as point sources of secondary wavelets, as illustrated in Fig. 1.7.

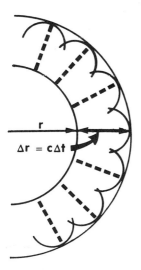

Figure 1.7 Huygens' principle for advancing spherical waves.

Therefore, as waves impinge on a hard, smooth surface, the waves are
reflected with shape and propagation characteristics preserved.
Hence, spherical waves are reflected as spherical waves, cylindrical
waves remain cylindrical, plane waves remain plane, etc. Of equal importance to the acoustical engineer are the laws of reflection which
deal with the direction of propagation of the reflected wave. To
understand the laws of reflection, consider the impingement of a series
of plane waves on a reflecting plane surface, as shown in Fig. 1.8. In
this construction, the arrows drawn normal to the wave fronts and in
the direction of propagation are called rays. From a direct application
of Huygens' principle and the geometry, the first law of reflection
follows; i.e., the angle of reflection is equal to the angle of incidence,
where the angles are defined between a line drawn normal to the reflecting plane and the incident and reflected rays, respectively. Of
less importance, but to complete the discussion, we also note that the
incident ray, the reflected ray, and the normal all lie in the same
plane.

In short summary, a wave front incident upon a plane surface
will be reflected at an angle equal to the angle of incidence, and with
waveshape preserved. The geometrical ray construction for a di-

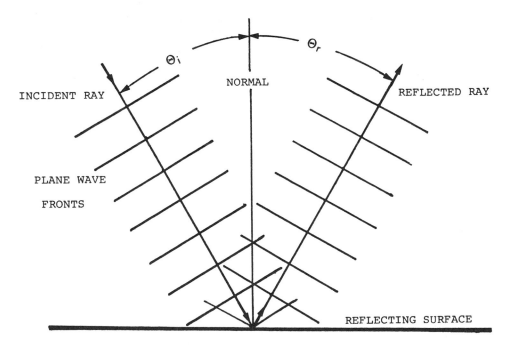

Figure 1.8 Geometrical illustration of plane wave reflection where θ_i
and θ_r are the angles of incidence and reflection, respectively.

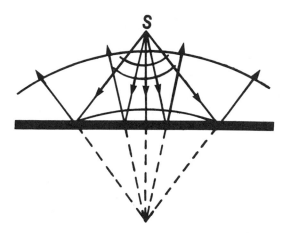

Figure 1.9 Geometrical illustration of spherical diverging sound waves reflected from a plane surface.

verging *spherical* wave incident on a plane surface is shown in Fig. 1.9. Here again, the direction of reflected sound can be at least qualitatively determined and, as we shall see in a later chapter, accounted for in many industrial measurement situations. Finally, it should be emphasized again that it is the *standing wave interference patterns*, resulting from the reflections, that are of most interest to the acoustical engineer. These patterns occur inherently in typically enclosed industrial environments.

Actually, in the previous section dealing with standing waves, we considered a special case of reflection when we had the superposition of two waves traveling in opposite directions. Specifically, the wave referred to as traveling to the left could have been considered without loss of generality a reflection of the original wave from a plane surface at 0° angle incidence.

Utilizing the laws of reflection, we can now extend these results to include the more general case where waves are incident at any oblique angle θ_i and determine the interesting properties of the resultant sound field.

Refer back to Fig. 1.8, which illustrates geometrically the superposition of two traveling plane waves at oblique angles θ_i and θ_r. Again, without loss of generality, let the sound waves be sinusoidal in nature. As the incident or reflected wave front intersects any line normal to the reflecting plane, there is a scissorlike effect which is often observed when ocean waves break obliquely along a beach. The intersection of these waves along the normal is in a sense a *projection* of the incident and reflected waves. From the original concept of wave motion, then, the distance between crests along the

normal can be considered a projected wavelength λ' and from the construction (Figure 1.8) is related to the incident wave as follows:

$$\lambda' = \frac{\lambda}{\cos \theta_i} = \lambda \sec \theta_i \qquad [m] \qquad\qquad (1.17)$$

From the laws of reflection, the reflected wave also scissors back
along the normal in an opposite direction, producing a traveling wave
with a projected wavelength also equal to λ'. We therefore have along
any normal line the superposition of two waves traveling in opposite
directions with wavelength λ'.

From the analysis of the preceding section concerning standing
waves, the presence of standing waves can similarly be anticipated
as one moves along a normal line, and, further, the spacing between
nodes and antinodes needs only to be modified by the factor $\sec \theta_i$
as given in Eq. (1.17).

In summary, then, when a complex periodic wave is incident upon
a plane or nearly plane reflecting surface, a standing wave sound
field will be present. Second, the distance d' between peaks along
the normal follows from Eq. (1.15) and is given as:

$$d' = \frac{\lambda'}{2} = \lambda_n \sec \theta_i \qquad [m] \qquad\qquad (1.18)$$

where λ_n is the wavelength associated with the nth harmonic and θ_i
is the angle of incidence of the propagating plane wave fronts.

Examining Eq. (1.18), one notes that for the special case $\theta_i = 0$,
i.e., normal incidence, the nodal spacing reduces to $\lambda/2$ as given by
Eq. (1.15). As the angle of incidence increases, the spacing between
nodes increases, and in the limit $\theta_i = 90$, i.e., $\pi/2$, there is no reflected wave, and naturally the standing wave field vanishes.

The practical justification for the depth of this discussion can be
seen from the following example.

Example
   The noise from a reduction gearbox is dominantly a pure tone at
   1000 Hz. An adjacent masonry wall extends alongside the gearbox as shown in the plan view sketch of Fig. 1.10. What is the
   character of the standing wave field for sound pressure measurements which are to be made along an envelope* around the gearbox? In particular, what is the spacing between nodes at (1) location A, $\theta_i = 0$, and (2) along the line segment BC at $\theta_i \cong 60$?

---

*Measurements on an envelope around machine tools, compressors,
etc., are common in industrial association measurement procedures.

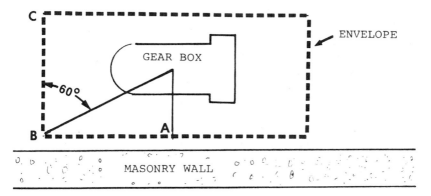

Figure 1.10 Plan view of gearbox installation.

Solution

Since the angle of incidence at location A is 0, the standing wave pattern will have nodes (minimal sound pressure) every one-half wavelength, as given by Eq. (1.15). Hence, for f = 1000 Hz,

$$\lambda = \frac{1100}{1000} = 1.1 \text{ ft}$$

and the distance d between nodes, from Eq. (1.18), is

$$d = \frac{\lambda}{2} = 0.55 \text{ ft} \cong 6 \text{ in.}$$

Considering the line segment BC, a normal to the reflecting plane, the angle of incidence θ_i is approximately 60°, and the distance d' between nodes along BC is then given from Eq. (1.18) as

$$d' = \frac{\lambda}{2} \sec \theta_i = \frac{1.1}{2} \sec(60) = 1.1 \text{ ft}$$

Therefore, the distance between *peak* and *dwell* in the standing wave pattern at location A is approximately 3 in., whereas along the ends of the envelope it is a little over 5 in. Needless to say, if meaningful measurements are to be obtained, these variations in sound pressure due to reflection must be taken into account, or first-order errors and anomalies will surely result.

The phenomenon of sound wave reflection also has many useful applications. In particular, a sound wave pulse generated from the surface of the ocean propagates to the bottom and is generally reflected back. By measuring the time interval or elapsed time of the echo and knowing the speed of sound in water, the depth of the water can be computed. Further, by comparing the spectral character of

the generated wave to the reflected wave, a good measure of the
geological composition, i.e., mud, rock, sand, etc., can also be obtained. Analogously, the properties of reflected sound are used to
determine the depth and composition of stratified layers in the earth
crust, especially with respect to locating both oil and natural gas.

Finally, it is universally agreed that the control and balance of
reflected sound (reverberation) in music halls are essential for good
listening. In fact, the "proper" balance has been the source of controversy in many recently built prestigious concert halls. This is
a subjective area in acoustics, but most people notice an overall im-
provement in tone quality when an aria or popular melody is sung in
the highly reverberant shower room or stall. We shall discuss reverberation in a later chapter.

1.1.5 Refraction

Generally, when sound waves pass from one medium to another, the
direction of the advancing wave front is *bent* or, better, refracted.
The change in direction is due to the difference in the speed of sound
generally found between different mediums. Shown in Fig. 1.11 is a
geometrical ray construction which illustrates the refraction of sound
waves passing from one medium to another. By again applying
Huygens' principle, the basic laws of refraction are obtained, the most
useful of which is

$$\frac{\sin \theta_i}{v_1} = \frac{\sin \theta_r}{v_2} \qquad\qquad (1.19)$$

where

θ_i = angle of incidence (rad)

θ_i = angle of refraction (rad)

v_1, v_2 = speed of sound in medium 1 and medium 2, respectively (m/s)

Note that the basic law of sound refraction is similar in form to
the well-known Snell law for light refraction. In fact, the classic example of refraction often given in optics is the apparent bending of a
stick partially immersed in a clear pool. Analogously, the direction of
sound is also bent at a medium interface, and again, as with the laws
of reflection, waveshape is preserved, and the incident ray, the refracted ray, and the normal lie in the same plane.

It is important to consider for a moment that zones of severe
temperature differences do commonly occur in both the atmosphere

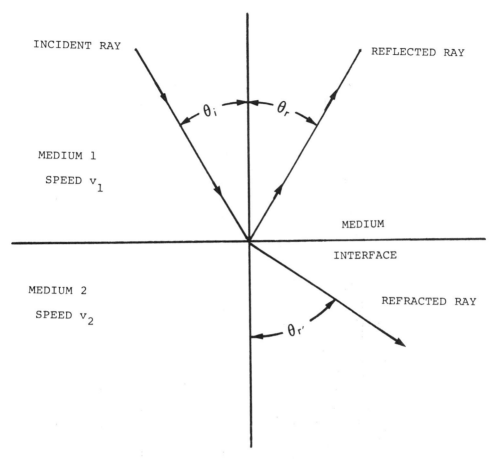

Figure 1.11 Refraction of sound rays passing from medium 1 to medium 2. Here the speed of sound is greater in medium 2 than in medium 1.

and oceans. As such, when sound travels from zone to zone across often sharp thermal gradients, the direction of propagation changes measurably and cannot be ignored. For example, on a sunny day, the surface of the earth heats up more rapidly than the atmosphere. Further, due primarily to heat conduction, the temperature of the air close to the surface rises correspondingly. Since the speed of sound is higher in the warmer lower layer, sound waves traveling along the surface are refracted, or bent, upward. Similarly, on a clear night, the earth's crust cools faster, and a cooler layer of air

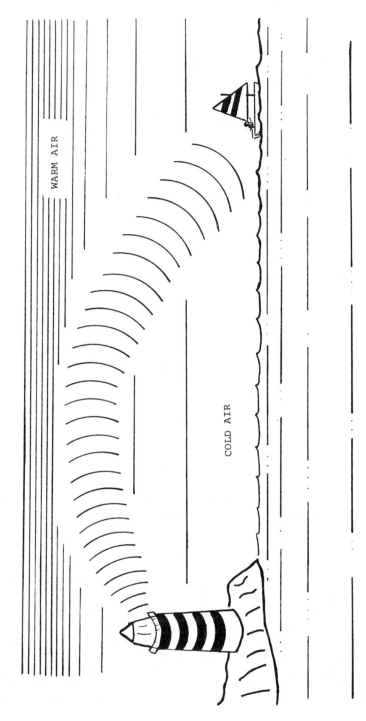

Figure 1.12 Refraction of sound due to temperature differences in the atmosphere.

24

forms and bends sound downward along the surface, as illustrated in Fig. 1.12. Thus to an observer the sound, for example, from an industrial plant would be refracted downward and would appear louder at night than during the day, which is often the situation.

In addition to thermal effects, vertical wind velocity gradients also provide a refractive medium. These effects will be treated in more detail in Chapter 4, Sound Propagation.

1.1.6 Diffraction

When sound waves impinge upon a partial barrier, as shown in Fig. 1.13, some sound is reflected back, some sound continues on without change, and some sound bends or diffracts over the top. As such, the barrier does not cast a sharp *acoustical shadow* as one might expect. The diffraction effect is common in industry and everyday living. A good everyday example is the diffraction of sound over a wall or around a building corner. The ease with which sound diffracts is evident when one recalls that one hardly has to raise the voice to be clearly understood over a wall 10 to 15 ft high.

We shall, in a later chapter, treat in analytical detail the sound reduction associated with diffraction as applied to partial barriers. These barriers are useful and a very important noise control measure.

It shall suffice at this time to say qualitatively that sound at lower frequencies tends to diffract over partial barriers better than sound at higher frequencies. In addition, the sharpness and extent of the shadow zone behind the barrier depends on the relative position of the source and receiver. Intuitively, the closer a source is to a barrier, the longer the shadow over the barrier, or, more important, the more sound reduction obtained.

1.1.7 Speed of Sound

As mentioned earlier, the speed of sound in a medium depends to the first order on the medium itself, but it is often essential in industrial noise control to account for the second-order effect of medium temperature. As such, we shall now consider in some detail the effect of temperature on wave velocity for air and exhaust gases common in gas turbine and aerospace applications.

Since the nature of sound is a compression-type wave motion, the speed of sound in a fluid medium follows from the general wave equation and is given as

$$c = \sqrt{\frac{B}{\rho}} \quad [m/s] \tag{1.20}$$

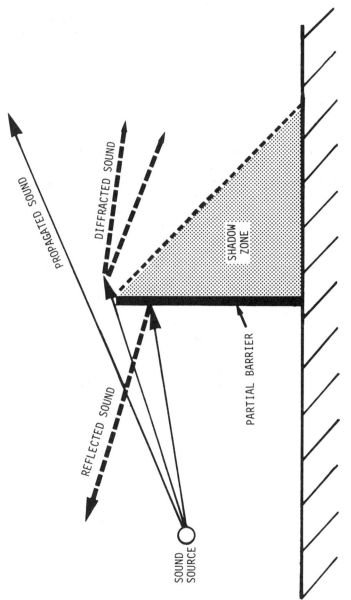

Figure 1.13   Illustration of sound diffraction over a partial barrier.

where

B = elastic bulk modulus (N/m^2)

ρ = density of medium (kg/m^3)

The bulk modulus is qualitatively a measure of how difficult it is to compress a material, and, from Eq. (1.20), it is easy to see that "hard" materials such as steel or water conduct sound more rapidly than, for example, air.

Under reasonable assumptions dealing with adiabatic heat transfere and ideal gas laws, Eq. (1.20) simplifies and can be rewritten in a form which is applicable for most gases under a wide range of conditions:

$$c = \sqrt{\frac{\gamma RT}{M}} \quad [m/s] \tag{1.21}$$

where

γ = (specific heat at constant pressure)/(specific heat at constant volume) = 1.4 for air and most diatomic gases

R = gas constant = 8317 $m^2/(s^2 \, ^\circ K)$

M = molecular weight of gases

T = temperature $(^\circ K)$

In short, the speed of a sound wave is nearly independent of ambient pressure but dependent to the first order on the square root of the absolute temperature.

Example
What is the speed of sound in hydrogen at 0°C? $M = 2.017$ for H_2.
Solution
From Eq. (1.21),

$$c = \sqrt{\frac{1.4 \times 8317 \times 273}{2.017}}$$

$$= 1255 \ m/s$$

Since we most often deal with air at atmospheric pressure, Eq. (1.21) reduces simply to

$$c = 49.03 \ \sqrt{^\circ R} \quad [ft/s] \tag{1.22}$$

where $^\circ R$ = Rankine temperature = 459.7 + $^\circ F$. In mks units, for air,

$$c = 20.05 \ \sqrt{^\circ T} \quad [m/s] \tag{1.23}$$

where $^\circ T = 273.2^\circ + ^\circ C$.

Example
What is the speed of sound in air at 1000 °F?

Solution
From Eq. (1.22),

$$c = 49.03 \sqrt{°R}$$

$$= 49.03 \sqrt{460° + 1000°}$$

$$= 1873 \text{ ft/s}$$

1.1.8. Impedance

To understand the absorption and transmission of sound, it is essential to consider the acoustical impedance of the medium. This important physical property is somewhat analogous to electrical impedance and is basically a measure of the resistance a fluid or material medium gives to the propagation of sound waves.

In its most common form, the *specific acoustic impedance* Z is defined as the complex ratio of the sound pressure to the particle velocity as follows:

$$Z = \frac{p}{u} \quad \text{[mks rayls]} \tag{1.24}$$

Just as in electronics, the impedance is in general complex and can be separated into a real and imaginary part:

$$Z = r + jx \tag{1.25}$$

where

r = specific acoustic resistance

x = specific acoustic reactance

$j = \sqrt{-1}$

The units of impedance are rayls (dyne-s/cm^3) or mks rayls (N-s/m^3), named after Lord Rayleigh, who is often credited with first putting acoustics on a rigorous mathematical basis.

The electrical analogy is now clearly evident from Eq. (1.24) by comparing pressure and particle velocity to voltage and current, respectively. For a sound wave propagating in a free field, say, outdoors, or in a nonreactive material, Z is real and reduces to

$$Z = \frac{p}{u} = \rho c \quad [\text{kg/m}^2\text{-s}] \tag{1.26}$$

where

ρ = density of the medium (kg/m^3)

c = speed of propagation (m/s)

In air, the product ρc is called the *characteristic impedance* of free air and is 40.7 rayls or 407 mks rayls at 22°C and 0.76 m Hg. Typical values for water and steel are $(\rho c)_{water} = 1.48 \times 10^6$ and $(\rho c)_{steel} = 4.0 \times 10^7$.

Consider an example that brings these concepts together.

Example

A piston in a long tube is driven in sinusoidal motion by a motor so that the displacement amplitude of the piston is $x_0 = 0.1$ cm. The speed of the motor is 6000 rpm; hence the frequency of oscillation of the piston is $f = 6000/60 = 100$ Hz. What is the amplitude of the sound pressure of the waves generated in the tube?

Solution

Since the motion is sinusoidal, the displacement x can be written as

$$x = x_0 \sin \omega t$$

where

x_0 = displacement amplitude

ω = angular frequency of the motion

$= 2\pi f$

$= 2\pi \times 100$ (rad/s)

The velocity u of the piston is then the time derivative of the displacement, or

$$\frac{dx}{dt} = u = \omega x_0 \cos \omega t$$

Hence, from the preceding expression, the velocity amplitude u_0 is then

$$u_0 = \omega x_0$$

Now, from Eq. (1.26), the sound pressure amplitude P_0 is

$$P_0 = \rho c u_0 = \rho c \omega x_0$$

$$= 407 \times 2\pi \times 100 \times 0.1 \times 10^{-2} = 256 \text{ Pa}$$

From Eq. (1.7) the effective or rms pressure is then,

$$p_{rms} = 0.707 P_0 = 181 \text{ Pa}$$

For absorbing materials, the normal resistive and reactive components of the impedance are obtained from measurements utilizing an acoustic impedance tube. These methods and applications will be discussed in detail in a later chapter. The importance of this discussion will also be seen later as one realizes that the acoustical performance of absorbing materials and barrier materials depends on the *match* or *mismatch*, respectively, of the impedance at the media interfaces.

1.2 MECHANICS OF HEARING

We shall complete this chapter by considering the basic physics or mechanics of hearing.

The mechanisms of the hearing phenomena are still being refined after centuries of study. To be sure, a complete understanding may well be a long time away when one realizes the complex interaction of the hearing mechanism and the auditory nervous system, including the brain. No small physical apparatus possesses any more remarkable properties than the ear, especially when one considers that sound pressures as small as 10^{-5} μPa can be detected. These pressure fluctuations produce displacements of the eardrum of the order of 10^{-9} cm for frequencies near 1000 Hz or roughly one-tenth the diameter of a hydrogen molecule [1].

From an energy standpoint, experiments prove that for a sensitive ear, sound at 2000 Hz is just audible if the rate at which energy supplied to the ear is about 10^{-12} W. If energy at this miniscule rate were converted to heat, it would take more than 300 million years to raise 1 g of water 1°C.

To further appreciate this remarkable device, let us consider the structure and basic function of each part. For convenience, the ear is usually divided into three parts (see Fig. 1.14): the outer ear, the middle ear, and the inner ear. The outer ear comprises the visible part, the pinna, and the passage leading down to the eardrum (tympanic membrane). The pinna is essentially a sound-gathering and -focusing device for the incident energy. Most animals can move the pinna, but humans have lost the ability and must now move the head to locate the direction of a sound source.

Leading from the pinna is a small passage about 1/4 in. in diameter and 1 in. in length terminating at the eardrum. It is the oscillating sound pressure in this cavity that excites the drumskin and ultimately gives rise to the sensation we call sound. The drumskin is a very delicate and sensitive membrane which is slightly conical and contains muscle tissue for tensioning.

The middle ear contains a chain of bones which act as a lever system to transmit the motion of the drumskin to the inner ear. Because of the shape of these bones, they are called the hammer (attached to the tympanic membrane), anvil, and stirrup. To equalize pressure on both sides of the drumskin, the eustachian tube connects the back of the throat to the middle ear. As such, the act of swallowing or yawn-

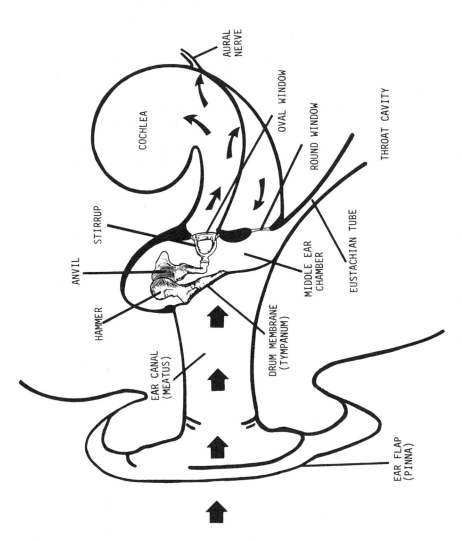

Figure 1.14 Human ear showing major anatomical parts and passages.

ing will often relieve the temporary hearing loss most people experience during a rapid change in altitude on airplanes or mountain
roads.

The inner ear is more complex and consists essentially of a bony
cavity containing a spirally wound passage called the cochlea. The
passage is filled with a colorless liquid and is divided down the middle
by the cochlear partition, which is part bone and partly gelatinous
membranes. These membranes contain hairlike cells and nerve endings
associated with the auditory system.

Further, as the fluid in the cochlea vibrates, the hairlike cells
move. This motion is sensed by the nerve cells and is processed by
the brain, giving rise to the perception of sound. As a matter of
interest, the loss of hearing due to long-term exposure to noise is
thought to be associated with stiffening of the hairlike cells in the
inner ear.

Naturally, the mechanical model just described is considerably
oversimplified and incomplete. One area of omission which must be
mentioned because of its relevance in noise control is other transmission paths of sound to the inner ear. In our mechanical model, we
discussed only the transmission path from the outer ear to the eardrum, then on to the bones of the middle ear, etc. Recent experiments have shown that sound can also be transmitted to the inner ear
by paths involving the skull or cranium area and similarly from the
jawbone. As such, it is important to note that personal hearing protection such as plugs and muffs, which do indeed reduce the transmission of sound from the outer ear, is limited to the extent of these
and other *flanking* transmission paths.

1.3 AUDIBLE, INFRASONIC, AND ULTRASONIC SOUND

A relatively narrow frequency range, from about 30 to 17,000 Hz,
makes up the audible spectrum. It is these sets of acoustic waves that
stimulate the human ear and brain to the sensation of hearing. Sound
waves with frequencies below 30 Hz, called infrasonic waves, impress
the ear as a flutter or a series of separate puffs. With sound waves
above 17,000 Hz, called ultrasonic waves, there is no sensation of
hearing at all, more or less depending on the individual. Many animals
have wider ranges of hearing than humans. Bats, for example, seem
to have the widest range and hear sounds up to 100,000 Hz or more.

In the past, control of noise in either the infrasonic or ultrasonic
regions was given little attention. However, with the development of
ultrasonic welders, ultrasonic medical and dental procedures, and so
forth, the potential for health hazards has been recognized. Thus,
guidelines for exposure [2] have been established (Appendix E) and
noise control in this region can no longer be ignored.

REFERENCES

1. L. L. Beranek. *Acoustics*. McGraw-Hill, New York, 1954.
2. *Threshold Limit Values and Biological Exposure Indices for 1985–
   1986*. American Conference of Governmental Industrial Hygienists,
   Cincinnati, Ohio, 1985.

BIBLIOGRAPHY

Acoustic Handbook. Hewlett-Packard Company, Palo Alto, California,
   1968.
Bell, L. H. *Fundamentals of Industrial Noise Control*. Harmony
   Publications, Trumbull, Connecticut, 1973.
Beranek, L. L. *Noise and Vibration Control*. McGraw-Hill, New
   York, 1971.
Faulkner, L. *Handbook of Industrial Noise Control*. Industrial
   Press, New York, 1975.
Harris, C. M. *Handbook of Acoustical Measurements and Noise Con-
   trol*. McGraw-Hill, New York, 1991.
Ingard, K. U., and G. C. Maling, Jr. *Physical Principles of Noise
   Reduction: Properties of Sound Sources and Their Fields*.
   Swedish Academy of Engineering, 1973.
Jens, Trampe Broch. *Acoustic Noise Measurements*. Brüel and Kjaer,
   Marlborough, Massachusetts, 1971.
Kinsler, L. E., and A. R. Frey. *Fundamentals of Acoustics*. Wiley,
   New York, 1950.
Lawrence, K., D. P. Lewis, and R. C. Bryant. *Noise Control in
   the Workplace*. Aspen Systems Corporation, 1978.
Lipscomb, D. M., and A. C. Taylor, Jr. *Noise Control Handbook
   of Principles and Practices*. Van Nostrand Reinhold, New York,
   1978.
Morse, P. M. *Vibration and Sound*. McGraw-Hill, New York, 1948.
Morse, P. M., and K. U. Ingard. *Theoretical Acoustics*. McGraw-
   Hill, New York, 1968.
Olishifski, J. B., and E. R. Harford. *Industrial Noise and Hearing
   Conservation*. National Safety Council, 1978.

EXERCISES

1.1 The fundamental frequency of a note played from an organ was
256 Hz with a sound pressure amplitude of 400×10^{-5} Pa. What is

a. The wavelength of the tone in air?
b. The effective or rms value?
Assume that c = 330 m/s.

> *Answer:* (a) 1.29 m;
> (b) 282.8 × 10^{-5} Pa

1.2 What is the wavelength of the standard tone A, 440 Hz, in
a. Water?
b. Steel?
c. Glass?

> *Answer:* (a) 3.18 m;
> (b) 11.36 m;
> (c) 12.50 m

1.3 Utilizing Eq. (1.6), show that the effective or rms pressure of
a simple sinusoidal pressure wave is $P_0/\sqrt{2}$, where P_0 is the amplitude
of the pressure wave. [Hint: Let p(t) = P_0 sin ωt, and note that
∫ $\sin^2 u$ du = 1/2(u − sin u cos u) + C.

1.4 The rotational speed of a twin-engine aircraft differs by 120
rpm. What is the beat frequency of the noise heard in the cockpit?
> *Answer:* 2 Hz

1.5 The noise from a diesel engine which drives a generator is
periodic with a fundamental tone at 240 Hz and numerous higher harmonics. The engine-generator set is located outdoors with a steel
roof for protection from rain and snow. Due to sound reflection from
the roof, what is the spacing of standing wave nodes anticipated for
a. The fundamental?
b. The third harmonic?
Use 1110 ft/s for the speed of sound.

> *Answer:* (a) 2.29 ft;
> (b) 0.76 ft

1.6 The echo of a gunshot is heard 4.0 s after the gun is fired by
the hunter. How far from the hunter is the cliff that reflects the
sound? Use c = 330 m/s.
> *Answer:* · 660 m

1.7 What is the speed of sound in nitrogen (N_2) at 20°C? The
molecular weight of Nitrogen is 28.0.
> *Answer:* 349 m/s

1.8 What is the angle of refraction of a plane wave moving from a
layer of air at 22°C to a layer at 20°C with the angle of incidence of
85°?
> *Answer:* 83.1°

1.9 The exhaust gases from a jet engine are 1700°F. What is the approximate speed of sound in the exhaust stream assuming the gases are dominantly air?

Answer: 2279 ft/s

1.10 What is the characteristic impedance of air at 20°C given that the density of air is 1.21 kg/m^3 and the speed of sound is 343 m/s?

Answer: 415 mks rayls

2
LEVELS AND SPECTRA

The first step in any systematic approach to noise control requires
that the spectral character of the sound or noise be determined. The
two physical properties which are easily measured and will suffice in
most cases of industrial noise control are sound pressure and frequency. However, because of the wide range of variation of both,
logarithmic and geometric scales were created to condense the range
and to generally simplify computations.

We shall, then, in this chapter develop a working knowledge of
these scales and bring together several other physical parameters
essential to characterizing sound and noise sources.

2.1 THE DECIBEL SCALE

The range of sound pressure of most interest in noise control varies
from about 1×10^{-9} psi, the threshold of hearing, to about 1 atm, approximately 15 psi. This wide range of variation represents 10 orders
of magnitude and is cumbersome to deal with quantitatively. Therefore, to condense this extremely wide range into a more manageable
scale, the concept of sound pressure level (L_p), a logarithmic scale,
has been adopted by acoustical engineers and is defined as

$$L_p = 10 \log_{10} \left(\frac{p^2}{p_{re}^2} \right) \qquad [dB] \qquad (2.1)$$

or

$$L_p = 20 \log_{10} \left(\frac{p}{p_{re}} \right) \qquad [dB] \qquad (2.2)$$

37

where

$\quad$ p = root-mean-square (rms) sound pressure (Pa or N/m^2)

$\quad$ p_{re} = international reference pressure of 2.0×10^{-5} Pa or N/m^2

In this form, the sound pressure level is expressed in decibels, commonly abbreviated dB, and is dimensionless.

$\quad$ For reference and comparison, presented in Fig. 2.1 are some examples of typical everyday noise levels given in decibels and the more common engineering unit, pounds per square inch. As a matter of interest, the decibel scale originated with electrical engineers, where electrical power ratios were defined in terms of a *bel* or 10 dB.

$\quad$ Returning to Eq. (2.2), it is noteworthy to mention that the international reference pressure 2.0×10^{-5} Pa is the average threshold of hearing for young adults when listening to a pure tone at 1000 Hz. From this one can again recognize that the human ear is an extremely sensitive instrument in that such miniscule pressure fluctuations can be detected. It is interesting, however, to note that the ear is not very sensitive when it comes to detecting differences in sound level. For example, under controlled conditions, the ear can barely detect a 3-dB difference in level, and with normal everyday background level present, a change of 5 dB requires careful listening. A drop of 10 or 20 dB can readily be sensed, and a 40-dB change is dramatic and can be likened as "going from a shout to a whisper."

$\quad$ By utilizing the algebraic properties of logarithms, Eq. (2.2) can be rewritten in the following simpler and, in some cases, more useful form:

$$L_p = 20 \log_{10} p + 94 \qquad [\mathrm{dB(re\ 2.0 \times 10^{-5}\ Pa)}] \qquad (2.3)$$

$\quad$ Consider now a few examples to clarify and illustrate these new concepts of sound pressure level.

Example

$\quad$ Calculate the sound pressure level L_p associated with an effective or rms sound pressure of 100 Pa.

$\quad$ *Solution*

$\quad$ From Eq. (2.3)

$$L_p = 20 \log_{10} (100) + 94$$

$$= 40 + 94$$

$$= 134 \qquad [\mathrm{dB(re\ 2.0 \times 10^{-5}\ Pa)}]$$

$\quad$ Because of the universal acceptance of the reference pressure 2.0×10^{-5} Pa, we shall hereafter omit the reference unless an ambiguity might arise. Likewise, the abbreviation dB will be conventionally used for decibel. Consider another example.

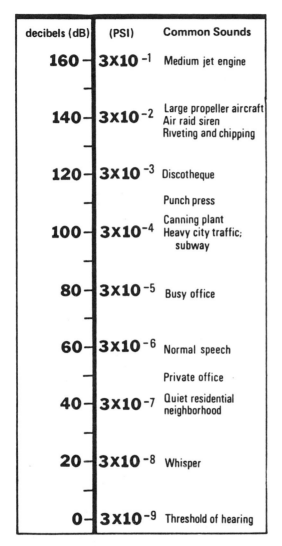

decibels (dB)	(PSI)	Common Sounds
160	3×10^{-1}	Medium jet engine
140	3×10^{-2}	Large propeller aircraft Air raid siren Riveting and chipping
120	3×10^{-3}	Discotheque
100	3×10^{-4}	Punch press Canning plant Heavy city traffic; subway
80	3×10^{-5}	Busy office
60	3×10^{-6}	Normal speech
40	3×10^{-7}	Private office Quiet residential neighborhood
20	3×10^{-8}	Whisper
0	3×10^{-9}	Threshold of hearing

Figure 2.1 Typical everyday sounds for comparison in sound pressure level (decibels) and pounds per square inch.

Example

The sound pressure level of a pure tone is 104 dB. What is the corresponding rms sound pressure?

Solution

From Eq. (2.3),

$$104 = 20 \log_{10}(p) + 94$$

$$20 \log_{10}(p) = 104 - 94 = 10$$

$$\log_{10}(p) = \left(\frac{1}{2}\right)$$

$$p = \text{antilog} \left(\frac{1}{2}\right) = 10^{1/2}$$

$$= 3.16 \text{ Pa}$$

It should be emphasized that the basic field sound-measuring instruments are calibrated to read sound pressure level in decibels directly as defined by Eq. (2.1) and (2.2). As such, the noise control engineer deals with the actual sound pressure infrequently and only in special situations.

2.2 SOUND POWER LEVEL

Another physical parameter of sound and wave motion which is fundamental and which must be thoroughly understood is the concept of sound power. By definition, sound power is the rate at which acoustical energy is radiated from a sound source, and its unit (from basic physics) is the watt. Students often find sound power a difficult physical concept to visualize, and there is also an inherent tendency to confuse sound power and sound pressure. This difficulty is unnecessary if one considers the common light bulb as an analogy.

Light bulbs are rated in terms of their power consumption, i.e., 100-W bulbs, 25-W bulbs, etc. From experience, however, the light intensity or illumination of a 100-W bulb is greater than a 25-W bulb at any given radius. Analogously, a sound source of 100 W produces a higher sound pressure level or is *louder* than a 25-W source at a given radius. In short, sound power is a single-number index that measures the flow rate of acoustical energy and is independent of any point in space or time. Sound pressure, on the other hand, is related to intensity or illumination as in the light bulb analogy and generally depends on how far you are from the source. As the reader might expect, and as we shall see later, when sound propagation is discussed, sound power and sound pressure are indeed functionally related. It is also noteworthy to point out that sound pressure is readily measured with simple hand-held instruments but that sound power cannot be measured directly.

Again, just as in the case of sound pressure, the range of interest over which sound power varies is extremely wide, over 10 orders of magnitude. Therefore, to condense the range of variation, acoustical engineers created a logarithmic scale where the sound power level L_W is defined as

$$L_W = 10 \log_{10} \left(\frac{W}{W_{re}} \right) \qquad [dB] \qquad\qquad (2.4)$$

where

W = acoustic power of interest (W)

W_{re} = internationally accepted reference = 10^{-12} W

Shown in Table 2.1 is a listing of the sound power and equivalent sound power levels for common acoustic sources. As can be seen, and to some extent unfortunately, sound power level L_W is expressed in decibels just as sound pressure level L_p, which is the origin of some confusion. Therefore, to avoid ambiguity, strict attention must be given to the proper terminology and symbolism. This is best accomplished by always including an antecedent reference to sound power level and sound pressure level, whichever is appropriate, in conjunction with decibel.

The reference power, 10^{-12} W, has been accepted almost universally. However, some literature still exists which uses 10^{-13} W as a reference. The relationship between the two references is of course 10 dB:

$$L_W(\text{re } 10^{-13} \text{ W}) = L_W (\text{re } 10^{-12} \text{ W}) + 10 \qquad [dB] \qquad (2.5)$$

Hereafter the reference power level 10^{-12} W will be omitted when the term sound power level or symbol L_W is used in accordance with convention.

Consider now an example to illustrate these concepts of power level.

Example

What is the sound power level (re 10^{-12} W) of a sound source radiating energy at a rate of 0.2 W? Re 10^{-13} W?

Solution

From Eq. (2.4),

$$L_W = 10 \log_{10}(0.2/10^{-12}) = 10 \log_{10}(2/10^{-11})$$

$$= 10 \log_{10}(2) + 110$$

$$= 3.01 + 110$$

$$\cong 113 \text{ dB}$$

For the reference 10^{-13}, using Eq. (2.5), we get

$$L_W (\text{re } 10^{-13}) = L_W (\text{re } 10^{-12}) + 10 \text{ dB}$$

$$= 113 + 10 = 123 \text{ dB}$$

Table 2.1 Typical Sound Power Levels and Equivalent Power for Common Everyday Sound Sources

Source	Typical Power Levels, L_W (dB re 10^{-12} W)	Power (W)
Saturn rocket	180	1,000,000
Turbojet engine with afterburner	170	100,000
Turbojet engine, 7000-lb thrust	160	10,000
4-propeller airliner	140	100
75-piece orchestra	130	10
Large chipping hammer	120	1
Auto horn	110	0.1
Radio hi-fi	100	0.01
Voice, shouting (average)	90	0.001
Office	80	0.0001
Voice, conversational level	70	0.00001
Bedroom	60	0.000001
Whisper	50	0.0000001

By again utilizing the properties of logarithms, Eq. (2.4) can be written in a simpler and often more useful form since the reference level is a constant:

$$L_W = 10 \log_{10} \left(\frac{W}{W_{re}}\right)$$

$$= 10 \log_{10}(W) - 10 \log_{10}(10^{-12})$$

$$= 10 \log_{10}(W) + 120 \qquad [dB] \qquad (2.6)$$

Often the absolute power of a sound source is required, given the sound power level. This can be accomplished easily by rearranging Eq. (2.6) and solving for W:

$$\log_{10}(W) = \frac{L_W - 120}{10}$$

$$W = \text{antilog} \frac{(L_W - 120)}{10} \qquad [W] \qquad (2.7a)$$

$$= 10^{(L_W - 120)/10} \qquad [W] \qquad (2.7b)$$

Example

What is the acoustical power of a 100-hp vane-type air compressor whose rated sound power level is 130 dB?

Solution

From Eq. (2.7a),

$$W = \text{antilog} \left(\frac{130 - 120}{10}\right)$$

$$= \text{antilog } 1$$

$$= 10 \text{ W}$$

Now, interestingly, the 10 W of radiated acoustical power represents only 0.013 hp or less than 0.1% of the rated power of the compressor (1 hp = 745 W). However, to those familiar with vane-type compressors, this small acoustical power conversion represents a great deal of noise.

2.3 SOUND INTENSITY LEVEL

The acoustical power passing through a unit area is defined as the sound intensity. Symbolically, the sound intensity is then

$$I = \frac{W}{A} \qquad [W/m^2] \qquad\qquad (2.8)$$

where

W = acoustical sound power of the source (W)

A = surface area (m^2)

To better understand this concept, consider a sound source radiating uniformly into free space. Now, with the source at the center, construct a spherical surface around the source at an arbitrary radius r. From Eq. (2.8) the sound intensity I on the surface at r is then

$$I = \frac{W}{4\pi r^2} \qquad [W/m^2]$$

where $4\pi r^2$ is the surface area of the sphere. Therefore, from Eq. (2.8), for a 2.0-W source the sound intensity at 10 m is

$$I = \frac{2.0}{4\pi(10)^2}$$

$$= 0.0016 \qquad [W/m^2]$$

At a radius of 20 m, the intensity is

$$I = \frac{2.0}{4\pi(20)^2}$$

$$= 0.0004 \qquad [W/m^2]$$

From this example, it is easy to see that the sound intensity depends generally on distance from the source, in a manner similar to sound pressure. In fact, it can be shown that for a freely propagating plane wave the average sound intensity is related simply to sound pressure as follows:

$$I = \frac{p_{rms}^2}{\rho c} \qquad [W/m^2] \qquad\qquad (2.9)$$

where

p_{rms} = rms sound pressure (Pa)

ρc = characteristic impedance of the medium (mks rayls)

Consider now an example of this concept for a better understanding.

Example

A sinusoidal plane wave is propagating outdoors with an amplitude of 10 Pa. What is the sound intensity?

Solution

From Eq. (1.7) the rms value of the amplitude is $0.707 \times 10 = 7.07$ Pa, and at sea level and 22°C, $\rho c \cong 415$ mks rayls. Therefore, upon substitution into Eq. (2.9),

$$I = \frac{(7.07)^2}{415} = 0.120 \text{ W/m}^2$$

Sound intensity also has a wide range of variation, and, as for
sound pressure and sound power, a more manageable logarithmic decibel scale has also been adopted. The sound intensity level L_I is
defined as

$$L_I = 10 \log_{10} \left(\frac{I}{I_{re}} \right) \qquad [\text{dB (re } 10^{-12} \text{ W/m}^2)] \qquad (2.10)$$

where

I = sound intensity of interest (W/m^2)

I_{re} = international reference for sound intensity = 10^{-12} W/m^2

With the definition of sound intensity level, we can now consider
an example which will bring together and clarify the relationship between the major physical parameters of wave motion that the acoustical
engineer deals with on a routine basis.

Example

The tone from a tuning fork is 440 Hz and has a measured sound
pressure level of 54 dB at a radial distance of 10 m. Assuming
the sound waves are radiating uniformly in all directions (spherically), what is the power level of the source? What is the sound
pressure level at 20 m?

Solution

To obtain the power level, we must first determine the sound
intensity at 10 m. The rms sound pressure at 10 m can be
calculated from Eq. (2.3):

$$p_{rms} = \text{antilog} \frac{(54 - 94)}{20}$$

$$= \text{antilog}(-2)$$

$$= 10^{-2} \text{ Pa} \qquad \text{at 10 m}$$

With the sound pressure, the sound intensity can now be calculated from Eq. (2.9) and taking ρc to be 415 mks rayls is

$$I = \frac{(10^{-2})^2}{415}$$

$$= 2.4 \times 10^{-7} \ W/m^2 \qquad \text{at 10 m}$$

or in terms of sound intensity level L_I, from Eq. (2.10),

$$L_I = 10 \log_{10} \left(\frac{2.4 \times 10^{-7}}{10^{-12}} \right)$$

$$= 10 \log_{10}(2.4) + 10 \log_{10}(10^5)$$

$$= 3.8 + 50$$

$$= 53.8 \ dB$$

Now from Eq. (2.8) the power radiated by the source is

$$W = IA$$

where in this example A is the area of a sphere of radius 10 m. A sphere was chosen since the source is radiating uniformly in all directions. Therefore, the required sound power of the source is

$$W = 2.4 \times 10^{-7} \times 4\pi(10)^2$$

$$= 30.16 \times 10^{-5} \ W$$

or in terms of sound power level, from Eq. (2.4),

$$L_W = 10 \log_{10} \left(\frac{30.16 \times 10^{-5}}{10^{-12}} \right)$$

$$= 10 \log_{10}(30.16) + 10 \log_{10}(10^7)$$

$$= 14.8 + 70$$

$$= 84.8 \ dB$$

For the second part of the example, the sound intensity at 20 m is given by Eq. (2.8):

$$I = \frac{W}{4\pi r^2} = \frac{30.16 \times 10^{-5}}{4\pi(20)^2}$$

$$= 6.0 \times 10^{-8} \ W/m^2$$

From Eq. (2.9) we again get the rms sound pressure at 20 m:

$$p_{rms} = \sqrt{\rho c I}$$

$$= \sqrt{415 \times 6.0 \times 10^{-8}}$$

$$= 49.9 \times 10^{-4} \text{ Pa}$$

and from Eq. (2.3) the required sound pressure level then is

$$L_p = 20 \log_{10}(49.9 \times 10^{-4}) + 94$$

$$= 48 \text{ dB}$$

Comparing this result with sound pressure levels at 10 m, we see a difference of $54 - 48 = 6$ dB. In short, in doubling the distance from the source, the sound level was reduced by 6 dB.

In this example, we have considered a special case of sound propagation, spherical divergence, where the sound intensity decreases inversely as the square of the distance from the source. This is, of course, the origin of the well-known *inverse square law*. As we shall see in a later chapter, not all sound radiates in this manner; however, the basic physical parameters presented here will be just as applicable in characterizing the sound fields.

2.4 SPECTRA: FREQUENCY

The second parameter, in addition to sound pressure level, required to characterize noise is frequency. As mentioned before, the frequency range or spectrum of most interest in noise control engineering varies from about 50 to 20,000 Hz. Here again, because of the wide range of variation, acoustical engineers convened for purposes of analysis to divide the spectrum into easy-to-use geometrically related bands. Shown in Table 2.2 are the center, upper, and lower band limit frequencies for octave and one-third octave bands as specified by the American National Standards Institute (ANSI). Although these band limits and center frequencies are well defined, it is important to consider the mathematical background in order to assure a complete understanding. The basic concept here lies in the definition of bandwidth (BW):

$$f_{n+1} - f_n = \text{bandwidth (BW)} \qquad [\text{Hz}] \qquad (2.11)$$

where f_{n+1} and f_n are successive frequency band limits. Referring to Table 2.2, in the common band number 5 the successive band limits are 707 and 1414 Hz, respectively, and from Eq. (2.11) the bandwidth is

$$1414 - 707 = 707 \text{ Hz}$$

Table 2.2 Octave and One-third Octave Center Frequencies and Corresponding Band Limits as Specified by American National Standards Institution (ANSI) SI.6

Common Band Number	Octave			One-third Octave		
	Lower Band Limit (Hz)	Center Frequency (Hz)	Upper Band Limit (Hz)	Lower Band Limit (Hz)	Center Frequency (Hz)	Upper Band Limit (Hz)
	22	31.5	44	22.4	25	28.2
				28.2	31.5	35.5
				35.5	40	44.7
1	44	63	88	44.7	50	56.2
				56.2	63	70.8
				70.8	80	89.1
2	88	125	177	89.1	100	112
				112	125	141
				141	160	178
3	177	250	354	178	200	224
				224	250	282
				282	315	354

4	354	500	707	354	400	447
				447	500	562
				562	630	707
5	707	1,000	1,414	707	800	891
				891	1,000	1,122
				1,122	1,250	1,414
6	1,414	2,000	2,828	1,414	1,600	1,778
				1,778	2,000	2,239
				2,239	2,500	2,828
7	2,828	4,000	5,656	2,828	3,150	3,548
				3,548	4,000	4,467
				4,467	5,000	5,656
8	5,656	8,000	11,312	5,656	6,300	7.079
				7,079	8,000	8,913
				8,913	10,000	11,220
	11,312	16,000	22,624	11,220	12,500	14,130
				14,130	16,000	17,780
				17,780	20,000	22,390

Now these bands are geometrically related by the following recursive relationship:

$$\frac{f_{n+1}}{f_n} = 2^k \qquad (2.12)$$

where f_n and f_{n+1} are successive band limits, i.e., the lower and upper bands, respectively.

The index k may be a positive integer or a fraction accordingly as whole octave or fractional octave band limits are being calculated. For example, if k = 1, the ratio between successive bands is 2, as is easily seen in Table 2.2. Correspondingly, if k = 1/3, then from Eq. (2.12) the ratio between the upper and lower band limits is $2^{1/3}$ or the cube root of 2 = 1.26. Referring again to Table 2.2, we see that for one-third octave bands, the ratio of successive band limits is indeed 1.26. It should be noted that in some cases the ratio is not exactly 1.26 due to some *rounding off*.

Finally, associated with each of the bands is a center frequency f_c, which is given by

$$f_c = \sqrt{f_n f_{n+1}} \qquad [Hz] \qquad (2.13)$$

where again f_n and f_{n+1} are the lower and upper band limits, respectively.

It should be noted that center frequency is a bit of a misnomer in that the "so-called" center frequency is actually the geometrical mean by definition. However, the term center frequency is so commonly used, we shall continue to use it throughout the text.

We can now derive a general expression to calculate the center frequencies for the commonly used octave bands. From Eq. (2.12), with k = 1 for octave bands

$$f_{n+1} = 2f_n$$

substituting into Eq. (2.13), we get

$$f_c = \sqrt{2f_n^2} = f_n \sqrt{2} \qquad [Hz] \qquad (2.14)$$

where f_n is the frequency of the lower band limit.

Example
  What is the center frequency of the common band number 6 whose lower and upper bands are 1414 and 2828 Hz, respectively?
  Solution
  From Eq. (2.13),

$$f_c = 1414\sqrt{2}$$

$$= 2000 \text{ Hz}$$

Similarly, for one-third octave bands from Eq. (2.12)

$$k = \frac{1}{3}$$

$$f_{n+1} = 2^{1/3} f_n$$

Then from Eq. (2.13), upon substitution we have

$$f_c = \sqrt{2^{1/3} f_n^2}$$

$$= 2^{1/6} f_n$$

$$= 1.12 f_n \qquad [Hz] \qquad\qquad\qquad (2.15)$$

Example

What is the center frequency for the one-third octave band whose lower and upper frequency limits are 7079 and 8913 Hz, respectively?

Solution

From Eq. (2.15),

$$f_c = 1.12 \times 7079$$

$$= 7928 \cong 8000 \text{ Hz}$$

It should not be inferred from this discussion that the octave and one-third octave bands discussed are the only geometric scales currently used by acoustical engineers. To be sure, the one-half octave and one-tenth octave geometric scales are used but have rather limited and special applications.

Before leaving the topic of frequency, it must be mentioned that any spectrum can always be divided into equal or constant bandwidths. As we shall see later, the use of recently developed constant (narrow) bandwidth analyzers is fundamental to characterizing industrial noise. However, since the bands are arbitrary, that is, no current standards exist specifying the bandwidths, we shall defer a detailed discussion until later.

It should also be emphasized that characterizing noise in octave bands is the crudest form of analysis and that often more spectral resolution is required. This is especially true when the noise is from numerous sources or contains a multiplicity of pure tones. In these cases, fractional octave analyses, obtained from more sophisticated instruments, are utilized. For example, shown in Fig. 2.2 is a one-third octave band analysis of the noise from a bank of transformers located near an electrical power station. Note here the obvious pure tone character of the radiated sound in the 63-, 125-, and 250-Hz bands. For comparison, shown in Figure 2.3, is an octave band analysis of the same transformer noise. Note that the pure tone characteristic in the low-frequency range is not discernable with the greater bandwidths.

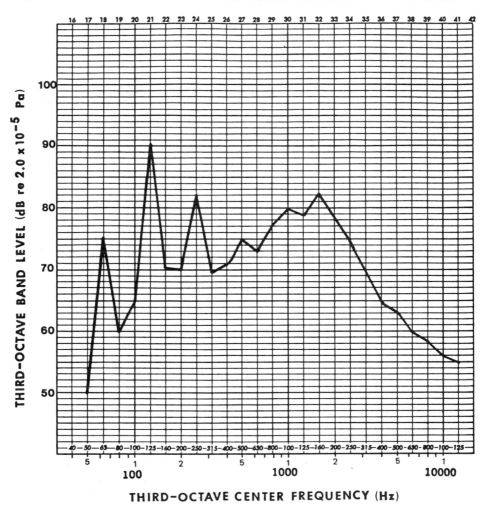

Figure 2.2 One-third octave band analysis of a bank of transformers located near an electrical power station.

From this discussion and examples, it should now be clear that determining the spectral character of the noise is essential in developing a systematic approach to noise control. We shall return to this topic in greater detail in a later chapter.

Finally, with the geometric scales to define frequency and the decibel scales to define sound pressure, the essential elements to characterize industrial noise in its most basic form are now present. To better illustrate this concept, shown in Fig. 2.4 is a typical octave

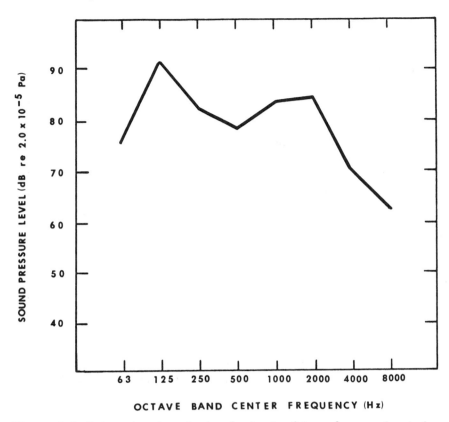

Figure 2.3 Octave band analysis of a bank of transformers located near an electrical power station.

band spectral analysis of a small gas turbine measured outdoors at 100 m. With the noise from the gas turbine presented in this format, the acoustical engineer concludes that peak noise levels are in the band whose center frequency is 2000 Hz, i.e., between 1414 and 2828 Hz, and that a secondary peak lies in the 500-Hz band. One could now compare these levels to specification criteria to determine the required noise reduction and, if familiar with the gas turbine, conclude that (1) the noise in the 2000-Hz band is probably originating in the compressor and (2) the noise in the 500-Hz band is the exhaust. In short, from this octave band analysis of the noise, the magnitude of the required noise reduction can be determined and the major sources identified. With this information, a systematic approach to control the noise can now be established. It should be emphasized that analyses, such as this example, are not difficult to obtain. In fact, portable instruments called octave band analyzers

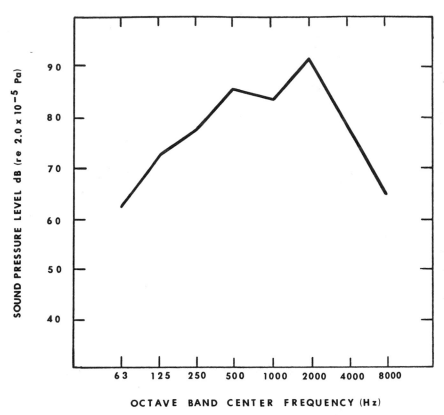

Figure 2.4 Typical octave spectral analysis of a small gas turbine.

will measure the sound pressure level in each octave band just as illustrated. We shall discuss these instruments in detail later.

2.5 COMBINING DECIBELS

In most industrial environments, sound is emitted from more than one source or at different frequencies, and it is necessary to calculate the cumulative or overall sound pressure level or in some cases the sound power level. Obviously, since the decibel scales are logarithmic, they cannot be added algebraically. For example, if the sound pressure level of a machine at a given point is 70 dB and a second machine is turned on, also producing a sound level of 70 dB, the combined sound level is *not* 140 dB, which is about the sound level within a few meters of a jet airliner during takeoff.

To develop a method to combine these level values, let L_{p1}, L_{p2}, L_{p3}, . . . , L_{pn} be n sound pressure levels to be combined to yield the cumulative or *overall* sound pressure level. Here, for example, each sound level to be combined might be the noise from n individual machines or the sound levels associated with n octave bands. Refering to the definition of sound pressure level [Eq. (2.1)], we have

$$L_p = 10 \log \left(\frac{p}{p_{re}} \right)^2 \quad [dB]$$

and solving for $(p/p_{re})^2$, we get

$$\left(\frac{p}{p_{re}} \right)^2 = \text{antilog} \left(\frac{L_p}{10} \right)$$

Now the mean-square values of sound pressure can be combined algebraically, assuming, of course, that the sound sources have a random phase relationship. If not random, standing waves would result, as we saw in Chap. 1. Hence, the total algebraic sum of the mean squares is

$$\frac{p_1^2 + p_2^2 + p_3^2 + \cdots + p_n^2}{p_{re}^2} = \sum_{i=1}^{n} \text{antilog} \left(\frac{L_{p,i}}{10} \right) \tag{2.16}$$

$$= \sum_{i=1}^{n} 10^{L_{p,i}/10} \tag{2.17}$$

Taking the $\log_{10}$ of each side and multiplying by 10, we get

$$10 \log \left[\sum_{i=1}^{n} \left(\frac{p_i}{p_{re}} \right)^2 \right] = 10 \log \left(\sum_{i=1}^{n} 10^{L_{p,i}/10} \right) \tag{2.18}$$

But the left-hand side is by definition the sound pressure level of the total mean-square pressure or the total sound pressure level $L_{p,t}$. Hence,

$$L_{p,t} = 10 \log \left(\sum_{i=1}^{n} 10^{L_{p,i}/10} \right) \quad [dB] \tag{2.19}$$

An example will illustrate the ease of using this expression.

Example
    Three machines produce at a given distance noise levels of 86, 84, and 89 dB when operated individually. What is the cumulative sound level at the given point if all are operating at the same time?

Solution

From Eq. (2.19) we have

$$L_{p,t} = 10 \log_{10}(10^{8.6} + 10^{8.4} + 10^{8.9})$$

$$= 10 \log_{10}[(10^{0.6} + 10^{0.4} + 10^{0.9})10^8]$$

$$= 10 \log_{10}(3.98 + 2.51 + 7.94) + 80$$

$$= 11.6 + 80$$

$$= 91.6 \text{ dB}$$

Another method which is popular, easy to use, but less accurate for combining decibel levels utilizes the chart shown in Fig. 2.5. An example will clearly illustrate the use of the chart.

Example

The noise level of a small sewing machine is 75 dB. The level of an adjacent larger machine is 81 dB. What is the combined level?

Solution

The difference between the two levels is 6 dB; i.e., 81 dB − 75 dB = 6 dB. Therefore, by entering the chart at 6, the ordinate intercept is 1 dB, which is to be added to the higher of the two or 81 dB + 1 dB = 82 dB, the combined level.

More than two unequal levels can be combined by taking the combinations in pairs, as illustrated in the following example.

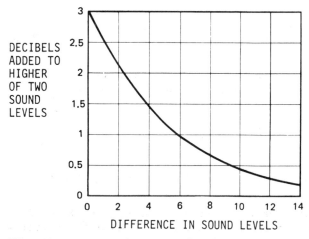

Figure 2.5 Chart for combining or adding sound levels.

Example

Four distinct sources have noise levels L_p of 81, 75, 75, and
73 dB, respectively. What is the overall noise level?

Solution

If we take the first pair, 81 dB and 75 dB, the combined level
is 82 dB, as seen from the previous example. For the second
pair, 75 dB and 73 dB, the difference is 2 dB, and from the
chart in Fig. 2.5 the number of decibels to be added to the
larger is approximately 2 dB. Therefore, 75 dB + 73 dB = 77 dB.
We now have the resultant combined levels, 82 dB and 77 dB, to
combine. The difference here is 5 dB, i.e., 82 dB − 77 dB,
and from the chart in Fig. 2.5 the number of decibels to be added
to the larger, corresponding to a 5-dB difference, is approximately 1.5 dB. Therefore, 82 dB + 77 dB = 82 dB + 1.5 dB =
83.5 dB. Shown in Fig. 2.6 is a recommended format for performing this recursive addition. Note that the differences in
levels are in parentheses and that the numbers to be added
from the chart are inserted as mnemonics.

From this example it is clear that the recursive method can be extended to any number of sound levels. There are also some time-
saving "tricks" worth mentioning. First, when two levels are equal,
their sum is just 3 dB higher, and second, when the difference exceeds 10 dB or more, the contribution of the smaller is less than 0.5
dB. The author does not want to encourage careless calculation
habits, but in most noise control problems, measured or combined
noise levels can be rounded off to the nearest 0.5 dB with negligible
error or design impact.

When rather accurate calculations are required and to avoid an
accumulation of rounding errors when numerous levels are to be combined, Table 2.3 can be most useful. This table is of course just a
list of selected coordinate pairs from the chart of Fig. 2.5. Consider
an example, and note the speed and ease of its use.

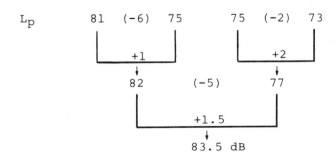

Figure 2.6 Recommended format for combining sound levels.

Table 2.3 Selected Integer Decibel Differences for Ease of
Combining Sound Levels

If the Difference in Decibels Between the Two Levels Being Added Is	Add to the Higher Level
0	3.0
1	2.6
2	2.2
3	1.8
4	1.4
5	1.2
6	1.0
7	0.8
8	0.6
9	0.5
10	0.4
13	0.2
16	0.1

Example
　　Frequently, from octave or fractional octave band level analyses,
　　the overall or cumulative sound pressure level is required.
　　Shown in Fig. 2.7 in tabular form are the measured octave band
　　sound levels from a noisy diesel engine at 200 ft. Compute the
　　overall cumulative sound pressure level.
　　Solution
　　As illustrated in Fig. 2.7 the pairs of sound levels are added
　　successively, and the overall cumulative sound level is 90.0 dB.
　　It should be noted that the differences between levels were also
　　rounded to the nearest whole integer for ease in entering the
　　table.

　　Another example occurring often in noise control involves the
combining of one-third octave levels to obtain octave levels. Consider the following example.

Example
　　The one-third octave sound pressure levels for the 1600-,
　　2000-, and 2500-Hz bands were, respectively 90, 92, and 93
　　dB, as shown in Fig. 2.8. What is the combined sound levels
　　for the octave band whose center frequency is 2000 Hz?

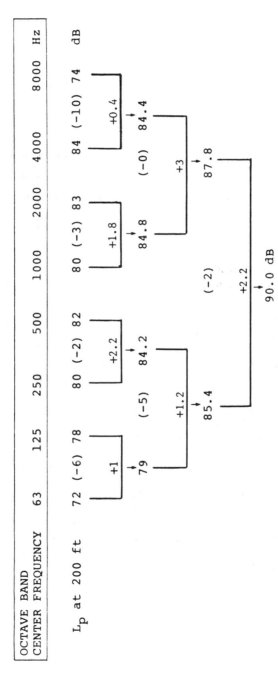

Figure 2.7 Combining octave band sound pressure levels. Note that differences were rounded to the nearest 0.5 dB.

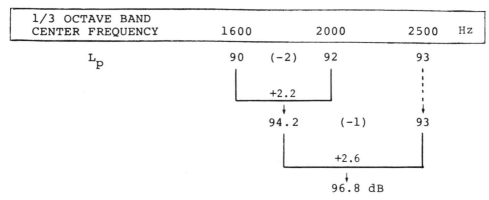

1/3 OCTAVE BAND CENTER FREQUENCY	1600	2000	2500	Hz

Figure 2.8 Combining one-third octave band sound levels to obtain octave band levels.

Solution

As illustrated in Fig. 2.8, the levels are combined from Table 2.3, and the overall octave band level in the 2000-Hz band is then 96.8 dB.

It is obvious from the preceding example that if sound levels for all 24 one-third octave bands from 50 to 10,000 Hz were combined in triplets about the octave band center frequencies, an octave band analysis would be obtained. This is a common requirement in noise control.

It must be emphasized again that the preceding examples utilizing the chart in Figure 2.5 and the values in Table 2.3 should be used only when first-order approximations are acceptable. For precise calculations, Eq. (2.19) can be easily programmed into a hand-held calculator or a personal computer. With this approach, rounding errors can be easily controlled and time-consuming arithmetic effort is sharply reduced.

Shown in Figure 2.9 is an example of a typical spreadsheet used to calculate overall sound levels from octave band sound levels. Note that column J is the logarithmic sum of user-entered data in column B through I. Shown also in Figure 2.9 are the "cell" formulae, or algebraic commands, found in column J, which are used to calculate the overall sound level.

In a similar fashion, the cumulative total or combined sound power level L_{W_t} of n sound power levels is

$$L_{W,t} = 10 \log \left(\sum_{i=1}^{n} 10^{L_{W,i}/10} \right) \quad \text{[dB]} \quad (2.20)$$

	A	B	C	D	E	F	G	H	I	J
1		Octave Band Center Frequency (Hz)								
2	Source	63	125	250	500	1000	2000	4000	8000	Overall
3										
4	Vacuum Cleaner 3'	48	66	69	73	79	73	73	72	82
5	Large Truck at 50'	83	85	83	85	81	76	72	65	91
6	Passenger Car at 50'	72	70	67	66	67	66	59	54	77
7	Window Air Conditioner	64	64	65	56	53	48	44	37	69
8										
9										
10										

Where the equation for cell J4 = 10*@LOG(10^(B4/10)+10^(C4/10)+10^(D4/10)+10^(E4/10)
+10^(F4/10)+10^(G4/10)+10^(H4/10)+10^(I4/10))

Where the equation for cell J5 = 10*@LOG(10^(B5/10)+10^(C5/10)+10^(D5/10)+10^(E5/10)
+10^(F5/10)+10^(G5/10)+10^(H5/10)+10^(I5/10))

⋮

Where the equation for cell J# = 10*@LOG(10^(B#/10)+10^(C#/10)+10^(D#/10)+10^(E#/10)
+10^(F#/10)+10^(G#/10)+10^(H#/10)+10^(I#/10))

Figure 2.9 Example of a spreadsheet format for calculating overall sound levels from octave band sound levels. Also shown are the formulae for the cells in column J.

where $L_{W,i}$ is the ith sound power level. Here again, noting the similarity with the Eq. (2.19), the chart method just discussed is equally applicable for combining sound power levels.

Example

What is the combined sound power level of three sources whose individual sound power levels are 80, 90, and 96 dB?

Solution

Starting with the larger levels 90 and 96 dB, the difference is 6 dB; therefore from the chart in Fig. 2.5, 1.0 dB is to be added to the larger; i.e., 96 dB + 1 dB = 97 dB. The combination of the 80-dB source can now be ignored since the difference is 17.5 dB, i.e., 97 dB − 80 dB = 17 dB, and from the chart yields less than 0.25 dB. Therefore, the combined power level of the sources is, for all practical purposes, 97 dB.

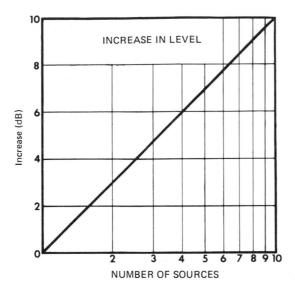

Figure 2.10 Chart for combining equal sound levels.

Frequently, it is desirable to combine sound levels which are equal. For this special case, the chart shown in Fig. 2.10 can be very useful. This chart can be very valuable in assessing the impact of operating several identical machines in the same area. To illustrate the use of the chart, consider the following example.

Example
The noise level of a small sewing machine at 1 m is 88 dB. What is the combined noise level at 1 m if four machines are operated in close proximity?
Solution
The number of sources is four; therefore, from the chart in Fig. 2.10, the increase in level will be 6 dB, or the combined level is 88 dB + 6 dB = 94 dB.

The chart in Fig. 2.10 was constructed from the following equation over the range of 0 to 10:

$$\text{Increase in dB level} = 10 \log_{10}(N) \qquad [\text{dB}] \qquad (2.21)$$

where N is the number of sources. Hence, for combining more than 10 sources or levels, Eq. (2.21) can be used directly, or, with a little cleverness, the combined levels of more than 10 sources can also often be determined from the chart in Fig. 2.10. An example will illustrate both methods.

Example

In the ceiling of a "clean room," 24 small fans are to be installed as part of a dust filtering system. Estimate the combined levels (worst case) of the blowers if each one produces 75 dB at ear level below.

Solution

From Eq. (2.21)

Increase in dB = $10 \log_{10} 24$

= 13.8 dB

or since one produces 75 dB, the combined level is 75 dB + 13.8 dB = 88.8 dB.

Alternatively, 24 sources can be considered as four sets of 6 fans. The level increase is then, from the chart, 7.8 dB for 6 fans, or the combined level 75 dB + 7.8 dB = 82.8 dB.

Since we have four sets of 6 fans, the additional increase is, from the chart, 6 dB more, or 82.8 dB + 6 dB = 88.8 dB.

In short, if the number of sources N has factors less than 10, the chart in Fig. 2.10 can also be used in a *bootstrap*-like recursive manner where the contribution of each factor is determined and combined logarithmically.

In final summary, in this chapter we have defined the basic physical parameters which, in most cases, will suffice to characterize industrial noise. In addition, simple analytical methods were developed to deal with these parameters, and they were illustrated with numerous commonly encountered examples. With this background, we can now proceed directly to more practical applications.

BIBLIOGRAPHY

Acoustic Handbook. Hewlett-Packard Company, Palo Alto, California, 1968.

Bell, L. H. *Fundamentals of Industrial Noise Control*. Harmony Publications, Trumbull, Connecticut, 1973.

Beranek, L. L. *Noise and Vibration Control*. McGraw-Hill, New York, 1971.

Faulkner, L. *Handbook of Industrial Noise Control*. Industrial Press, New York, 1975.

Harris, C. M. *Handbook of Accoustical Measurements and Noise Control*. McGraw-Hill, New York, 1991.

Ingard, K. U., and G. C. Maling, Jr. *Physical Principles of Noise*

Reduction: Properties of Sound Sources and Their Fields.
Swedish Academy of Engineering, 1973.

Irwin, J. D., and E. R. Graf. *Industrial Noise and Vibration
Control.* Prentice-Hall, Englewood Cliffs, New Jersey, 1979.

Jens, Trampe Broch. *Acoustic Noise Measurements.* Brüel and
Kjaer, Marlborough, Massachusetts, 1971.

Kinsler, L. E., and A. R. Frey. *Fundamentals of Acoustics.* Wiley,
New York, 1950.

Lawrence, K., D. P. Lewis, and R. C. Bryant. *Noise Control in
the Workplace.* Aspen Systems Corporation, 1978.

Lipscomb, D. M., and A. C. Taylor, Jr. *Noise Control Handbook of
Principles and Practices.* Van Nostrand Reinhold, New York,
1978.

Morse, P. M. *Vibration and Sound.* McGraw-Hill, New York, 1948.

Morse, P. M., and K. U. Ingard. *Theoretical Acoustics.* McGraw-
Hill, New York, 1968.

Peterson, A. P., and E. E. Gross, Jr. *Handbook of Noise Measurement.* GenRad, Concord, Massachusetts, 1972.

EXERCISES

2.1 Calculate the sound pressure level associated with the
following rms sound pressures:
a. 0.0002 Pa
b. 0.106 Pa
c. 10 Pa
d. 20 Pa
e. 10^4 Pa

Answer: (a) 20 dB;
(b) 74.5 dB;
(c) 114 dB;
(d) 120 dB;
(e) 174 dB

2.2 What is the rms sound pressure of noise with sound pressure
levels of
a. 110 dB?
b. 116 dB?
c. 74 dB?
d. 90 dB?
e. 10 dB?
f. 0 dB?

Answer: (a) 6.32 Pa;
(b) 12.62 Pa;
(c) 0.1 Pa;
(d) 0.63 Pa;
(e) 0.000063;
(f) 0.00002 Pa

2.3 The maximum rms sound pressure near the muzzle of a cannon was measured to be 4.38×10^4 Pa. What is the maximum sound pressure level?

> *Answer:* 187 dB

2.4 The sound pressure level measured at 30 ft from a transformer was 92 dB for the dominant 120-Hz tone. What was the rms sound pressure at the measurement location?

> *Answer:* 0.80 Pa

2.5 Three different pumps are installed in a room of a power station. The noise levels from each, operating individually, were measured at the operator's station and found to be 91, 88, and 96 dB, respectively. What is the combined overall noise level at the operator's station with all three pumps operating?

> *Answer:* 97.7 dB

2.6 Calculate the overall noise level L_p of the following octave band levels: 63 Hz, 72 dB; 125 Hz, 78 dB; 250 Hz, 80 dB; 500 Hz, 82 dB; 1000 Hz, 80 dB; 2000 Hz, 83 dB; 4000 Hz, 84 dB; and 8000 Hz, 74 dB.

> *Answer:* 90 dB

2.7 The noise level at 25 ft from one small axial fan is 83 dB. What is the combined overall noise level at 25 ft if 12 of the fans were operating?

> *Answer:* 94 dB

2.8 What is the sound power level L_W of an acoustic source that produces
a. 50 W?
b. 3 W?
c. 1.5 W?
d. 0.003 W?

> *Answer:* (a) 137 dB;
> (b) 124.8 dB;
> (c) 121.8 dB;
> (d) 94.8 dB

2.9 How many watts are represented by a sound power level of (a) 120 dB; (b) 123 dB; (c) 110 dB; (d) 90 dB; and (e) 45 dB?

> *Answer:* (a) 1 W;
> (b) 2 W;
> (c) 0.1 W;
> (d) 0.001 W;
> (e) 3.16×10^{-8} W

2.10 The sound power level (re 10^{-12}) of a small motor is 92 dB.
What is the combined power level L_W of
a. Four motors?
b. Sixteen motors?

> *Answer:* (a) 98 dB;
>   (b) 104 dB

2.11 What are the upper and lower band limits for the octave bands
whose center frequency is
a. 500 Hz?
b. 8000 Hz?

> *Answer:* (a) 354 and 707 Hz;
>   (b) 5656 and 11,312 Hz

2.12 What are the upper and lower band limits for the one-third
octave bands whose center frequency is
a. 160 Hz?
b. 6300 Hz?

> *Answer:* (a) 141 and 178 Hz;
>   (b) 5656 and 7079 Hz

2.13 What is the ratio of lower and upper band limits for
a. one-half octave bands?
b. One-tenth octave bands?

> *Answer:* (a) $2^{1/2}$;
>   (b) $2^{1/10}$

2.14 Show that for one-half octave bands the center frequencies
$f_c = 2^{1/4} \times f_n$.

2.15 What the the upper and lower band limits for a center frequency of 1000 Hz in one-half octave bands?

> *Answer:* 841 and 1189 Hz

2.16 The sound pressure level measured at 100 m from a military
rocket was 154 dB. Assuming a uniform spherical radiation pattern,
what is the sound intensity at (a) 100 m; (b) 1000 m?

> *Answer:* (a) 2421 W/m^2;
>   (b) 24.21 W/m^2

2.17 Using the results of Exercise 2.16, what is the sound pressure
level at 1000 m?

> *Answer:* 134 dB

2.18 Again, using the results of Exercise 2.16, what is the acoustic
power of the rocket?

> *Answer:* 3.0×10^8 W

3
CHARACTER OF NOISE

In Chap. 2 we saw how the overall character of noise can be presented in terms of sound pressure level and frequency. In this
chapter we shall see that, generally, the spectral character of noise
can be further divided into three basic and distinctly different types:

1. Discrete frequency noise (pure tones)
2. Broadband noise (random)
3. Impulsive noise (impact)

Furthermore, most noise sources will take on one or more of these sub-
characteristics and as such will possess a unique acoustical signature.
It is this signature, along with access to modern instrumentation,
that allows the acoustical engineer to identify and rank in order the
major sources of noise in a complex acoustical environment.

In this chapter we shall therefore discuss the analytical representation of the basic spectral characteristics of sound and essential
noise-producing mechanisms. Examples common in industry will be
used to illustrate these powerful diagnostic methods. More detailed
discussions of the noise-generating mechanisms of major industrial
sources will follow in later chapters.

In addition, for a variety of reasons such as human response and
regulatory criteria, the spectral character of noise is often altered
by application of weighting scales. In the latter section of this chapter we shall therefore present these weighting concepts as a natural
follow-on to the overall characterization of noise.

3.1 DISCRETE FREQUENCY NOISE

The most common type of noise found in industry is discrete in spectral character. Discrete frequency noise has the characteristic of pure

tones and is generated mainly from rotating machinery. Classic
sources of discrete frequency noise are

1. Fans, rotary positive displacement blowers
2. Compressors, pumps
3. Internal combustion engines
4. Gears, timing belts
5. Transformers
6. Planars, routers, saws
7. Jar-to-jar and can-to-can impact (packaging lines, etc.)

   A typical discrete frequency spectrum from a refrigeration compressor is shown in Fig. 3.1. Note that the spectrum is composed
essentially of pure tones (often referred to as a *line spectrum*) and
that in this case they are harmonically related.

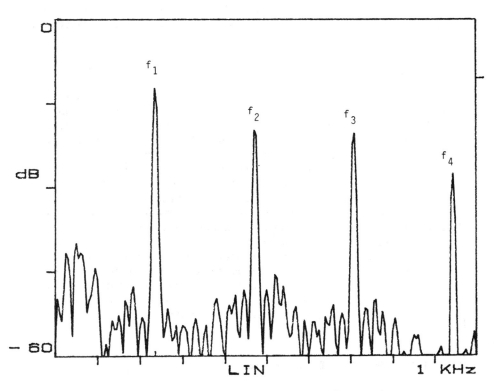

Figure 3.1 Narrow-band spectral analysis of a refrigeration compressor showing discrete frequency pure tones.

The frequencies of discrete tones from fans, blowers, gears, and rotating machinery are usually rather easy to predict, since they are related to shaft rotational speed and the number of fan blades, gear teeth, pistons, vanes, etc. For example, the fundamental blade-passing frequency of axial fans is given by

$$f_1 = Nk \qquad [Hz] \tag{3.1}$$

where

N = shaft rotational speed (r/s)

k = number of blades

As discussed in previous chapters, the higher harmonics are at integer multiples of the fundamental frequency, i.e., $2Nk$, $3Nk$, $4Nk$, ... Hence, the discrete frequency character of axial fan noise is well defined and easy to predict, given shaft speed and some basic knowledge of the mechanical components of the fan. Consider an example.

Example

What is the spectral character of the noise from a four-bladed axial fan rotating at 100 r/s?
Solution
From Eq. (3.1),

$$f_1 = 100 \times 4 = 400 \text{ Hz}$$

The fundamental blade-passing frequency is then 400 Hz. However, the spectrum of axial fans generally contains integer-order higher harmonics of the fundamental tone. Hence, one would expect additional tones at

$$f_2 = 2 \times 400 = 800 \text{ Hz}$$

$$f_3 = 3 \times 400 = 1200 \text{ Hz}$$

.
.
.

$$f_n = n \times 400$$

with generally diminished sound pressure amplitude.

The basic form of Eq. (3.1) needs to be modified only slightly to predict the fundamental tones from other common rotating machinery by letting k represent the number of gear teeth, cylinders, impeller lobes, etc.

For example, consider the noise from the meshing of a pair of simple spur gears. Intuitively, for each revolution of the pinion gear there are k metal-to-metal interactions, where k is the number of gear teeth. Therefore, it follows that the spectral character of the resultant noise would be discrete at the gear meshing frequency Nk and higher harmonics. Consider the following example.

Example

Calculate the fundamental gear meshing frequency and higher harmonics for a reduction gear rotating at 3600 rpm with 32 teeth.

Solution

From Eq. (3.1) with k = 32,

$$f_1 = \frac{3600}{60} \times 32$$

$$= 1920 \text{ Hz}$$

For the higher harmonics,

$$f_2 = 2 \times 60 \times 32 = 3840 \text{ Hz}$$

$$f_3 = 3 \times 60 \times 32 = 5760 \text{ Hz}$$

.
.
.

$$f_n = n \times 60 \times 32 = n \times 1920 \text{ Hz}$$

We should therefore expect the noise from the gear box to contain discrete tones at 1920, 3840, 5760 Hz, etc.

Shown in Fig. 3.2 is an actual spectral analysis of the noise from a pair of spur gears as considered in the preceding example.

The discrete character of many piston-type compressors, pumps, and internal combustion engines can also be predicted from Eq. (3.1), where k in this case represents the number of cylinders. Another example will illustrate this concept.

Example

What is the spectral character of the exhaust noise from a small six-cylinder diesel engine at 1800 rpm?

Solution

The fundamental piston firing frequency is given from Eq. (3.1), where k = 6 = the number of cylinders:

$$f_1 = \frac{1800}{60} \times 6 = 30 \times 6$$

$$= 180 \text{ Hz}$$

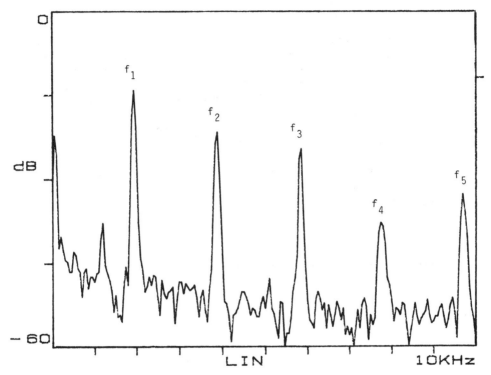

Figure 3.2 Narrow-band spectral analysis of the noise from a gearbox showing the fundamental (f_1) tooth meshing tone and higher harmonics.

The higher harmonics of the exhaust noise are then

$$f_2 = 2 \times 180 = 360 \text{ Hz}$$

$$f_3 = 3 \times 180 = 540 \text{ Hz}$$

.

.

.

$$f_n = n \times 180 \text{ Hz}$$

Consideration naturally must be given here as to whether the engine or compressor operates on a two- or four-cycle basis, has in-line cylinders or V configuration, etc. Regardless, the point to be emphasized is that for most rotating machines, the spectral character of the noise will be discrete, and predicting the spectral character of the noise often follows rather easily.

Another common example of dominantly discrete tones
generated as empty jars or cans impact in food packaging
the loudest areas are accumulation points just before the fi
Here the jars, for example, moving along a conveyor from
loader or cleaner impact the jars waiting to be filled. The
noise is like the ringing of dozens of bells. In short, at ja
accumulation points, there is a noise source resembling the
of bells, all about the same size, i.e., similar to sleigh bel
spectral character of the radiated noise is discrete at the
bell tone frequencies of the jars or cans. Shown in Fig. 3
narrow-band analysis taken near the filling machine of a ja
line. Note the discrete bell tones at approximately 2800 an
It should be emphasized that generally the bell tones are no
monically related but depend on the shape, stiffness, or g
struction of the jar or can.

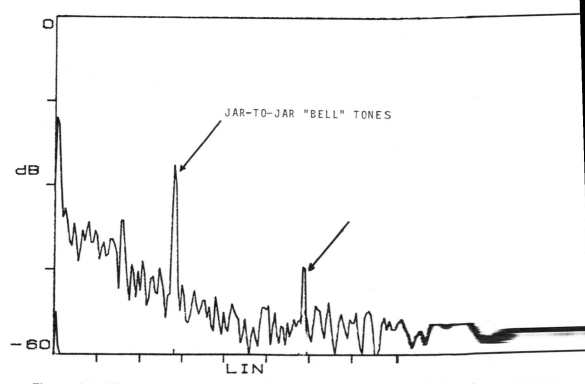

Figure 3.3 Narrow-band spectral analysis of the noise from
packaging line. Note the discrete frequency bell tones from
to-jar (empty) impact.

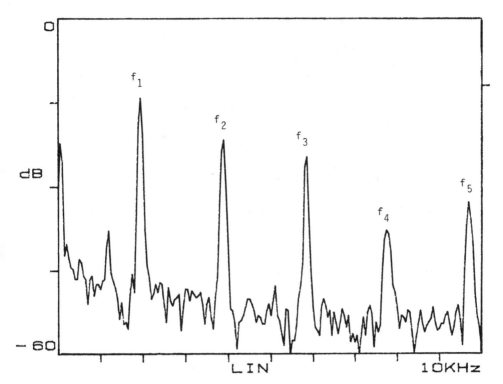

Figure 3.2 Narrow-band spectral analysis of the noise from a gearbox showing the fundamental (f_1) tooth meshing tone and higher harmonics.

The higher harmonics of the exhaust noise are then

$$f_2 = 2 \times 180 = 360 \text{ Hz}$$

$$f_3 = 3 \times 180 = 540 \text{ Hz}$$

.

.

.

$$f_n = n \times 180 \text{ Hz}$$

Consideration naturally must be given here as to whether the engine or compressor operates on a two- or four-cycle basis, has in-line cylinders or V configuration, etc. Regardless, the point to be emphasized is that for most rotating machines, the spectral character of the noise will be discrete, and predicting the spectral character of the noise often follows rather easily.

   Another common example of dominantly discrete tones is the noise
generated as empty jars or cans impact in food packaging lines. Often
the loudest areas are accumulation points just before the filling machine.
Here the jars, for example, moving along a conveyor from the case un-
loader or cleaner impact the jars waiting to be filled. The resultant
noise is like the ringing of dozens of bells. In short, at jar or can
accumulation points, there is a noise source resembling the ringing
of bells, all about the same size, i.e., similar to sleigh bells. The
spectral character of the radiated noise is discrete at the natural
bell tone frequencies of the jars or cans. Shown in Fig. 3.3 is a
narrow-band analysis taken near the filling machine of a jar packaging
line. Note the discrete bell tones at approximately 2800 and 5900 Hz.
It should be emphasized that generally the bell tones are not harmonically related but depend on the shape, stiffness, or general construction of the jar or can.

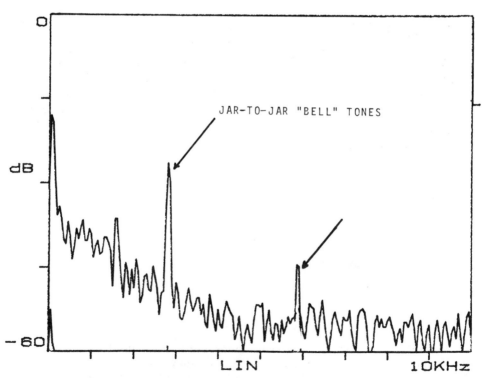

Figure 3.3 Narrow-band spectral analysis of the noise from a food
packaging line. Note the discrete frequency bell tones from the jar-
to-jar (empty) impact.

Other examples of discrete noise, which are often not so simply related to rotational speeds but yet are easy to identify, are tire *whine*, metal cutting tool *screech*, saw blade *howl* (not cutting), etc.

In summary, discrete frequency noise and its simple association to noise mechanisms give the acoustical engineer immediate insight into the identification and origin of probably the majority of industrial noise sources. We shall return in more detail to specific cases as major noise sources are considered individually in later chapters.

3.2 BROADBAND NOISE

Broadband noise is the second most common type of industrial noise and is best characterized as a *rumble*, *roar*, or *hiss*. By far the

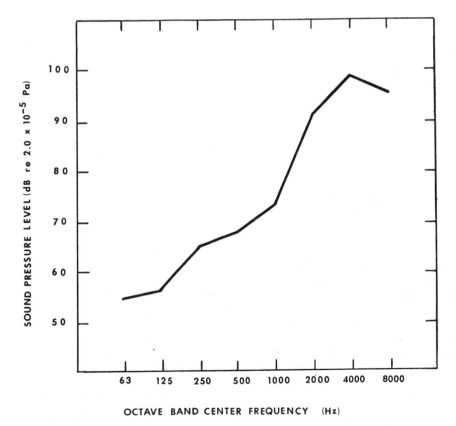

Figure 3.4 Broadband noise levels at 3 ft from a 1/4 in. diameter shop air blowoff nozzle.

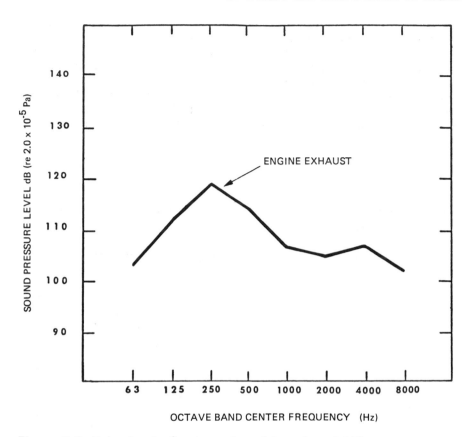

Figure 3.5 Noise levels from a modern jet engine at 100 m.

major source of broadband noise in industry is the noise associated with discharging high-velocity *shop* air. Here the broadband noise originates from the aerodynamic shearing of the ambient air by the high-velocity jet. The shearing action creates swirling eddylike turbulence with corresponding dynamic sound pressure fluctuations. Since the turbulent mixing is, in general, nonperiodic, the resulting sound pressure is random in both phase and amplitude. As such, there are no discrete frequency tones present, but the acoustical energy may be heavily concentrated in one or more areas of the spectrum.

    For example, shown in Fig. 3.4 is an octave band spectral analysis of the noise from a *shop air* blowoff nozzle. Here a 1/4 in. diameter copper tube (connected to the shop air supply) is used to blow small metal parts out of the die area of a punch press into a collector bin. Note that the peak levels are in the range of 2000 to 8000 Hz.

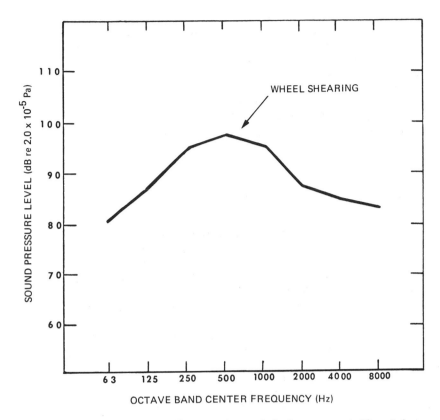

Figure 3.6 Broadband noise levels at 6 ft from a centrifugal fan.

The signature of these nozzles is so typical that one can often look at the octave band spectral analysis from a press room and determine the air noise contribution at a glance.

Other examples of high-frequency broadband noise sources are

1. Steam leaks in power plants (boilers), food processing kitchens, petrochemical plants, etc.
2. The impingement of grain or plastic pellets on conveying pipes
3. Plastic granulizers
4. Hammer mills
5. Pneumatic chipping, grinding, or burring tools, etc.

As mentioned before, subjectively, the high-frequency broadband noise is best described as a hiss.

The best examples of low-frequency broadband noise are the noise from the exhaust of a jet engine or large gas turbine. Shown in Fig.

3.5 is an octave band analysis of the noise at 100 m from a modern jet engine. Note that the peak levels, dominantly due to the exhaust, are in the range 125 to 500 Hz. Subjectively, this noise is often described as a *rumble*. Another example of low-frequency broadband noise, common to almost everybody, is the rumble associated with oil- or gas-fired burners.

Shown in Fig. 3.6 is an octave band analysis of broadband *wheel shearing* noise inherent in most centrifugal fans. Here the peak noise levels are concentrated in the middle of the audio spectrum and are often subjectively described as a roar.

3.3 DISCRETE AND BROADBAND NOISE

Finally, shown in Fig. 3.7 is the narrow-band spectral analysis of the noise from a high-speed (3400 rpm), 24 in. diameter, carbon-tipped wood cutoff saw at idle, i.e., not cutting. Note that here we have a combination of discrete tones and broadband noise. The discrete tone at approximately 2800 Hz is due to self-excited resonant vibration of the blade and is neither harmonic nor directly related to shaft speed. The broadband noise with maximum levels in the range of 1200 to 6000 Hz is aerodynamic in origin as the carbon tips *shear* the air. Interestingly, the amplitude of the shearing noise depends to the first order on the size of the tips or the *set* of the blades. That is, the larger the tips or the greater the set, the more noise at a given speed. As a matter of further interest, during the cutting cycle, the resonant discrete tones often vanish but are replaced by discrete tones at the sawtooth-to-workpiece impact frequency and integer harmonics thereof.

From these examples, it is easy to see that broadband noise is a principal contributor to both community and industrial noise and must be recognized and dealt with on a routine basis. Because of its random character, geometrically related octave or one-third octave band analyses are usually the best measurement approaches and presentation formats.

3.4 IMPULSE-IMPACT NOISE

Impulsive or impact noise is best characterized as a transient acoustical event of short duration, usually less than 0.5 s. Examples include a hand clap, gunshot, stamping machine (repeated), etc. The impulsive character can be further broken down into two types, type A and type B. Type A is illustrated in Fig. 3.8a and is best described as a rapid rise in sound pressure followed by uniform decay

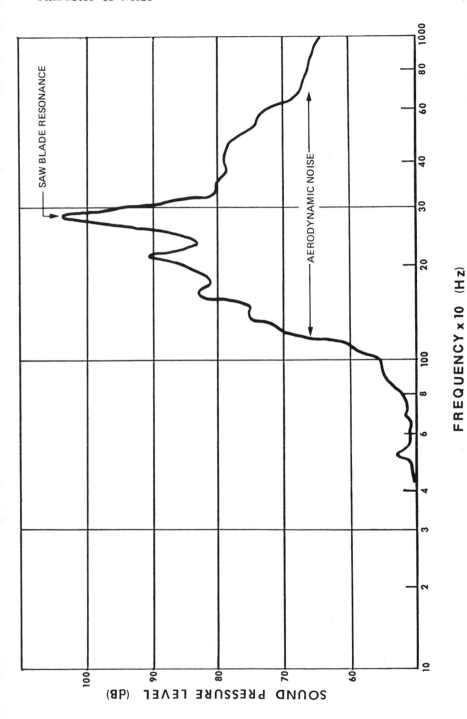

Figure 3.7 Narrow-band spectral analysis of noise from a cutoff saw at idle (not cutting).

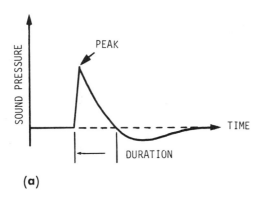

(a)

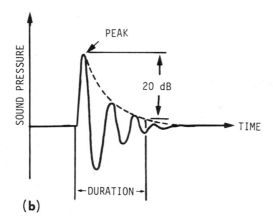

(b)

Figure 3.8 Two principal types of impulse sounds.

to a negligible amplitude. Type B impulsive noise also possesses a rapid rise in sound pressure, but the decay is oscillatory in nature, as illustrated in Fig. 3.8b.

The parameters common to both types A and B which are used to characterize impulsive noise, and also illustrated in Fig. 3.8, are

1. *Peak:* the maximum sound pressure amplitude reached in the event.
2. *Rise time:* the time from the start of the impulse to when the sound pressure reaches peak value.
3. *Duration:* usually the time from the start of the impulse to a specified decay level. In the case of type A, the duration is essentially the time for the peak sound level to decay to the initial level or 40 dB *down*. For type B, the duration is usually

taken as the time for the envelope of the oscillation to decay
20 dB down.

It should be noted that the parameters describing impulse noise
lack some technical sophistication and are in some ways arbitrary.
However, if one imagines that a simple impulse is a periodic function
with a large or, in the limiting case, infinite period, then one might
expect that the pulse also has a spectral character. This is indeed
the case, and it can be shown that an impulse has a unique continuous spectrum, much like random noise. The shape or density of
the spectrum is determined to the first order by the time history of
the impulse, or, qualitatively, it is dependent on the parameters just
presented, i.e., the rise time, peak, and duration. For example,
shown in Fig. 3.9 is the octave band spectral analysis of a shotgun
report and a power press punch impact. Note that the shotgun re-

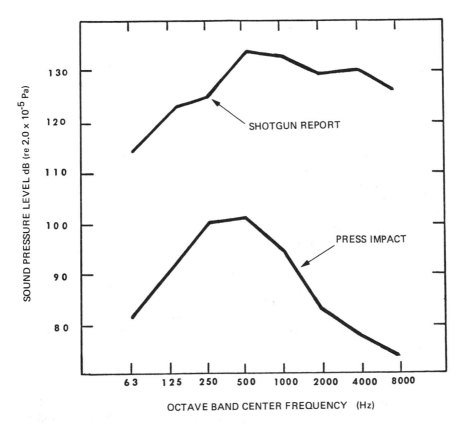

Figure 3.9 Octave band spectral analysis of punch press die impact
and shotgun report.

port contains considerably more high-frequency energy than the
press impact. Now the shotgun report has a very fast rise time
and short duration. It is these qualities that usually produce a spectrum rich in high-frequency energy. The press impact, on the other
hand, is dominantly low frequency in spectral character and, compared
to a shotgun report, is more like a dull thud.

In summary, impulsive or impact noise can be described in terms
of a well-defined spectral character. As such, the acoustical engineer can again characterize impulsive noise in terms of the same
basic parameters as the discrete or broadband noise, that is, sound
pressure level and frequency. Octave band analysis is usually the
preferred presentation format.

3.5 FREQUENCY WEIGHTING SCALES

The acoustical engineer often must alter or *weight* the spectral character of noise for comparison to subjective, psychoacoustic, or regulatory criteria. Over the years, dozens of frequency weighting
scales have been developed for a variety of reasons. Many are obsolete, and most have undergone what appears to be a never-ending
process of revision, modification, updating, etc. As such, we shall
focus here only on those weighting scales which have to date "stood
the test of time" and are currently applicable to industrial and community noise.

By definition, a weighted frequency scale is simply a table of corrections which is applied to sound pressure levels on a power basis
as a function of frequency. Shown in Table 3.1 are the corrections
for the A and C weighting scales at each of the one-third octave center
frequencies over the range of 20 to 20,000 Hz. Note that the A scale
corrections increase sharply below 1000 Hz, while the C scale is relatively "flat" over the range of most interest in noise control, i.e.,
50 to 10,000 Hz.

Statistically, the A scale follows closely the frequency response
of the human ear to sound; that is, the ear is less sensitive to frequencies below 1000 Hz. In addition, recent studies show that the
rate of hearing loss tends to follow the A scale in that one can tolerate
higher levels of low-frequency noise for a longer period without hearing impairment. It is for these reasons, primarily, that the A scale
plays such a prominent role in noise control.

The C scale, with its relative flatness, is useful since it provides
a baseline reference for comparison. It should also be noted that
basic sound-measuring instruments have built-in electronic networks
which can be easily selected to apply these weighting scales. We shall
deal with this aspect in detail in a later chapter.

Table 3.1 Weighting Corrections
for the A and C Scales

	Weighting Scales	
Frequency (Hz)	A Scale	C Scale
20	−50.5	−6.2
25	−44.7	−4.4
31.5	−39.4	−3.0
40	−34.6	−2.0
50	−30.2	−1.3
63	−26.2	−0.8
80	−22.5	−0.5
100	−19.1	−0.3
125	−16.1	−0.2
160	−13.4	−0.1
200	−10.9	0
250	−8.6	0
315	−6.6	0
400	−4.8	0
500	−3.2	0
630	−1.9	0
800	−0.8	0
1000	0	0
1250	0.6	0
1600	1.0	−0.1
2000	1.2	−0.2
2500	1.3	−0.3
3250	1.2	−0.5
4000	1.0	−0.8
5000	0.5	−1.3
6300	−0.1	−2.0
8000	−1.1	−3.0
10,000	−2.5	−4.4
12,500	−4.3	−6.2

An example of spectral weighting will clarify this concept.

Example
The octave band sound pressure levels of a centrifugal fan measured at 3 m is shown in line 1 in Table 3.2. Apply the A and C
weighting corrections to the measured values, and then calculate
the overall A and C weighted noise levels.
Solution
Referring to line 2 of Table 3.2, one sees that the weighting corrections for the C scale are −0.8 dB at 63 Hz, −0.2 dB at 125 Hz,
−0.2 dB at 2000 Hz, −0.8 dB at 4000 Hz, and −3.0 dB at 8000
Hz. To find the C weighted equivalents, these corrections are
added algebraically (i.e., noting the sign + or −) to the measured
values illustrated in line 3. The overall sound level equivalent
for the C scale is then obtained by combining or adding logarithmically each C weighted value. Specifically, taking the rounded
C weighted octave band levels, the overall C weighted level is
then calculated as shown in Table 3.3. The overall C weighted
noise level is then 105 dB at 3 m.

For the A weighted equivalent levels, the corrections are more
significant. From Table 3.1, the A weighting corrections are
−26.2 dB at 63 Hz, −16.1 dB at 125 Hz, −8.6 dB at 250 Hz, −3.2
dB at 500 Hz, 0 dB at 1000 Hz, +1.2 dB at 2000 Hz, +1.0 dB at
4000 Hz, and −1.1 dB at 8000 Hz. These corrections are shown
in line 2 of Table 3.4. Again adding the corrections algebraically
to the measured values at the corresponding center frequencies,
we obtain the A weighted equivalent levels shown in line 3 of
Table 3.4.

After rounding the A weighted octave band values, the important overall A weighted sound pressure level is then obtained by
combining each weighted level logarithmically as shown in Table
3.5. The overall A weighted sound level is then 101 dBA at 3 m.

Comparing the results of the calculations in the example, one notes
that the overall A scale level is 4 dB lower than the overall C scale
level. This relationship is typical of most industrial and community
environments, and unless the noise is exclusively in the range of
1000 to 4000 Hz, the overall A scale levels will always be less than
those of the C scale.

Note that the symbol A was added to the abbreviation of decibel,
i.e., dBA, to denote A weighting. Further, the symbol L_A or L_{pA}
symbolically denotes A weighted sound level. However, since the C
scale corrections are usually negligible over the frequency range of
interest, the letter C is not conventionally added to the decibel abbreviation to denote C weighting. Therefore, with little risk of confusion, dB thereafter will refer to dBC. The reader should not infer
that the A and C scales are the only weighting scales. On the con-

Table 3.2 Computation Table

	Center Frequency (Hz)							
	63	125	250	500	1000	2000	4000	8000
1. L_p at 3 m (dB)	91	95	101	100	96	90	84	75
2. C weighting corrections (dB)	-0.8	-0.2	0	0	0	-0.2	-0.8	-3.0
3. C weighted equivalent (dB)	90.2	94.8	101	100	96	89.8	83.2	72

Table 3.3 Overall C Weighted Sound Pressure Level

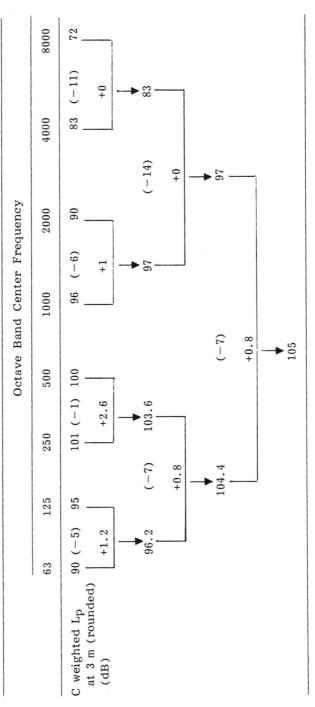

Table 3.4 Computation Table

	63	125	250	500	1000	2000	4000	8000
1. L_p at 3 m (dB)	91	95	101	100	96	90	84	75
2. A weighting corrections (dB)	−26.2	−16.1	−8.2	−3.2	0	+1.2	+1.0	−1.1
3. A weighted equivalent (dB)	64.8	78.9	92.4	96.8	96	91.2	85	73.9

Table 3.5 Overall A Weighted Sound Pressure Level

Octave Band Center Frequency

	63	125	250	500	1000	2000	4000	8000
A weighted L_p at 3 m (rounded) dB	65 (−14)	79	92 (−5)	97	96 (−5)	91	85 (−11)	74

+0 → 79

(insignificant)

+1.2 → 98.2

+1.2 → 97.2

+0 → 85

(insignificant)

(−1) +2.6 → 101

trary, and especially in the area of community noise, as will be shown later, Listed in App. D are a reasonably complete list of descriptors currently being used, and we shall deal with most in later chapters.

It must be emphasized again that for precise calculations, a direct application of Eq. (2.19), utilizing a programmable computer, is recommended. Shown in Fig. 3.10 is an example of a spreadsheet which calculates the overall A weighted sound level from octave band levels. In this instance, the A scale corrections from Table 3.1 are subtracted from each octave band level and the logarithmic sum, or overall A weighted level, is computed from Eq. (2.19). Shown also in Fig. 3.10 are the cell formulae for column J.

	A	B	C	D	E	F	G	H	I	J
1					Octave Band Center Frequency (Hz)					
2	Source	63	125	250	500	1000	2000	4000	8000	A-Weight
3										
4	Vacuum Cleaner 3'	48	66	69	73	79	73	73	72	82
5	Large Truck at 50'	83	85	83	85	81	76	72	65	86
6	Passenger Car at 50'	72	70	67	66	67	66	59	54	72
7	Window Air Conditioner	64	64	65	56	53	48	44	37	60
8										
9										
10										

Where the equation for cell J4 = 10*@LOG(10^((B4-26.2)/10)+10^((C4-16.1)/10)+10^((D4-8.6)/10)+10^((E4-3.2)/10) +10^(F4/10)+10^((G4+1.2)/10)+10^((H4+1.0)/10)+10^((I4-1.1)/10))

Where the equation for cell J5 = 10*@LOG(10^((B5-26.2)/10)+10^((C5-16.1)/10)+10^((D5-8.6)/10)+10^((E5-3.2)/10) +10^(F5/10)+10^((G5+1.2)/10)+10^((H5+1.0)/10)+10^((I5-1.1)/10))

Where the equation for cell J# = 10*@LOG(10^((B#-26.2)/10)+10^((C#-16.1)/10)+10^((D#-8.6)/10)+10^((E#-3.2)/10) +10^(F#/10)+10^((G#+1.2)/10)+10^((H#+1.0)/10)+10^((I#-1.1)/10))

Figure 3.10 Example of a spreadsheet format for calculating overall A weighted sound levels from linear octave band levels. Also shown are the cell formulae for column J.

To see the effect of the A weighting process, compare the overall A weighted levels in Fig. 3.10 with the overall levels shown in
Fig. 2.9 (Chapter 2).

3.6 LOUDNESS

One other characteristic of sound that needs to be mentioned is the
subjective attribute loudness. Because loudness is highly subjective,
i.e., ranges from loud to soft and varies dynamically from person
to person, it does not lend itself to direct measurement. However,
over the years statistical methods have been developed which predict rather reliably the response of human beings to sound in terms
of loudness.

Since loudness plays a rather small role in noise control, we shall
not spend a great deal of time with the subject. We shall, however,
present a calculation method for estimating loudness which has been
accepted by both national and international standard committees. The
method presented was developed primarily by Stevens and is based on
relatively easy-to-measure one octave, one-third octave, and one-
half octave band sound pressure levels.

The step-by-step calculation procedure is as follows [1,2]: First,
for each of the band sound levels, determine the loudness index I.
The loudness index is obtained from the equal loudness index contours, as presented in Fig. 3.11. Next, the total loudness S_t is calculated from

$$S_t = I_m (1 - K) + K \sum_{i=1}^{n} I_i \qquad [sones] \qquad (3.2)$$

where

  I_m = maximum loudness index

  I_i = each of the loudness indices, including I_m

  K = weighting factor for the bands chosen: K = 0.3 for one
      octave bands, K = 0.2 for one-half octave bands, K = 0.15
      for one-third octave bands

Note that the basic unit of loudness is the sone, which is also defined
as the loudness of a pure tone at 1000 Hz with a sound pressure level
of 40 dB. Consider an example of a loudness calculation.

Example
  Presented in line 1 of Table 3.6 are the measured octave band sound
  levels at the operator's station 1 m from a noisy ventilating hood.
  Calculate the loudness at the operator's station.

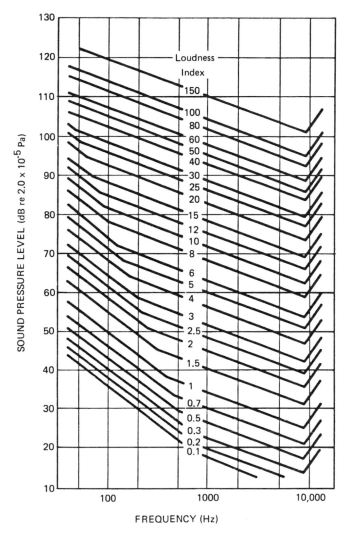

Figure 3.11 Equal loudness index curves.

Solution
See Table 3.6. Substitution in Eq. (3.2) yields

$$S_t = 18(1 - 0.3) + 0.3 \; \Sigma \; I_i$$

$$= 12.6 + 0.3 \times 73.1$$

$$= 12.6 + 21.9$$

$$= 34.5 \text{ sones}$$

Table 3.6 Computation Table

	Octave Band Center Frequency (Hz)							
	63	125	250	500	1000	2000	4000	8000
1. L_p at 1 m (dB)	56	62	65	70	81	75	73	62
2. Loudness index (from Fig. 3.11) sones	1.2	2.7	4.7	7.5	18	15	15	9

3. ΣI_i = sum of line 2 = 73.1 sones

4. I_m = maximum loudness index = 18 sones

5. For octave bands, K = 0.3

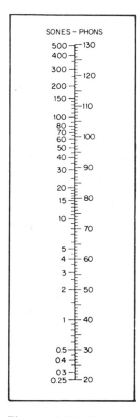

Figure 3.12 Nomogram showing the relationship between loudness (sones) and loudness level (phons).

The total loudness is then 34.5 sones at a distance of 1 m from the hood.

It should be emphasized that the method presented here is for calculating loudness of steady, rather constant in level, complex sound. As such, if the noise levels are fluctuating dynamically or are impulsive in character, the method is inappropriate. There are other more applicable methods which will be considered later.

3.7 LOUDNESS LEVEL

Since loudness is related to hearing, it follows that perhaps a log-arithmic scale would better relate to a hearing response, and indeed

Table 3.7 Conversion of Loudness Level in Phons to Loudness in Sones[a]

					Sones				
Phons	0	+1	+2	+3	+4	+5	+6	+7	+8	+9
40	1	1.07	1.15	1.23	1.32	1.41	1.51	1.62	1.74	1.87
50	2	2.14	2.30	2.46	2.64	2.83	3.03	3.25	3.48	3.73
60	4	4.29	4.59	4.92	5.28	5.66	6.06	6.50	6.96	7.46
70	8	8.57	9.20	9.85	10.6	11.3	12.1	13.0	13.9	14.9
80	16	17.1	18.4	19.7	21.1	22.6	24.3	26.0	27.9	29.9
90	32	34.3	36.8	39.4	42.2	45.3	48.5	52.0	55.7	59.7
100	64	68.6	73.5	78.8	84.4	90.5	97.0	104	111	119
110	128	137	147	158	169	181	194	208	223	239
120	256	274	294	315	338	362	388	416	446	478
130	512	549	588	630	676	724	776	832	891	955

[a] To find the number of sones corresponding to 76 phons, find the intercept of 70 phons and +6. The correct value is 12.1 sones.

this is the case. The loudness level L_L in phons is defined in terms of the total loudness S_t as follows:

$$S_t = 2^{(L_L - 40)/10} \qquad \text{[sones]} \qquad (3.3)$$

where L_L is the loudness level (phons). Hence, it can be shown (see the Exercises) that the loudness level L_L in terms of loudness is

$$L_L = 33.2 \log_{10} S_t + 40 \qquad \text{[phons*]} \qquad (3.4)$$

For ease of calculation a nomogram for converting sones to phons or the converse is given in Fig. 3.12 [4].

It is interesting to note that the equations relating loudness and loudness level were developed empirically from observations that a majority of listeners would judge a doubling of loudness to correspond to an increase of about 10 phons in loudness level—hence, the geometric factor 2 in Eq. (3.3). Consider an example.

---

*Because the argument of a logarithm must be dimensionless, S_t is actually the ratio S_t/S_{40}, where $S_{40} = 1$ sone.

Example

In the previous example, we found that the total loudness 1 m from the ventilating hood was 34.5 sones. What is the corresponding loudness level?

Solution

Utilizing Eq. (3.4), we get

$$L_L = 33.2 \log_{10}(34.5) + 40$$

$$= 51.2 + 40$$

$$= 91.1 \text{ phons}$$

Alternatively, utilizing Table 3.7, we see also that 34.5 sones is correspondingly 91.1 phons.

As mentioned before, the concept of loudness rarely enters the realm of industrial noise control because of the subjective nature. However, loudness is basic in evaluating human response to noise, and, as such, research continues in this area. In particular, refinements to the method presented here have been incorporated into both the ANSI S3.4 and ISO R-532 Standards [1,2]. These refinements address the presence of discrete tones, irregular spectra, diffuse sound fields, and so forth. For further study, the reader is referred to the References and Bibliography at the end of this chapter.

REFERENCES

1. Procedure for the Computation of Loudness of Noise. *American National Standards Institute USAS S3.4*—1980 (R 1986). American National Institute, New York.
2. Method for Calculating Loudness Level. *ISO Recommendation R-532*, 1975.
3. J. D. Irwin and E. R. Graf. *Industrial Noise and Vibration Control*. Prentice-Hall, Englewood Cliffs, New Jersey, 1979.
4. *Acoustic Handbook*. Hewlett-Packard Company, Palo Alto, California, 1968.

BIBLIOGRAPHY

Bell, L. H. *Fundamentals of Industrial Noise Control*. Harmony Publications, Trumbull, Connecticut, 1973.

Beranek, L. L. *Noise and Vibration Control, Revised Edition*. INCE, Washington, D.C., 1988.

Diehl, *Machinery Acoustics*. Wiley, New York, 1975.

Faulkner, L. *Handbook of Industrial Noise Control*. Industrial Press, New York, 1975.

Harris, C. M. *Handbook of Acoustic Measurements and Noise Control*. McGraw-Hill, New York, 1991.

EXERCISES

3.1 A single-stage axial compressor has 20 blades. What are the frequencies of the fundamental and second harmonics if the compressor is operating at 6000 rpm?

Answer: f_1 = 2000 Hz;
f_2 = 4000 Hz

3.2 A pinion gear has 36 teeth. What is the gear meshing frequency if the gear is driven at
a.   1740 rpm?
b.   3450 rpm?

Answer: (a) 1044 Hz;
(b) 2070 Hz

3.3 A saw blade has 64 teeth and turns at 3400 rpm. What are the sawtooth impact frequency and next two higher harmonics?

Answer: f_1 = 3627 Hz;
f_2 = 7253 Hz;
f_3 = 10,880 Hz

3.4 The sprocket that drives a timing belt has 12 teeth and runs at 1740 rpm. What is the fundamental sprocket to timing belt impact frequency?

Answer: 348 Hz

3.5 Calculate the fundamental piston firing frequency for an in-line eight-cylinder, four-cycle diesel engine driving a generator at 3600 rpm?

Answer: 240 Hz

3.6 A four-cylinder hydraulic pump on a plastic molding machine is operated at 3450 rpm. Calculate the fundamental and three higher harmonics of the *compression* frequency.

Answer: f_1 = 230 Hz;
f_2 = 460 Hz;
f_3 = 690 Hz;
f_4 = 920 Hz

3.7 The octave band levels measured 500 ft from an electrical power plant (at the boundary line) were as follows: 63 Hz, 74 dB; 125 Hz, 63 dB; 250 Hz, 50 dB; 500 Hz, 48 dB; 1000 Hz, 46 dB; 2000 Hz, 40 dB; 4000 Hz, 35 dB; and 8000 Hz, 30 dB. What are the overall A and C weighted noise levels?

Answer: 53 dBA;
            74 dB

3.8 Octave band noise levels as follows were measured in an office adjacent to a machine shop. Calculate
a. The loudness
b. The loudness level in the office

Center Frequency (Hz)	Band Pressure Level (dB)
63	66
125	63
250	65
500	70
1000	73
2000	76
4000	81
8000	79

Answer: (a) 47 sones;
            (b) 96 phons

3.9   Show that if the sound level in an area increases by 10 dB, the loudness is doubled.

3.10   Given Eq. (3.3), derive Eq. (3.4). [Hint: Take the logarithm of both sides of Eq. (3.3).]

3.11   The sum of loudness indices for a one-third octave analysis was 105, with a maximum $I_m = 35$. Calculate
a. The loudness
b. The loudness level

Answer: (a) 46 sones;
            (b) 95 phons

4
SOUND PROPAGATION

One of the most frequently asked questions in noise control is, Given a sound source at point A, what is the sound level at point B? To answer this question and others of a related nature, one must have an understanding of the basic properties of sound propagation. The basic term used here should not infer simplicity. To be sure, the radiation characteristics of most sound sources are generally complex, and the myriads of reflecting surfaces, especially in the industrial and community environments, further add to the complexity of the sound field. In this chapter we shall, however, develop analytical methods for predicting the propagation characteristics of noise both in outdoor and indoor environments. In addition, methods to account for the influence of reflecting surfaces, partial barriers, and meteorological and topological conditions will be presented.

4.1 POINT SOURCE

The most basic sound source is called the point, simple, or monopole source. We shall use the term point source hereafter, since it will be consistent with geometrical descriptions used for other source types. In its most elementary form, the point source is often likened to a pulsating sphere. Here the rapid pulsation produces a displacement of molecules and a corresponding dynamic pressure fluctuation. Since the wave fronts generated with each pulsation are always in phase, the resultant wave motion diverges uniformly in a spherical manner. Now, as we saw previously for uniform spherical divergence, the sound intensity I at a distance r is given by

$$I = \frac{W}{4\pi r^2} \qquad [W/m^2] \tag{4.1}$$

where

$\quad$ W = acoustical power of the radiating source (W)

$\quad$ $4\pi r^2$ = surface area of sphere of radius r (m^2)

In addition, it can be shown that for freely propagating plane waves, the intensity is related to the rms sound pressure as follows:

$$I = \frac{p^2}{\rho c} \quad [W/m^2] \tag{4.2}$$

where

$\quad$ p = rms sound pressure (Pa)

$\quad$ ρc = characteristic impedance of medium (mks rayls)

At radial distances greater than a single wavelength, spherically diverging waves from a point source are nearly plane. Thus we can combine Eqs. (4.1) and (4.2) to obtain the relationship between sound pressure and sound power.

$$\frac{p^2}{\rho c} = \frac{W}{4\pi r^2} \tag{4.3}$$

With a little algebra (see the Exercises), the more useful relationship between sound pressure level L_p and sound power level L_W is obtained:

$$L_p = L_W - 20 \log_{10}(r) - 11 \quad [dB] \tag{4.4}$$

where

$\quad$ L_W = sound power level of the point source (re 10^{-12} W)

$\quad$ r = radial distance from source (m)

The constant term was rounded to the nearest decibel.

$\quad$ Equation (4.4) is one of the most important equations in acoustics for it brings together the concepts of sound power level and sound pressure level. Consider an example of its application.

Example

$\quad$ A small source whose sound power level L_W is 110 dB is hanging freely outdoors. What is the sound pressure level at 20 m from the source?

Solution

By utilizing Eq. (4.4), the sound pressure level is given by

$$L_p = 110 - 20 \log(20) - 11$$

$$= 110 - 26 - 11$$

$$= 73 \text{ dB}$$

The sound pressure level is then 73 dB at 20 m.

Equation (4.4) can be modified for common usage in British units by simply noting that 1 m = 3.28 ft (Appendix B):

$$L_p = L_W - 20 \log_{10}(r \times 3.28) - 11$$

$$= L_W - 20 \log_{10}(r') - 0.5 \text{ dB} \qquad (4.5)$$

where r' is the radial distance (ft).

Now it must be emphasized that in deriving Eqs. (4.4) and (4.5) it was assumed that the radiation was uniform, nondirectional, and freely propagating plane waves. Most sources do not radiate uniformly, and therefore care must be taken in applying Eq. (4.4). We shall develop methods to deal with the directivity later in this chapter. However, at large radii (large compared to a wavelength and the dimensions of the source), spherically diverging waves are nearly plane, and Eqs. (4.4) and (4.5) can be used with negligible error. This region of applicability is commonly called the *far field* and, closer to the source, the *near field*.

Equations (4.4) and (4.5) can be put into another useful form as follows: Let $L_{p,1}$ and $L_{p,2}$ be the sound pressure levels at radial distance r_1 and r_2, respectively, as shown in Fig. 4.1. Now from Eq. (4.4), the sound pressure levels at r_1 and r_2, respectively, are

$$L_{p,1} = L_W - 20 \log(r_1) - 11 \text{ dB}$$

$$L_{p,2} = L_W - 20 \log(r_2) - 11 \text{ dB}$$

By subtracting the two equations and noting that the acoustical power level L_W for the source is the same, we get

$$L_{p,1} - L_{p,2} = 20 \log(r_2) - 20 \log(r_1)$$

and, from the property of logarithms,

$$L_{p,1} - L_{p,2} = 20 \log\left(\frac{r_2}{r_1}\right)$$

or, rewriting,

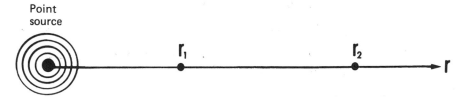

Point source

Figure 4.1 Radiating point source.

$$L_{p,2} = L_{p,1} - 20 \log\left(\frac{r_2}{r_1}\right) \quad [dB] \tag{4.6}$$

In this form, the dependence of the hard-to-measure sound power level is eliminated. Equation (4.6) then yields the sound pressure level $L_{p,2}$ at r_2 in terms of sound pressure level $L_{p,1}$ at r_1 and the ratio of the radial distances. In short, if we measure the sound pressure level $L_{p,1}$ at r_1, we can calculate the sound pressure level $L_{p,2}$ at any distance r_2 on the same radial line. Consider the following example.

Example

The measured sound pressure level at 5 ft from a radiating sound source is 98 dB. What is the sound pressure level at 20 ft along the same radial line?

Solution

From Eq. (4.6),

$$L_p \text{ (at 20 ft)} = L_p \text{ (at 5 ft)} - 20 \log_{10}\left(\frac{20}{5}\right)$$

$$= 98 - 12$$

$$= 86 \text{ dB}$$

It is interesting to note that if we take the special case of $r_2/r_1 = 2$ and rewrite Eq. (4.6) in terms of the difference between the sound pressure levels, we get

$$L_{p,2} - L_{p,1} = 20 \log_{10}(2) \cong 6 \text{ dB}$$

Since $r_2/r_1 = 2$ amounts to doubling the distance from the source, we have the origin of the often quoted rule "6 dB for doubling the distance." It is worth noting, however, that for sufficiently large distances most sources can be considered a point source and spherical radiation assumed for first-order results. As mentioned earlier, most sound sources encountered in industrial and community environments are not nondirectional. For example, shown in Fig. 4.2 is a plane polar plot of the radiation pattern of a small axial fan showing acute directional radiation characteristics. Therefore, to account for the directionality, a term DI_θ must be added to the basic propagation equation (4.4) which adds considerably to its generality:

$$L_p = L_W + DI_\theta - 20 \log_{10}(r) - 11 \quad [dB] \tag{4.7}$$

where DI_θ is called the directivity index. More specifically, the directivity index is usually defined as follows:

$$DI_\theta = L_{p,\theta} - L_{p,re} \quad [dB] \tag{4.8}$$

where

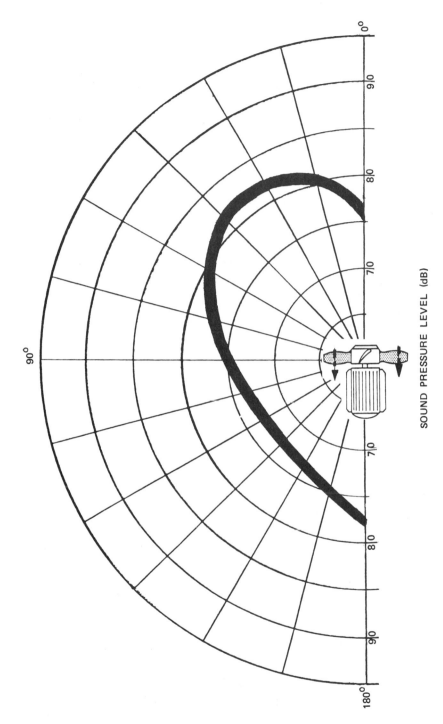

SOUND PRESSURE LEVEL (dB)

Figure 4.2 Typical radiation pattern of a small axial fan.

101

$L_{p,\theta}$ = sound pressure level measured at distance r and angle θ from a source of sound power W radiating into free space

$L_{p,re}$ = sound pressure level measured at distance r from a non-directive point source of power W radiating into free space

In short, the directivity index is just the difference between the sound pressure level one measures at radius r, angle θ, and the level at radius r of a nondirective reference point source of the same power. It should be noted that the directivity index is algebraic, that is, positive or negative. Consider the following example.

Example

A source of sound power level 100 dB radiates into free space. The sound pressure level at 10 m and angle 30° is 75 dB. What is the directivity index for 30°?

Solution

Here $L_{p,30°}$ = 75 dB, and, from Eq. (4.4), assuming uniform radiation, we have

$$L_{p,re} = 100 - 20 \log_{10}(10) - 11$$

$$= 110 - 20 - 11$$

$$= 79 \text{ dB}$$

From Eq. (4.8) the directivity index is then

$$DI_{30°} = 75 - 79 = -4 \text{ dB}$$

If the sound pressure level at 30° and radius 10 m had been 85 dB, the index would be

$$DI_{30°} = 85 - 79 = +6 \text{ dB}$$

If the radiation patterns are not extremely directional, say less than ±6 dB, a simplified, more useful form of Eq. (4.7) can be written to obtain a good first-order approximation of the sound field:

$$L_p = L_W - 20 \log_{10}(r) + (L_{p,\theta} - \overline{L}_p) - 11 \qquad [\text{dB}] \qquad (4.9)$$

where

$L_{p,\theta}$ = measured sound pressure level at a radial distance r and angle θ

$\overline{L}_p$ = average of the sound pressure levels measured at 12 or more equidistant points circumferentially around the source

To illustrate these concepts, consider the following example.

Example

   The acoustical power level of an axial fan is 112 dB. The directionality factor at an angle of 40° off the centerline, obtained
by 12 circumferential measurements, is +4 dB. What is the sound
pressure level at a distance of 10 m at the same angle?

Solution

From Eq. (4.9)

$$L_p = 112 - 20 \log_{10}(10) + 4 - 11$$

$$= 112 - 20 + 4 - 11$$

$$= 85 \text{ dB}$$

   Again, it should be emphasized that in Eq. (4.9) the directionality term $(L_{p,\theta} - \overline{L_p})$ can be either positive or negative.

   It should also be noted here that much more sophisticated measurement procedures and standards exist to obtain the sound power
level. These procedures involve the use of large rooms of either
anechoic or reverberant types. In the case of anechoic rooms, the
walls, floors, and ceilings are heavily treated with wedge-shaped
acoustical absorbing material. As such, the rooms are considered
acoustically *dead* since there are little or no sound reflections. These
rooms can be as large as a gymnasium or the size of a breadbox, as
illustrated in Fig. 4.3, depending on the space required for the source.

   In the case of reverberation rooms, they are always large, and
the walls, floors, and ceilings are highly reflective, usually of heavy
masonry construction. The resultant sound field in the reverberation
room is very *diffuse;* that is, the sound is a complex combination of
reflected sound. As such, the sound level in the far field is, in the
ideal case, uniform in level and nondirectional. To obtain this diffuse
sound field, great care must be taken to avoid parallel walls which
could, by direct reflection, produce standing waves. In many reverberation rooms, large rotating *wing*like diffuser panels are also used
to further minimize the buildup of standing wave fields. Shown in
Fig. 4.4 is a reverberation room used for acoustical testing. Note
the size and bare character of the interior.

   In addition to elaborate facilities as just described, dozens of
sound level measurements on the faces of 8- to 20-sided polyhedrons
surrounding the source are required. Further, some standards or
procedures insist that the microphones orbit and/or rotate to improve
the statistical properties of the measurements. It is because of these
involved measurement requirements that reliable sound power data
are difficult to obtain or unavailable for most noise sources. Notable
exceptions are found in the heating, refrigeration, ventilating, air-
conditioning, electrical motor, and gas product industries. Here the
trouble and expense are usually taken to provide the users with
octave band acoustical power levels for most product lines. For ex-

Figure 4.3 Large and small anechoic chambers for sound power determination. (Courtesy of Eckel Industries Inc., Cambridge, Mass.)

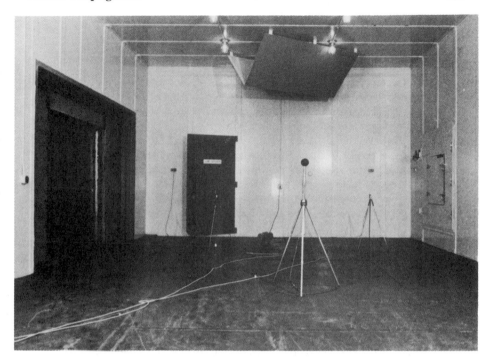

Figure 4.4 Reverberation room for acoustical measurements and test-ing. (Courtesy of Industrial Acoustics Company, Inc., Bronx, New York.)

ample, most fan manufacturers will provide octave band sound power level data for each fan model at representative operating and per-formance conditions.

We shall close the subject of the measurement of sound power at this time and summarize the discussion with the following comments. With a few exceptions, reliable sound power data do not exist for most sources of noise found in industry. However, with 10 or 12 measure-ments taken on an envelope around the periphery of a machine, say at 1 m from major machine surfaces, the directivity index can be ap-proximated to the first order. Thus Eq. (4.9) can then be used to estimate propagation characteristics and the corresponding noise impact of the equipment when installed in a similar environment. In fact, some industrial associations such as the National Machine Tool Builders have written standards or procedures for in situ measure-ments, as just described, and they have been enthusiastically ac-cepted by their customers. A partial list of these standards cur-rently in use is found in Appendix C.

Another factor which can strongly influence the directivity of a radiating point source and must be taken into account in applying Eqs. (4.4) and (4.5) is the presence of reflecting surfaces. For example, if a point sound source of acoustical power W were placed on a hard reflecting surface, say in the center of a parking lot, twice as much energy than predicted by Eqs. (4.4) and (4.5) would be radiated in any given direction. That is, the sound field would contain sound energy reflected from the asphalt and also the sound energy radiated directly to an observer. As such, on a hypothetical hemisphere above the reflecting plnae, the sound intensity would be doubled, and to an observer it would appear that the acoustical power of the source has also doubled.

Therefore, to account for the presence of reflecting surfaces, a directivity factor generally symbolized as Q is defined in terms of the directivity index as follows:

$$DI = 10 \log_{10}(Q) \qquad [dB] \tag{4.10}$$

or

$$Q = \text{antilog } \frac{DI}{10} \qquad [\text{unitless}] \tag{4.11}$$

Illustrated in Fig. 4.5 are the values for the directivity factor Q and corresponding directivity indices DI for some common sound source locations. An example will clarify these concepts.

Example

The overall sound power level of an air conditioner is 110 dB. What is the sound pressure level at 2 m if the air conditioner is installed (1) in a wall, (2) in a wall near the floor, and (3) at the base of the wall near the corner of a room?

Solution

For the midwall installation, we have hemispherical radiation (Fig. 4.5b), Q = 2, and from Eq. (4.10) the DI is

$$DI = 10 \log_{10}(2)$$

$$\cong 3 \text{ dB}$$

From Eq. (4.7) the sound pressure level then at 2 m is,

$$L_p = 110 + 3 - 20 \log_{10}(2) - 11$$

$$= 102 - 20 \log_{10}(2)$$

$$= 102 - 6$$

$$= 96 \text{ dB}$$

With the air conditioner near the floor, i.e., the junction of two planes (Fig. 4.5c), Q = 4, and the DI is

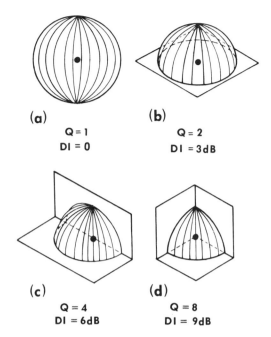

(a)
$Q = 1$
$DI = 0$

(b)
$Q = 2$
$DI = 3dB$

(c)
$Q = 4$
$DI = 6dB$

(d)
$Q = 8$
$DI = 9dB$

Figure 4.5 Examples of reflecting planes which strongly influence the radiation patterns of sound sources.

$$DI = 10 \log_{10}(4)$$
$$= 6 \text{ dB}$$

Again from Eq. (4.7), the sound pressure level at 2 m is

$$L_p = 110 + 6 - 20 \log_{10}(2) - 11$$
$$= 105 - 6$$
$$= 99 \text{ dB}$$

Note that the placement near the floor increased the noise level by 3 dB, i.e., 96 to 99 dB. From sound power considerations this is what one should expect, since the intensity has doubled.

With respect to locating the air conditioner in a corner, the directivity factor would be $Q = 8$, as illustrated in Fig. 4.5d. Hence, the $DI = 9$ dB, and the resultant level at 2 m is

$$L_p = 110 + 9 - 20 \log_{10}(2) - 11$$
$$= 108 - 6$$
$$= 102 \text{ dB}$$

Note again that the sound pressure level increased 3 dB over the floor installation and 6 dB over the midwall.

From this discussion it is easy to see that the propagation characteristics of a point source can be strongly influenced by reflecting elements common in both industrial and community environments. However, if the directivity of the source and the location relative to reflecting surfaces are accounted for, a useful and rather accurate approximation of the propagation properties can be obtained. Let us consider some other types of sources and their propagating characteristics.

4.2 LINE SOURCE

Another type of source common in industry is the line source. Shown in Fig. 4.6 is an illustration of a radiating line source and associated corresponding cylindrical wave front patterns. Two common examples of line sources are a busy highway and the noise from a long pipe filled with high-velocity steam. To determine the propagation characteristics of a line source, refer again to Fig. 4.6 and consider the sound intensity at a distance r on a cylindrical surface. Here the intensity I on the surface is

$$I = \frac{W}{2\pi rh} \qquad [W/m^2] \tag{4.12}$$

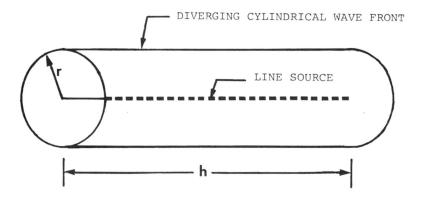

Figure 4.6 Diverging cylindrical wave front from line source.

where

W = acoustical power of the source (W)

$2\pi rh$ = surface area of a cylinder of radius r and length h (m^2)

Comparing the intensity of a line source, Eq. (4.12), to the intensity of a point source, Eq. (4.1), one notes that for the line source the inverse radial divergence is *linear*, whereas for a point source the divergence depends on the *square* of the radial distance r. In short, the radial divergence rate of a line source is less than the rate for a point source. In the previous section we saw that the divergence rate for a point source was 6 dB per doubling the distance. For a line source, the sound pressure level drops only 3 dB per doubling the radial distance or one-half the rate of a point source.

A more general and useful expression for the sound pressure level divergence of a line source can be derived in a manner similar to the development of Eq. (4.6) and is

$$L_{p,2} = L_{p,1} - 10 \log_{10} \left(\frac{r_2}{r_1}\right) \qquad (r_1 < r_2) \qquad [\text{dB}] \qquad (4.13)$$

where

$L_{p,2}$ = sound pressure level at radial distance r_2 (dB)

$L_{p,1}$ = sound pressure level at radial distance r_1 (dB)

An example will illustrate this concept.

Example

The noise level at 10 m from a long pipe carrying high-velocity steam was 95 dBA. What is the noise level at 100 m?

Solution

From Eq. (4.13),

$$L_{p,2} = 95 - 10 \log_{10} \left(\frac{100}{10}\right)$$

$$= 95 - 10$$

$$= 85 \text{ dBA}$$

Here again, given an easy-to-measure sound pressure level at a given radial distance, the sound level at any other distance can be calculated. It should be noted also that in the example the overall noise level was A weighted. This can usually be done with negligible error provided that there are no influences in the medium which would alter the propagation spectrally, i.e., with respect to frequency. We shall see examples that do selectively alter the propagation when we consider propagation outdoors.

4.3 SOURCES ON A LINE

Another common acoustic source type which has interesting propagating
sound characteristics is the distributed line source. The distributed
line source is an array of point sources equally spaced on a line of
nearly equal sound power, as illustrated in Fig. 4.7. The array
clearly approximates a row of machines such as looms or weaving
machines often found in the textile industry. Other examples include
rows of automatic bar, milling, or boring machines, as frequently installed in the machine shop, and rows of *eyelet* machines (transfer
presses), as found in the metal forming industry. Now, Rathé [1]
has shown that for radial distances $r < b/\pi$, where b is the spacing
between sources, sound propagates from the array like a point
source, and the sound pressure level decays 6 dB every time the
distance is doubled. However, for values $r > b/\pi$, the propagation
resembles a line source with a decay rate of 3 dB per doubling the
distance. The character of the radiation divergence is illustrated in
Fig. 4.8. Intuitively, one should expect this form of divergence
when one considers that in the near field or close to any one source
the contribution of adjacent sources is small, and we would expect
point source spherical divergence. However, as one moves away
along the normal, the contribution of adjacent sources is more and
more significant, since the radial distances to the observer approach
each other in magnitude.

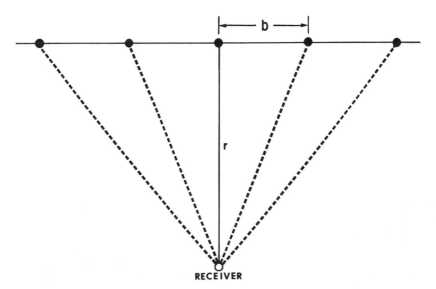

Figure 4.7 Array of point sources on a straight line.

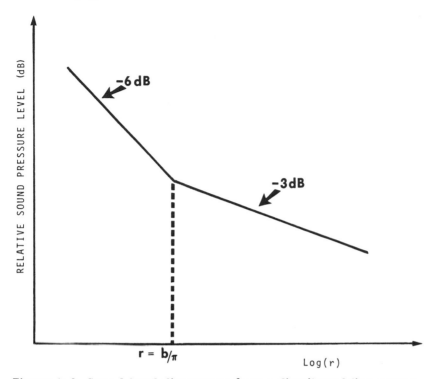

Figure 4.8 Sound level divergence from a distributed line source.

To compute the sound pressure level, then, at a given distance r_2 from an array of point sources of equal power, the following equations apply over the ranges as given:

$$L_{p,2} = L_{p,1} - 20 \log_{10}\left(\frac{r_2}{r_1}\right) \quad \left(r_1 \text{ and } r_2 < \frac{b}{\pi}\right) \quad [\text{dB}] \quad (4.14)$$

$$L_{p,2} = L_{p,1} - 10 \log_{10}\left(\frac{r_2}{r_1}\right) \quad \left(r_1 \text{ and } r_2 > \frac{b}{\pi}\right) \quad [\text{dB}] \quad (4.15)$$

It should also be noted that the sound from each source must be incoherent. That is, unrelated, otherwise, upon superposition we would expect to find standing wave interference patterns, with corresponding peaks and dwells in the sound field. Consider an example.

Example

The noise level measured 10 ft from a row of looms spaced 8 ft apart was 105 dBA. Estimate the noise level at an inspection station located (1) 20 ft away and (2) 40 ft away.

Solution

With a row of looms we have a distributed line source, and the transition from point to line source divergence occurs at

$$r_0 = \frac{b}{\pi}$$

where b is the nominal spacing between the sources. In this case, b = 8, and

$$r_0 = \frac{8}{\pi}$$

$$= 2.55 \text{ ft}$$

Now the sound level measurements were made 10 ft from the looms, and since 10 ft > 2.55 ft, we are in the region where the propagation behaves like a line source, and Eq. (4.15) is applicable. Therefore,

$$L_p(\text{at } 20 \text{ ft}) = L_p(\text{at } 10 \text{ ft}) - 10 \log_{10} \left(\frac{20}{10}\right)$$

$$= 105 - 3$$

$$= 102 \text{ dBA}$$

The A weighted sound level, then, at the inspection station 20 ft away is 102 dBA.

For the inspection station at 40 ft,

$$L_p(\text{at } 40 \text{ ft}) = L_p(\text{at } 10 \text{ ft}) - 10 \log_{10} \left(\frac{40}{10}\right)$$

$$= 105 - 6$$

$$= 99 \text{ dBA}$$

or 3 dB less than at 20 ft, as would be expected.

4.4 FINITE PLANE SOURCE

Another source with interesting propagation characteristics is the finite plane source, as illustrated in Fig. 4.9. Rathé [1] has also calculated the divergence properties of a finite plane source with respect to a normal radial distance, and the relative response is shown graphically in Fig. 4.10. Note that as one moves from the plane there is no noise reduction until the radial distance $r = b/\pi$ is reached, where b is the width of the radiating plane. At this point and further, the sound pressure level diverges at the same rate as a line source, i.e., −3 dB per doubling the distance. However, at radial distance

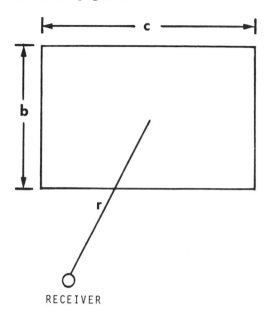

Figure 4.9 Finite plane source.

$r = c/\pi$, where c is the length of the radiating plane, the divergence rate again changes, resembling a point source, i.e., −6 dB per doubling the distance. In short, as one moves away from a radiating plane source, nothing happens for a while, and then suddenly the plane behaves like a line source and finally a point source. The plausibility of the final transition follows intuitively, since if you are far enough from any source of finite dimension, the source would appear small compared to the radial distance and behave to the first order like a point source.

Equations for governing the divergence of sound from a radiating finite plane are presented below; however, greater attention must be given to the regions of applicability as defined by the inequalities for radial distances r_1 and r_2:

1. Region I: $0 < r_1$ and $r_2 < b/\pi$:

    $$L_p = \text{constant level} \quad [\text{dB}] \quad\quad\quad (4.16)$$

where

   b = width of radiating plane surface (m)

   r_1, r_2 = radial distances ($r_1 < r_2$)

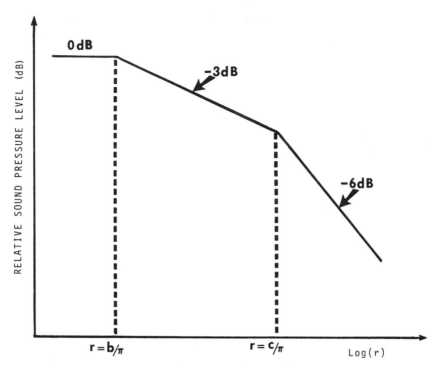

Figure 4.10 Sound level divergence from a finite plane source.

2. Region II: $b/\pi < r_1$ and $r_2 < c/\pi$:

$$L_{p,2} = L_{p,1} - 10 \log_{10}\left(\frac{r_2}{r_1}\right) \qquad [\text{dB}] \qquad (4.17)$$

where

$L_{p,2}$ = sound pressure level at r_2 (dB)

$L_{p,1}$ = sound pressure level at r_1 (dB)

b,c = width and length, respectively, of radiating plane (b < c) (m)

3. Region III: $c/\pi < r_1$ and r_2:

$$L_{p,2} = L_{p,1} - 20 \log_{10}\left(\frac{r_2}{r_1}\right) \qquad [\text{dB}] \qquad (4.18)$$

where

$L_{p,2}$ = sound pressure level at r_2 (dB)

$L_{p,1}$ = sound pressure level at r_1 (dB)

Consider an example.

Example

Describe the diverging sound field perpendicular to a large open delivery door whose dimensions are 30 × 10 ft. Noise levels in the door opening are 95 dB.

Solution

The open door is a good example of a radiating finite plane surface, and from the dimensions given, let b = 10 ft and c = 30 ft. Now there will be no noise level reduction in region I where the level is a constant of 95 dB up to

$$r = \frac{b}{\pi} = \frac{10}{\pi} = 3.18 \text{ ft}$$

The noise levels at this point now diverge at a rate of −3 dBA per doubling the distance to

$$r = \frac{c}{\pi} = \frac{30}{\pi} = 9.55 \text{ ft}$$

By applying Eq. (4.17), the sound level L_p at 9.55 ft given a sound level of 95 dB at 3.18 ft, is

$$L_p = 95 - 10 \log_{10} \left(\frac{9.55}{3.18} \right)$$

$$= 95 - 4.78$$

$$= 90.2 \text{ dB}$$

From this second transition point (9.55 ft), the noise levels diverge as a simple source with −6 dB per doubling the distance, and Eq. (4.18) applies.

In summary, the noise is constant at 95 dB from the door opening up to about 3 ft. At this the first transition point, the noise level decays at a rate of −3 dB per doubling the distance to about 90 dB at 9.5 ft. At about 9.5 ft from the door, and thereafter, the noise level decays at a rate of −6 dB per doubling the distance.

From this example it is clear that the first step to characterizing the propagation of a finite plane source is to determine the transition points and corresponding regions of constant divergence rate. With these regions determined, reliable first-order estimates of sound pressure levels at any radial distance can be calculated. Other examples of finite plane sources common in industry are open windows, thin walls, machine panels, etc.

A note of caution: Ellis [2] has shown that serious calculation errors arise if the radial divergence is taken near the edge of the plane or at angles off to the side. Therefore, the use of the equations given should be restricted to radial divergence near the center of the plane.

In final summary, we have considered here only the most fundamental types of sound sources and their basic propagation characteristics. There are, however, more complex source types, and in industrial environments there is often a combination of source types present which confound the rather straightforward analytical methods presented here. In either case, with a little cleverness, such as turning machines off or using a temporary isolation barrier and giving special attention to the spectral character of the noise, the acoustical engineer can often reduce complex situations into more simple, easy-to-analyze individual cases.

Let us now consider other influences that may alter the propagation of sound from the idealized cases considered thus far.

4.5 PARTIAL BARRIERS

Partial barriers, as illustrated in Fig. 4.11, occur frequently in both indoor and outdoor environments. As shown in Chap. 1, sound waves propagating over a partial barrier are diffracted, producing a *shadow zone* of generally diminished sound amplitude. From the standpoint of sound propagation, it is this excess attenuation that is of interest to the acoustical engineer as a noise reduction measure. We shall therefore develop analytical methods to estimate the attenuation for partial barriers.

The problem of semi-infinite (long compared to height) barrier attenuation was studied extensively by Maekawa in terms of source-to-barrier and receiver-to-barrier path length distances. Referring again to Fig. 4.11, the attenuation A_b due to the barrier at the receiver is given by

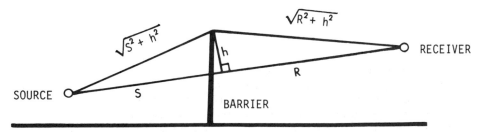

Figure 4.11 Partial barrier and relevent parameters to determine attenuation.

$$A_b = 20 \log_{10} \left(\frac{\sqrt{2\pi N}}{\tanh \sqrt{2\pi N}} \right) + 5 \qquad (N > 0) \qquad [dB] \qquad (4.19)$$

where tanh is the hyperbolic tangent and

N = Fresnel number

$$= \frac{2\delta}{\lambda} \qquad (4.20)$$

Now the factor δ is the difference between the diffracted path and the direct path, and λ is the wavelength of the sound approaching the barrier. In terms of the barrier geometry we have

$$\delta = \sqrt{S^2 + h^2} - S + \sqrt{R^2 + h^2} - R \qquad [m] \qquad (4.21)$$

where

S = distance from the source to the barrier along the line of sight (m)

R = distance of the receiver from the barrier along the line of sight (m)

h = effective barrier height (projected height of the barrier above the line of sight) (m)

For ease of calculation, Eq. (4.19) is shown graphically in Fig. 4.12 over the range of most interest in noise control. Consider now a practical example.

Example

An 8-ft barrier is to be installed 3 ft from a punch press. Calculate the attenuation due to the barrier at an inspector's station 10 ft away for the center frequencies of the octave bands.

Solution

Assuming the die area is a point source and is at the same level as the inspector's ear (see Fig. 4.13), from Eq. (4.21) the path length difference δ is

$$\delta = \sqrt{10^2 + 3^2} - 10 + \sqrt{3^2 + 3^2} - 3$$

$$= 1.7 \text{ ft}$$

From Eq. (4.20) the Fresnel numbers N for each of the octave band center frequencies N_{cf} are

$$N_{cf} = \frac{2 \times 1.7}{\lambda_{cf}}$$

where λ_{cf} is the wavelength for each octave band center frequency (i.e., 63, 125, 250, etc.) (ft). The calculated wavelengths are shown in line 1 of Table 4.1. In line 2 are the corresponding

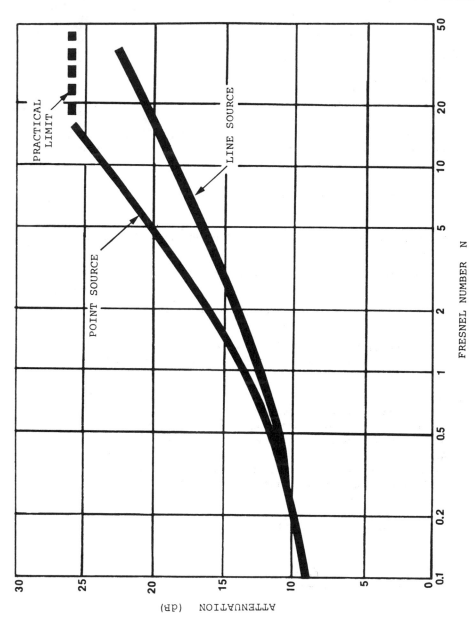

Figure 4.12 Attenuation due to a partial barrier for point and line sources.

Table 4.1 Computation Table

	Octave Band Center Frequencies (Hz)							
	63	125	250	500	1000	2000	4000	8000
1. λ_{cf} (ft) c = 1100 ft/s	17.5	8.7	4.4	2.2	1.1	0.55	0.27	0.14
2. N_{cf} = Fresnel number	0.2	0.4	0.8	1.5	3.1	6.2	12.4	24.7
3. Attenuation A_b at inspection station (dB)	8	10	12	15	18	21	24	27*

*Exceeds practical limit.

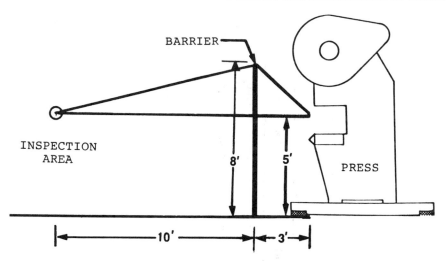

Figure 4.13 Press and inspection area and partial barrier.

Fresnel numbers, and in line 3 is the required attenuation A_b at the inspection station from Fig. 4.12.

In summary, the spectral attenuation of the barrier ranges from 8 to 27 dB. Note that as a noise reduction measure the acoustical performance is much better at higher frequencies, as described qualitatively in Chap. 1.

It must be emphasized that the attenuation calculated from Eq. (4.19) is for a point source with a semi-infinite barrier and the receiver is in the shadow zone. In practice, the performance of the barriers is usually somewhat less. This is especially true indoors where the barriers are generally finite and the noise sources are not point sources. In addition, the source noise will usually be reflected from the ceiling, walls, or equipment adjacent to the receiver. These effects generally reduce barrier effectiveness but can be accounted for and will be discussed in a later chapter when considering room acoustics.

We can now add the influence of a partial barrier to Eq. (4.7) to obtain an even more general expression for the sound field of a point source:

$$L_p = L_W + DI - 20 \log_{10}(r) + A_b - 11 \qquad [dB] \qquad (4.22)$$

where A_b is the attenuation due to a partial barrier (dB). Maekawa, Kurze, Anderson, etc. (see the Bibliography) have extended their analyses to include the attenuation of a barrier for a radiating incoherent line source. Results show that the excess barrier attenuation for a line source is 1 to 5 dB less than for a point source. This difference, as a function of Fresnel numbers, can be seen in Fig. 4.12.

Note that the attenuation is essentially the same at small Fresnel numbers (low frequencies) but that the line source attenuation is about 5 dB lower for Fresnel numbers above 10. Consider an example.

Example

>   Using the barrier configuration of the previous example and considering the die area of the press a line source, calculate the attenuation at 1000 and 8000 Hz.
>
>   *Solution*
>
>   The Fresnel numbers for 1000 and 8000 Hz are, from Table 4.1, 3.1 and 24.7, respectively. The attenuation for a line source from Fig. 4.12 is, then, approximately 15 dB for N = 3.1 or 1000 Hz and 21 dB for N = 24.7 or 8000 Hz.

Comparing these results to the attenuation for a point source, we see that the attenuation for a line source is 2 to 5 dB lower over the given frequency range, as mentioned. Here again, for a line source, the attenuation of a partial barrier can be considered excess attenuation and included as an additional logarithmic term in calculating the sound pressure level at a given radial distance.

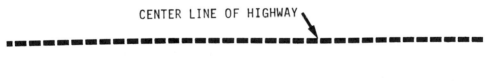

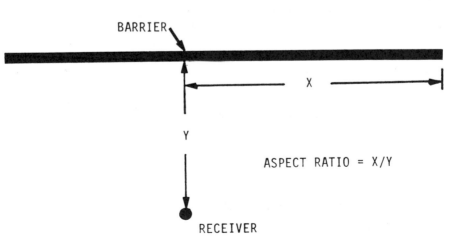

Figure 4.14 Aspect ratio for partial barriers.

It must be emphasized that the analysis just presented requires the barrier to be semi-infinite, that is, very long compared to the height. Intuitively, if the partial barrier, as commonly seen along major highways, is of finite length, some noise will diffract, or "flank," around the ends. As such, some compromise in acoustical performance can be expected

As a design guideline, it can be shown [5] that if the aspect ratio of the barrier is greater than 5, the performance degradation is negligible. (See Fig. 4.14, illustrating the aspect ratio of partial barriers.)

4.6 PROPAGATION OF SOUND OUTDOORS

The propagation of sound outdoors, at distances of several hundred feet or more, is often significantly affected by atmospheric, meteorological, and ground cover conditions. We shall, in this section, account for each of these effects, which in some instances are cumulative.

4.6.1 Atmospheric Absorption

In the case of the atmosphere, sound energy is gradually converted into heat by a variety of molecular processes called atmospheric absorption. This absorption process is dependent to the first order on air temperature and relative humidity (RH). Here again, as with partial barriers, the reduction of the sound pressure level due to atmospheric absorption can be considered excess attenuation with respect to divergence. There have been numerous attempts to develop analytical methods to describe this phenomenon, but the best approach to estimating the attenuation comes from using generally empirical data. In particular, the excess attenuation due to the atmosphere A_a is given empirically to the first order [3].

$$A_a = \frac{\alpha d}{100} \quad [dB] \tag{4.23}$$

where

α = atmospheric attenuation coefficient (dB/100 m)

d = distance between source and receiver (m)

Now, as mentioned, the attenuation depends primarily on the temperature and relative humidity, and shown in Table 4.2 are the attenuation coefficients for a wide range of both temperature and humidity. Note

Table 4.2 Atmosphere Attenuation Coefficient (dB/100 m) at
One Standard Sea-level Atmosphere

Temperature	Relative Humidity (%)	Frequency (Hz)					
		125	250	500	1000	2000	4000
30°C	20	0.06	0.18	0.37	0.64	1.4	4.4
(86°F)	50	0.03	0.10	0.33	0.75	1.3	2.5
	90	0.02	0.06	0.24	0.70	1.5	2.6
20°C	20	0.07	0.15	0.27	0.62	1.9	6.7
(68°F)	50	0.04	0.12	0.28	0.50	1.0	2.8
	90	0.02	0.08	0.26	0.56	0.99	2.1
10°C	20	0.06	0.11	0.29	0.94	·3.2	9.0
(50°F)	50	0.04	0.11	0.20	0.41	1.2	4.2
	90	0.03	'0.10	0.21	0.38	0.81	2.5
0°C	20	0.05	0.15	0.50	1.6	3.7	5.7
(32°F)	50	0.04	0.08	0.19	0.60	2.1	6.7
	90	0.03	0.08	0.15	0.36	1.1	4.1

Source: Ref. 6.

that the attenuation rises sharply with increasing frequency for a
given meteorological condition. The excess attenuation, as given by
Eq. (4.23), can also be included as an additional term in the general
propagation equation for the sources discussed in previous sections.
Consider an example.

Example
  The noise levels at 50 m from the inlet of a large forced draft
  fan are 110 dB in the 500-Hz band and 90 dB in the 4000-Hz
  band. What is the noise level in each band at a distance of 500
  m on a nice summer evening with the temperature at 20°C and
  the relative humidity 50%?
  Solution
  Assuming that the fan inlet is a point source, from Eq. (4.6)
  the noise levels at 500 m due to divergence are

$$L_p(\text{at } 500 \text{ m}) = L_p(\text{at } 50 \text{ m}) - 20 \log \frac{500}{50}$$

$$= L_p(\text{at } 50 \text{ m}) - 20 \text{ dB}$$

For the 500-Hz band, L_p at 50 m = 110, and

$$L_p(\text{at } 500 \text{ m}) = 110 - 20$$

$$= 90 \text{ dB}$$

For the 4000-Hz band, L_p at 50 m = 90, and

$$L_p(\text{at } 500 \text{ m}) = 90 - 20$$

$$= 70 \text{ dB}$$

From Table 4.2, the attenuation coefficient is 0.28 for a temperature of 20°C, RH = 50%, and a frequency of 500 Hz. Thus upon substitution in Eq. (4.23), we get

$$A_a = \frac{0.28 \times 450}{100} = 1.3 \text{ dB}$$

where

$$d = 500 - 50$$

$$= 450 \text{ m}$$

The noise level L_p at 500 m for the 500-Hz band, considering both divergence and atmospheric attenuation, is then

$$L_p = 90 - 1.3 = 88.7 \text{ dB}$$

Considering now the 4000-Hz band, from Table 4.2, $\alpha = 2.8$, and, upon substitution, the atmospheric attenuation is

$$A_a = \frac{2.8 \times 450}{100}$$

$$= 13 \text{ dB}$$

The noise level L_p at 500 m for the 4000-Hz band, considering both divergence and atmospheric attenuation, is then

$$L_p = 70 - 13 = 57 \text{ dB}$$

Upon comparison, one sees the strong influence of frequency; that is, for the 500-Hz band the attenuation is trivial, only 1.3 dB compared to the 13 dB at 4000 Hz.

Here again the excess attenuation due to absorption can be included as an algebraic term in the general expressions for sound propagation of both line and plane sources in the same manner as the partial barrier. In short, the attenuation due to the atmosphere is just subtracted from the resultant level due to divergence, DI, barriers, etc.

It should also be mentioned that the absorption rate depends on altitude, i.e., atmospheric pressure, but for most practical applications these corrections are negligible.

4.6.2 Ground Effects

Excess attenuation is also readily observed as sound propagates over large distances parallel to or in close proximity to the ground.

The magnitude of the attenuation is governed primarily by the type
of ground cover present, i.e., grass, shrubs, trees, etc. Numerous
studies and experimental measurement programs have produced some
basic guidelines for predicting the magnitude of the attenuation.
There is, however, some wide disagreement in the measured data despite well-qualified technical observers. As such, the data and
methods presented here are, in general, statistical condensations of
these results.

4.6.3 Sound Propagation Through Trees

With respect to trees, or, better, forests, the excess attenuation A_t
per 100 m for various tree types [4,7-9] is given from empirical
data as

$$A_t \cong f^{1/3} \qquad [dB/100\ m] \tag{4.24}$$

where f is the octave band center frequency. Equation (4.24) is
shown graphically in Fig. 4.15 along with measured results for
bare trees. Consider an example.

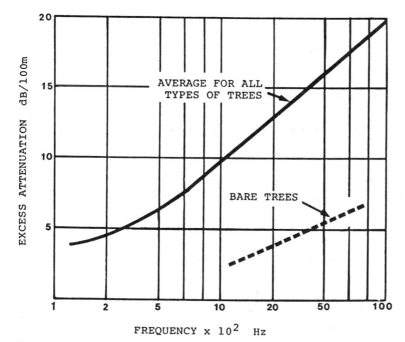

Figure 4.15 Excess attenuation A_t for sound propagation through
trees.

Example

Calculate the excess attenuation for a grove of deciduous trees 200 m deep along a highway at a frequency of 125, 1000, and 4000 Hz. What is the attenuation in the wintertime?

Solution

From Eq. (4.24), the attenuation at 125 Hz is

$$A_t = 125^{1/3}$$

$$= 5 \text{ dB}/100 \text{ m}$$

Therefore, the total attenuation for 200 m of a forest would be

$$5 \times 2 = 10 \text{ dB}$$

Similarly, at 1000 Hz,

$$A_t = 1000^{1/3}$$

$$= 10 \text{ dB}/100 \text{ m}$$

and the total attenuation for 200 m is

$$10 \times 2 = 20 \text{ dB}$$

For 4000 Hz,

$$A_t = 4000^{1/3}$$

$$= 15.9 \text{ dB}/100 \text{ m}$$

or for 200 m the total attenuation is

$$15.9 \times 2 = 31.8 \text{ dB}$$

During the winter, with the trees bare, the attenuation, from Fig. 4.15, is negligible at 125 Hz, about 6 dB at 1000 Hz, and about 10 dB at 4000 Hz.

From these results, one can conclude that a dense stand of trees with leaves will provide significant noise reduction. If, however, in the example we had used a foliage depth of 15 m (50 ft), the attenuation would have been negligible and, as such, would have only reduced the highway visibility.

4.6.4 Sound Propagation Through Shrubbery and Tall Grass

With respect to tall grass and shrubs, the average excess attenuation A_g is, again from empirical data [7],

$$A_g = 18 \log_{10}(f) - 31 \qquad [\text{dB}/100 \text{ m}] \qquad (4.25)$$

where f is the frequency of sound (Hz). Here again the attenuation through grass and shrubs is considerably more for higher frequency sound energy. Consider an example.

Example

What is the excess attenuation of sound through a dense cornfield 50 m deep at 125, 1000, and 4000 Hz?

Solution

From Eq. (4.25), the average excess attenuation rate for 125 Hz is

$$A_g = 18 \log_{10}(125) - 31$$

$$= 37.7 - 31$$

$$= 6.7 \text{ dB}/100 \text{ m}$$

For the cornfield 50 m deep, the total attenuation is

$$0.5 \times 6.7 = 3.4 \text{ dB}$$

Similarly, for 1000 Hz,

$$A_g = 18 \log_{10}(1000) - 31$$

$$= 54 - 31$$

$$= 23 \text{ dB}/100 \text{ m}$$

and for the 50-m field, the total attenuation is

$$0.5 \times 23 = 11.5 \text{ dB}$$

And finally for 4000 Hz,

$$A_g = 18 \log_{10}(4000) - 31$$

$$= 18 \times 3.6 - 31$$

$$= 64.8 - 31$$

$$= 33.8 \text{ dB}/100 \text{ m}$$

and for the 50-m field, the total attenuation is

$$0.5 \times 33.8 = 16.9 \text{ dB}$$

Again, from the example it is clear that sizable expanses of thick grass or shrub growth is required for significant noise reduction.

4.6.5 Wind and Meteorological Effects

Vertical differences in temperature and wind velocity are generally present in the lower boundary of the earth's atmosphere. With these zonal differences, changes in the speed of sound can be anticipated, and thus, as we saw in Chapter 1, so can the corresponding refractive effects. More simply, sound waves are refracted, or bent, as they propagate through thermal or wind velocity gradients in the atmosphere.

Consider first the effect of sound propagation across thermal gradients. Generally, there is a decrease in temperature with altitude. The decrease (positive lapse rate) is generally not constant. In fact, often at night or early morning hours, the lapse may turn negative and the phenomenon known as *inversion* occurs. In the resulting layer of warmer air, the speed of sound is higher than in the air below. Thus, as sound waves pass through the thermal gradient, they are refracted downward. Conversely, during the daytime hours, heat conduction from the earth's surface warms the air near the ground and sound waves propagating parallel to the ground are refracted upward. In either case, shadow or reinforcement zones of excess attenuation are produced. Measured differences of up to 20 dB between sunrise and sunset on the same day have been reported [10] over distances ranging from 200–700 meters. It should be noted that these shadow or reinforcement zones are generally symmetrical about the source.

With respect to the wind, generally wind velocity increases with altitude. Thus, as sound waves propagate downwind from a source, they are refracted toward the ground. Conversely, sound waves propagating upwind, say toward an observer, are refracted upward as they diverge from the source. Again, shadow or reinforcement zones are produced with corresponding excess attenuation. Measured variation of excess attenuation, both positive and negative, due to wind velocity gradients have been reported [11] in the range of 10–20 dB at distances of 110–615 meters, respectively. In addition, from the authors' own experience, measured overall level variations of 6–10 dBA have been frequently observed [12] at distances of 2100 or more feet from large petrochemical facilities, with strong correlation to wind direction.

Theoretical estimates [13, 14] confirm excess attenuation in sound level amplitude of this order or higher. Unfortunatley, a dilemma presents itself in using this background of information. Because of the obvious difficulty in measuring the ever-changing meteorological conditions, it is not likely that reliable estimates of the refractive effects can be predicted. Suffice it to say, however, that an awareness of these effects will provide the noise control engineer with at least plausible arguments to unsnarl measurement anomalies. This is especially true for measuring sound levels around petrochemical and electircal power plants, waste treatment plants, gas turbine installations, and so forth. In these situations, measured changes in sound level of 10–20 dB can be extremely confounding while one tries to assess the status of regulatory compliance or to evaluate the performance of installed noise reduction measures.

In final summary, the excess attenuation due to the atmosphere or ground effects is a first-order propagation influence only when distances are relatively large, i.e., greater than 100 m. This is often the case, and the effect must be accounted for in practical industrial situations.

REFERENCES

1. E. J. Rathé. Note on two common problems of sound propagation. *J. Sound Vib.* *10* (1969).
2. R. M. Ellis. The sound pressure of a uniform finite plane source. *J. Sound Vib.* *13*(4) (Dec. 1970).
3. C. M. Harris. *Handbook of Noise Control.* McGraw-Hill, New York, 1979.
4. F. J. Meister and W. Ruhrberg. The influence of green areas on the propagation of noise. *Lärmbekämpfung 3* (1959).
5. T. D. Northwood, J. D. Quirt, and R. E. Halliwell. Residential planning with respect to road and rail noise. *Noise Control Eng.* *13*(2) (Sept.–Oct. 1979).
6. Method for the Calculation of the Absorption of Sound by the Atmosphere. *Proposed American National Standards Institute* S1.26/ASA23, American National Standards Institute, New York, 1978.
7. L. L. Beranek. *Noise and Vibration Control.* McGraw-Hill, New York, 1971.
8. T. F. W. Embleton. Sound propagation in homogeneous deciduous and evergreen woods. *J. Acoust. Soc. Am.* *35* (1963).
9. F. M. Weiner. Sound propagation outdoors. *Noise Control 4* (July 1958).
10. S. Canard-Caruana, J. Vermorel, S. Lewy, and G. Parmentier. Long range sound propagation near the ground. *Noise Control Eng. J.* (May–June 1990).
11. P. H. Parkin, W. E. Scholes. The horizontal propagation of sound from a jet engine close to the ground, at Hatfield. *J. Sound Vibr.* *2*(4) (1965).
12. L. H. Bell. Unpublished data.
13. T. F. W. Embleton. Sound propagation outdoors—improved prediction schemes for the 80's. *Noise Control Eng. J.* (Jan.–Feb. 1982).
14. S. R. Solomon. Sound propagation in the atmosphere: Analysis of signal amplitude and phase temporal and spatial variations. Thesis, Pennsylvania State University Graduate School (May 1989).

BIBLIOGRAPHY

Bell, L. H. *Fundamentals of Industrial Noise Control.* Harmony
    Publications, Trumbull, Connecticut, 1973.
Daigle, G. A., J. E. Piercy, and T. F. W. Embleton. Line-of-sight
    propagation through atmospheric turbulence near the ground. *J.
    Acoust. Soc. Am.* *74*(5) (1983).
Eyring, C. F. Jungle acoustics. *J. Acoust. Soc. Am.* *18* (1946).
Faulkner, L. *Handbook of Industrial Noise Control.* Industrial
    Press, New York, 1975.
Harris, C. M. Absorption of sound in air in the audio-frequency
    range. *J. Acoust. Soc. Am.* *5* (1933).
Harris, C. M. Absorption of sound in air versus humidity and
    temperature. *J. Acoust. Soc. Am.* *40* (1966).
Hoover, R. M. Tree Zones as Barriers for the Control of Noise Due
    to Aircraft Operation. *Bolt, Beranek and Newman, Inc. Rept.
    844.* Bolt, Beranek and Newman, Inc., Feb 1961.
Ingard, U. A review of the influence of meteorological conditions on
    sound propagation. *J. Acoust. Soc. Am.* *25* (1953).
Ingard, U. On sound-transmission anomalies in the atmosphere.
    J. Acoust. Soc. Am. *25* (1953).
Kurze, U. J., and G. S. Anderson. Sound attenuation by barriers.
    Appl. Acoust. *4* (1971).
Maekawa, Z. Noise reduction by screens. *Mem. Fac. Eng. Kobe
    Univ.* *11* (1965).
Maekawa, Z. Noise reduction by screens of finite size. *Mem. Fac.
    Eng. Kobe Univ.* *12* (1966).
Nyborg, W. L., and D. Mintzer. Review of Sound Propagation in the
    Lower Atmosphere. *Wright Air Develop. Center, Ohio, Tech.
    Rept.* *54-602,* May 1955, ASTIA No. AD-67880.
Piercy, J. E., T. F. W. Embleton, and L. C. Sutherland. Review of
    noise propagation in the atmosphere. *J. Acoust Soc. Am.* *61*(6)
    (1977).
Piercy, J. E., T. F. W. Embleton, and G. A. Diagle. Atmospheric
    propagation of sound: Progress in understanding basic mechanisms.
    11th International Congress on Acoustics (Paris, 1983), Vol. 8,
    pp. 37–46.
Standard Values of Atmospheric Absorption as a Function of Tempera-
    ture and Humidity for Use in Evaluating Aircraft Flyover Noise.
    Aerospace Recommended Practice, ARP 866. Society of Automotive
    Engineers, New York, Aug. 31, 1964.
Weiner, F. M., and D. E. Keast. An experimental study of the propa-
    gation of sound over ground. *J. Acoust. Soc. Am.* *31* (1959).

EXERCISES

4.1 Given the relationship between sound pressure and sound power,

$$\frac{p^2}{\rho c} = \frac{W}{4 \pi r^2}$$

show that the relationship between sound pressure level and sound power level is

$$L_p = L_W - 20 \log(r) - 10.8 \qquad [dB]$$

(Hint: Take 10 times the logarithm of both sides, and use $\rho c = 415$ mks rayls.)

4.2 A siren has an overall acoustical power level of 140 dB. Assuming a point source and a directivity factor of Q = 1, calculate the sound pressure level L_p at
a. 1 mi
b. 2 mi   *Answer:* (a) 65 dB;
                                                           (b) 59 dB

4.3 The overall noise level measured 500 m from a jet engine test cell was 60 dBA. What is the noise level at 1500 m along the same radial line?

                                                Answer: 50.5 dBA

4.4 An electrical motor with an acoustical power level of 92 dB radiates into free space. The directivity index at 90° to the centerline is +5 dB and at 180° is −2 dB. Calculate the sound pressure level at a radial distance of 5 m.
a. 90° to the centerline
b. 180° to the centerline

                                                Answer: (a) 72 dB;
                                                           (b) 65 dB

4.5 The average sound pressure level for 12 sound level measurements taken at a radial distance of 10 m from a diesel engine was 98.3 dBA. At 10 m and 90° to the engine, the measured sound level was 103.5 dBA. What is the directivity index at 90°?

                                                Answer: 5.2 dB

4.6 The directivity index for the intersection of two planes is 6 dB. What is the directivity factor?

                                                Answer: Q = 4.0

4.7 The directivity factor is often defined more generally as

$$Q = \text{antilog } \frac{L_{p,\theta} - \overline{L}_p}{10}$$

If the average sound pressure $\overline{L}_p$ 10 ft from a source is 102 dBA and a measured sound level $L_{p,\theta}$ at $30°$ is 104 dB, what is the directivity factor for 30°?

Answer: Q = 1.58

4.8 The overall sound power level of a small air compressor is 103 dBA. If the compressor is installed in the corner of a mechanical room, what is the sound level at 7 m?

Answer: 84 dBA

4.9 Noise levels measured 200 m from a busy highway were 82 dBA. Assuming a line source radiation characteristic, what is the noise level at
a.   400 m?
b.   600 m?

Answer: (a) 79 dBA;
(b) 77 dBA

4.10 In a row, milling machines making identical parts are spaced 5 ft apart. The noise level measured 10 ft from the row was 96 dBA.
a. Calculate the distance to the transition point.
b. Estimate the noise levels 20 ft away at a maintenance bench.

Answer: (a) 1.6 ft;
(b) 93 dBA

4.11 A large window, 4 × 18 ft, on the first floor of a can manufacturing plant is left open in the summertime. Noise levels in the window are typically 92 dBA. Estimate the noise levels
a. On the sidewalk 4 ft from the window
b. At the nearest neighbor across the street, 30 ft from the window

Answer: (a) 87 dBA;
(b) 71 dBA

4.12 An 8-ft-high barrier is to be placed 6 ft from a small compressor located outdoors. What is the excess attenuation due to the barrier at the adjacent property line 60 ft away? Peak sound levels of the compressor are at 500 Hz, and the speed of sound is 1100 ft/s.

Answer: 19 dB

4.13 A barrier wall is to extend parallel along the *runs* of a dog
kennel. The barrier wall, 100 m long, will be 5 m high and will be installed 3 m from the run. Assuming a line source, i.e., numerous
dogs, estimate the attenuation due to the wall at a house 100 m away.
Peak noise levels of the barking dogs are at 1000 Hz.

Answer: 20 dB

4.14 Partial barriers 10 ft high are being installed parallel to a highway at 50 ft from the centerline. Estimate the attenuation due to the
barrier at a condominium complex 100 ft from the wall for tire noise in
the range of 2000 Hz.

Answer: 16 dB

4.15 A gas turbine generator is to be cited adjacent to a park. How
much attenuation will a stand of oak trees 100 m deep offer
a. In the summertime?
b. In the wintertime?
Dominant noise levels are in the 250-Hz octave band.

Answer: (a) 6 dB;
(b) nil

4.16 A jet engine test cell is to be installed near the corner of an
airport. A field of thick evergreen shrubs extends from the airport
property line 500 yd to the nearest neighbors. Calculate the excess
attenuation for the shrubbery for
a. The 125-Hz octave band
b. The 250-Hz octave band

Answer: (a) 30.8 dB:
(b) 55.6 dB

5
SOUND MEASUREMENT AND ANALYSIS

The measurement of sound and its characteristics plays an important role in the development of a systematic approach to noise control. In particular, the measurement of overall sound levels can be used to determine compliance with regulations or pertinent criteria. These measurements can also be used to assess the effectiveness of various control methods and to establish realistic goals. Although measuring sound level is an essential aspect of characterizing noise, other parameters associated with noise, such as duration and spectral content, are often equally as important. Today, sound measurement instrumentation embodies a wide range of complexity and sophistication. The need to understand the capabilities and limitations of this equipment is critical in order to properly interpret measurement results. The purpose of this chapter is to introduce the various types of available sound measuring and analysis tools, and to discuss the basic applications of these instruments.

5.1 SOUND LEVEL METER

The primary instrument for field sound measurement is the sound level meter (SLM). Figure 5.1 presents a range of commercially available portable sound level meters. The principal components of any sound level meter are illustrated in the block diagram of Fig. 5.2 and consist of the following:

1. *Microphone.* The microphone senses the sound pressure fluctuations and converts them to an analog electrical signal.
2. *Preamplifier.* The preamplifier is used for impedance matching and can sometimes provide a DC polarization voltage to the microphone.

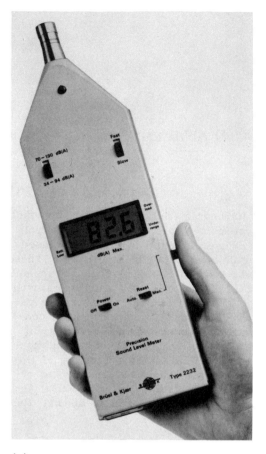

(a)

Figure 5.1 Commercially available sound level meters. [(a) Courtesy of Brüel and Kjær Instruments, Inc., Decatur, Ga.; (b) courtesy of Scantek Inc., Silver Spring, Md.; (c) courtesy of Quest Electronics, Oconomowoc, Wis.]

3. *Frequency Weighting Network.* This stage provides a system of networks, generally A, C, and linear, which are used to modify the frequency response charageristics of the measurement instrument. The selection of the appropriate frequency weighting characteristic is dependent upon the type of measurement being made.

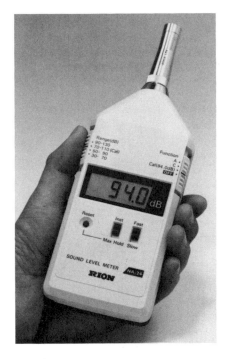

(b)

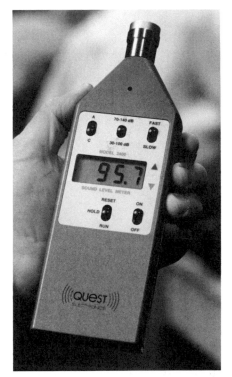

(c)

4. *Range Control Amplifier.* Most sound level meter detectors have a limited range of signal level for which they accurately operate. This amplifier is used to adjust the signal voltage to levels which are within this range.

5. *Detector.* This element is used to characterize the incoming signal amplitude. There are several types of detectors that are commonly used. They include RMS (root mean square), peak, and integrating.

6. *Display.* Once the amplitude of the signal has been detected, the display is used to indicate this level. Generally, SLM displays are scaled in decibels referenced to the international standard of 2×10^{-5} Pa.

7. *Outputs.* Sound level meters often provide signal outputs in order to obtain advanced analysis capabilities and graphical hard copy.

Each of these stages will be discussed in greater detail in the following sections.

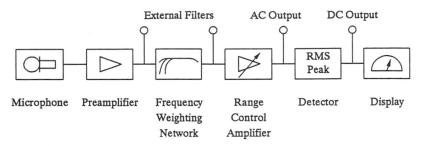

Figure 5.2   Block diagram showing the major components of a sound level meter.

5.1.1   Microphone

The most critical element of a sound level meter is the microphone, and it is here that the accuracy of the meter is usually determined. The purpose of the microphone is to convert the pressure fluctuations of sound waves into an electrical signal.   Ideally, these electrical signals will vary in a manner that is proportional to the measured pressure fluctuations.   Microphones that are of a sufficient quality to conform to the accuracy requirements of various international standards are called measurement microphones.   There are two basic types of measurement microphones: condenser and ceramic.

Condenser Microphones

The basic construction of a typical condenser microphone is presented in Fig. 5.3.   Functionally, an impinging sound pressure wave causes the thin metal diaphragm to deflect slightly and thus changes the capacitance between the diaphragm and the backplate. This change of capacitance is converted to an electrical signal by maintaining a constant polarizing voltage between the diaphragm and the backplate.   For a wide range of sound levels and frequencies, the generated electrical signal is proportional to sound pressure.   Condenser microphones have a long-term stability and are relatively insensitive to vibration or electromagnetic fields.   In addition, condenser microphones are relatively insensitive to the range of temperature or pressure excursions that one may encounter during field measurements.

The disadvantages of this type of microphone are its susceptibility to conditions of high humidity and its fragility.   With respect to humidity, electrical leakage results in excessive internal

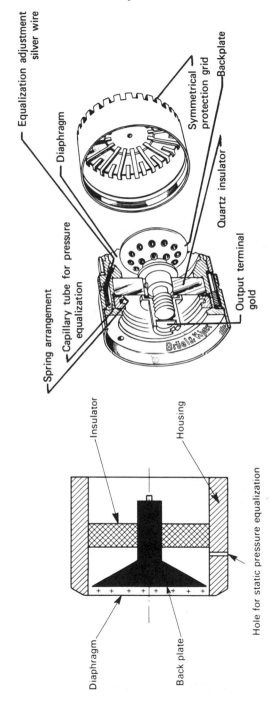

Figure 5.3 Schematic and cutaway views of a typical condenser microphone. (Courtesy of Brüel and Kjaer Instruments, Inc., Decatur, Ga.)

background noise and, under extreme conditions, failure. The delicate nature of the diaphragm and the tight manufacturing tolerances associated with construction require very careful handling.

The electret microphone is also a condenser microphone. However, this type of microphone requires no polarizing voltage supply. The major difference lies in the construction of the capacitor. Here a permanently polarized polymer film, called an electret, is sandwiched between the diaphragm and the backplate. A schematic illustration is presented in Fig. 5.4. Electret microphones are often preferred for field instrumentation due to their more rugged construction and their ability to operate without a polarizing voltage power supply.

Ceramic Microphones

Ceramic microphones utilize a piezoelectric crystal as a pressure-sensing element. In this type of microphone, the diaphragm is mechanically attached to the crystal, which in turn is attached to the backplate, as illustrated in Fig. 5.5. As the diaphragm deflects due to sound wave impingement, the crystal is strained or distorted. This distortion produces an electrical voltage (piezoelectric effect) across the crystal, which is proportional to the sound pressure. These ceramic microphones are more rugged than condenser microphones, are quite reliable, have

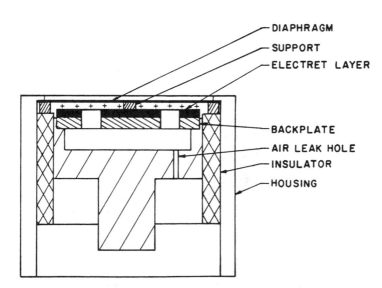

Figure 5.4 Schematic illustration of electret microphone. (Courtesy of GenRad, Concord, Mass.)

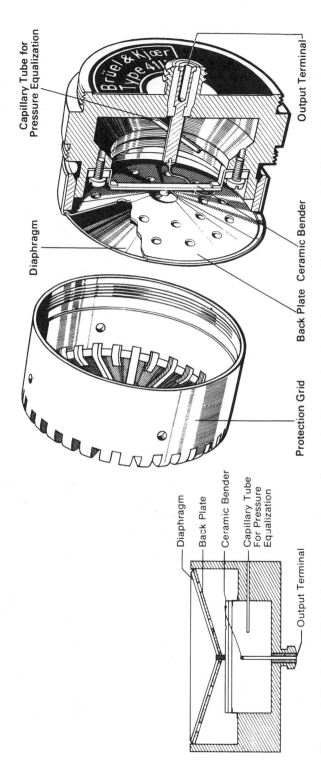

Figure 5.5 Schematic and cutaway views of a typical ceramic microphone. [Courtesy of Brüel and Kjaer Instruments, Inc., Decatur, Ga.]

141

excellent stability, and require no polarization voltage. They are, however, more susceptible to vibration than condenser microphones and are also more sensitive to temperature, particularly below 10°C. Finally, although the frequency response of ceramic microphones is good, they are generally considered inferior to condenser types.

Microphone Sensitivity

The two main parameters associated with microphone selection are sensitivity and frequency range. A microphone's sensitivity is the ratio of its electrical output to the sound pressure at the diaphragm of the microphone. Since a microphone output is usually measured in millivolts (mV) and sound pressure is measured in Pascals, the units of sensitivity are mV/Pa. Condenser microphone sensitivity is generally a function of diaphragm diameter. Typically, larger-diameter microphones have higher sensitivities and thus can usually measure lower level sounds. Microphone sensitivities usually range between 1 mV/Pa for 1/8" microphones, and 50 mV/Pa for 1" microphones.

Microphone Frequency Range

Ideally, one would like to have a uniform microphone sensitivity (flat frequency response) over a wide range of frequencies and microphone orientation. Considering these parameters, Fig. 5.6 [1] presents typical frequency response curves for condenser microphones at various angles of incidence. As you will note, the frequency response is relatively flat for each angle of incidence to about 1000 Hz. At higher frequencies, as the sensitivity begins to change, errors in sound level measurement can occur. For accurate sound measurements, microphones should be used only in the frequency range where their frequency response curves are flat. On the one hand, the accuracy of condenser microphones is not usually restricted by low frequency limits. On the other hand, the upper limiting frequency of a microphone can often provide measurement restrictions. As with sensitivity, a microphone's upper limiting frequency is directly related to its diameter. In this case, the smaller the microphone diameter, the higher the upper limiting frequency. A 1/8" microphone can extend to over 150 kHz, whereas a 1" microphone will often be limited to 10 kHz.

It is clear that, in selecting a microphone, a compromise between sensitivity and frequency range must be made. For most general purpose sound level measurements, a 1/2" microphone is adequate and can usually measure a level range between 20—140 dB and frequency range of 10—20 kHz. When selecting microphones for use outside of these ranges, care must be taken and consultation with the microphone manufacturer is recommended.

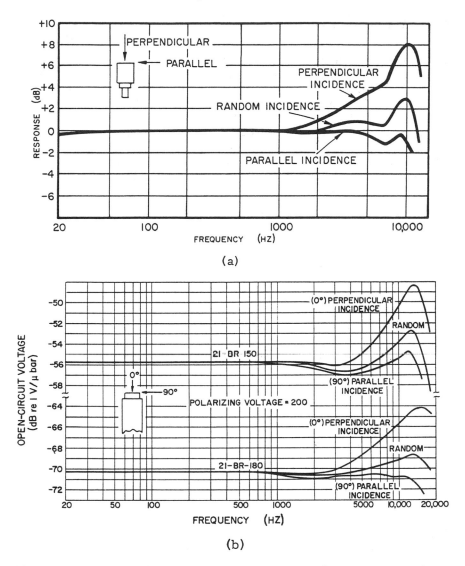

Figure 5.6 Typical frequency response curves for condenser microphones of various angles of incidence. Random incidence assumes
that the microphone is placed in a diffuse sound field. [Courtesy
of GenRad, Concord, Mass., from Ref. 1.]

5.1.2 Frequency Weighting Network

The frequency weighting network alters the frequency response of
the sound level meter. These frequency corrections are designed
to match the frequency weighting scales which have been discussed
in Section 3.5. Therefore, when selected, frequency band level
corrections need not be applied when measuring A- and C-weighted
values. However, when performing spectrum analysis or using a
sound level meter as a signal input for a tape recorder, it is recommended that *linear* weighting (no corrections) be selected. The
frequency response characteristics for the various weighting circuits are illustrated in Fig. 5.7.

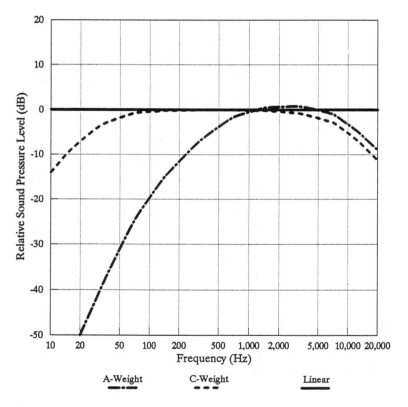

Figure 5.7 Frequency response characteristics for weighting networks of sound level meters.

5.1.3 Range Control Amplifier

Most sound level meter detectors can measure only signal levels that
vary over a range of 70 dB. Since the range of sound levels that
one might wish to measure spans from 10—140 dB (a 130 dB dynamic
range), a user-adjusted amplifier is employed. The range control
amplifier allows the user to select a maximum full-scale level. All
measurements that are below the full-scale sound level by less than
the dynamic range (typically 70 dB as noted for most detectors)
are valid. If measured levels fall outside of this range, then the
full-scale level can be adjusted appropriately. Proper measurement
procedure involves reducing the meter full-scale level until a signal
overload indication has occurred. The full-scale level should then
be increased one step to prevent overload. In this way the entire
detector dynamic range can be utilized.

5.1.4 Sound Level Meter Detectors

There are three basic signal detectors that are used in sound level
meters. These are: RMS (root mean square), peak, and integrating. Each type of detector has specific uses for sound measurements and will be discussed in the following sections.

RMS Detector

The RMS detector is the most common detector and is the one that
is used to measure sound pressure level (SPL). The RMS detector
provides a running time average of the square of the sound pressure signal. Two RMS detector modes of operation are usually
provided. These refer to the amount of *meter damping* in the detector circuit and are called *fast* and slow response. In the fast
mode, the meter responds quickly to rapidly changing sound levels, whereas in the slow mode the level variations appear to be
reduced. The significance of each mode is best illustrated in the
following examples. If one were measuring the noise level of a
passing truck, the fast mode would be used to obtain the maximum level during the passby. However, in a factory, where an
average level is often more useful, the slow mode would be selected to reduce rapidly changing readings.

It should not be inferred that these response features are
arbitrary with respect to meter manufacturers. In fact, the
ballistic characteristics are well controlled by international standards [2]. What is arbitrary, however, is which mode should be
used and when. This decision requires judgment, and only basic
guidelines can be given:

1. If sound levels are not changing too rapidly, or if the measurement of maximum levels is desired, use the fast response mode.
2. If the levels are varying dramatically, and average values are appropriate, use the slow mode.

Fortunately for measurements involving regulations, the response mode is often clearly stated.

Peak Detector

A peak detector indicates the highest instantaneous sound pressure during a measurement interval. Usually these detectors incorporate a hold circuit in order to continuously display the highest peak value until a higher peak level occurs or until the instrument is reset. One should be careful to recognize the difference between the *peak* level and *maximum* level. By definition the maximum level is the highest RMS value obtained during a measurement interval. When a measurement of maximum level is determined, the measurement mode (fast or slow) should always be reported, since the maximum level reached often depends upon the selected meter response time. However, the peak value is not derived from a RMS detector and is thus meter response time independent. In some cases, the peak sound pressure may exceed the maximum fast A-weighted sound level by as much as 20 dB [3].

Integrating Detector

It is often difficult to consistently estimate the average level of fluctuating noise. The equivalent-continuous sound level (L_{eq}) is often used to quantify varying sound pressure level. The L_{eq} is calculated using the following equation:

$$L_{eq} = 10 \, \log_{10} \left\{ \frac{(1/T) \int_0^T p^2(t)\,dt}{p_0^2} \right\} \tag{5.1}$$

where

  $p^2(t)$ = the square of the instantaneous sound pressure   [Pa]
  T = the measurement interval   [sec]
  p_0 = the reference sound pressure level   [2×10^{-5} Pa]

As can be seen from the above equation, the argument of the logarithm is simply the integrated average of the squared sound pressure. This is very similar to the RMS sound pressure measurement; however, L_{eq} is not a running average, and the measurement interval is unspecified. This interval can vary from 1 second to several hours. The L_{eq} measurement was not used until digital technology made it possible to compute Eq. (5.1) in a hand-held sound level meter. The most common application for L_{eq} measurement is in the assessment of environmental noise when the measurement interval is at least one hour.

5.1.5 Sound Level Meter Display

After the sound pressure signals have been detected, the results must be displayed. Signal levels are usually indicated in decibels (referenced to 2×10^{-5} Pa) and are displayed on either a digital or an analog meter. Digital meters provide a direct numerical readout of the measured sound pressure level, with the display updating at least once per second. When significant fluctuations in the sound level occur, large changes in the numerical display can often be difficult and confusing to interpret. Analog displays (traditional needle VU meters) provide easier observation of fluctuating levels but are generally less robust than digital displays. Some digital displays attempt to simulate analog displays using segmented bar graphs. These displays are called quasi-analog displays.

Although international standards require that a digital display indicate levels with a resolution of 0.1 dB, this does not mean that measurements will have an accuracy of 0.1 dB. In fact, as we will see, most field sound measurements are no more accurate than ±1 dB.

5.1.6 Sound Level Meter Outputs

Most sound level meters provide two different output signals. The first is the AC output, which is usually derived from the microphone after the frequency weighting circuits and after the range control amplifier. This signal can be used for tape recording or for detailed analysis utilizing more sophisticated instrumentation. It should be noted that any instrument overloads will result in distortions of the AC output signal and that the range control amplifier should be adjusted to avoid this occurrence.

The DC output is a voltage output that is proportional to the sound pressure level. This can be used to control the pen deflection on a strip chart recorder. In this way, a time history of the sound pressure level can be obtained. It should be noted that the range control amplifier affects both of these signal levels and care should be taken to record meter settings.

5.1.7 Environmental Influences

Sound level meters, and in particular their microphones, can be influenced by environmental conditions. Knowledge of instrumentation

limitations and manufacturer's specifications is a must for accurate
instrumentation usage. The following lists the most common environmental influences that affect the operation of sound level meters.

Temperature

The range of temperature excursions experienced by sound level
meters in most industrial environments is 0 to 30°C, and most manufacturers typically list operating conditions as −10 to 50°C. In
addition, temperature corrections provided by manufacturers are
usually less than 0.01 dB/°C, and over the entire operating range,
at most 1 dB. As such, corrections for temperature are generally
considered negligible under most outdoor or indoor situations. However, for specialized measurements under severe temperature ex-
tremes, the corrections for temperatures should be applied in accordance with the manufacturer's specifications.

Humidity

Extensive exposure of the microphone of a sound level meter to very
high humidity should be avoided. Ceramic microphones are highly
resistant to humidity, but condenser types are not. Excessive electrical leakage and possibly failure is likely if condenser microphones
are subjected to exposure involving rain or water condensation from
the atmosphere. As such, condenser microphones must be kept
covered in rain or snow conditions and kept warmer than the ambient air to avoid accumulations of moisture or condensation.

Wind

When a microphone is in the path of air currents of 6 knots or more,
a low-frequency broad band noise signal usually results. The magnitude of this *wind noise* signal is not negligible and usually oblit-
erates meaningful sound level measurements below 500 Hz. The
most common source of wind noise is of course wind experienced
in outdoor measurements, but high-velocity air currents are also
common indoors when fans and blowers are present. When it is
not possible to avoid these air currents, a windscreen which covers the microphone can be used, as shown in Fig. 5.8. These
screens can be obtained from the instrument manufacturers. With
these screens installed, the usefulness of the sound level meter is
extended to wind currents of up to 20−25 knots, with corrections
of less than 2 dB for frequencies up to 10 kHz, as can be seen in
Fig. 5.8.

Magnetic Fields

The effect of magnetic fields can usually be ignored for measurements in most industrial environments. However, most sound level

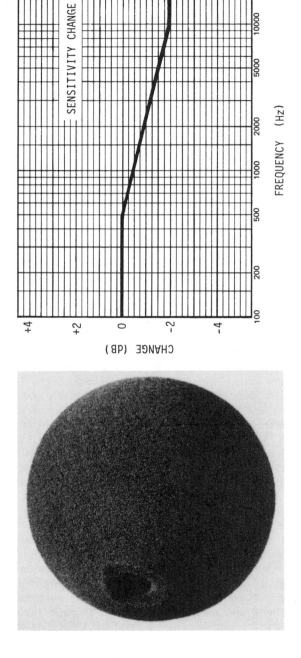

Figure 5.8 Foam-type windscreen which attaches over a microphone. Also shown is the typical sensitivity change due to the screen. (Courtesy of GenRad, Concord, Mass.)

meter manufacturers will specify the magnetic field effect, usually in terms of equivalent sound pressure level at a given magnetic field strength. For example, the effect of a magnetic field may be given as 50 dB for a 50-Oe field at 60 Hz. This may not be useful information to an engineer in the field, since the magnetic field strength of a motor may be harder to measure than the noise level itself. One method, however, that often suffices to establish whether magnetic fields are a serious influence is as follows:

1. Place the microphone dust cover (usually plastic) over the microphone, and note the sound level reading before and after.
2. If the sound level reading drops at least 10 dB, one can conclude that the influence of the magnetic field is negligible.

The conclusion here is based on the premise that the dust cover should reduce the incident sound upon the diaphragm 10 dB or more but will not significantly impede the magnetic field. The presence of magnetic fields is usually at line frequency, 50 or 60 Hz, and higher integer-ordered harmonics. Because of this hum-like character, magnetic fields can also be detected by listening to the sound level meter AC output through a pair of earphones.

Vibration

All microphones are highly susceptible to the influence of vibration. Typically, a sound level reading of 90 dB can be expected if the microphone is being vibrated at an acceleration rate of 1 g (9.81 m/s^2) perpendicular to the diaphragm. This effect is especially easy to see when one considers the basic construction of piezoelectric microphones. However, even for condenser types, the circumferentially supported diaphragms have mass and will flex and move with the vibratory motion just as with the incidence of sound. The microphone dust cover "trick" can also be used here to assess the presence and magnitude of vibratory motion. There are, however, no published corrections for dealing with meter or microphone vibration. Therefore, as a guideline, avoid vibration or, when unavoidable, provide a high level of vibration isolation.

Another problem called *microphonics*, which is related to vibration, also needs to be mentioned. Here, especially at high sound levels (120 dB or more), the sound field will acoustically *excite* the instrument case. This mechanical energy is then transmitted to the internal circuit elements, which often produce spurious unrelated signals called microphonics. Microphonic effects can often be minimized by using a microphone extension cable in order to remove the sound level meter from the sound level or by acoustically isolating the meter case by total enclosure or wrapping with a dense material.

Cables

Because of environmental conditions or constraints, microphone extension cables are often required for sound level measurements, and significant errors can be introduced if "off-the-shelf" cables are used indiscriminately. The problem with common cables is that they have a great deal of capacitance at high frequencies, and with typically low source impedance microphones, significant sound signal losses can be experienced. Therefore, it is strongly recommended that the sound level meter or microphone instruction manual be consulted for proper extension cable availability, possible additional amplification, or inclusion of matching devices.

Background Noise

Another environmental factor to be dealt with in most measurement situations is the presence of background or ambient noise. The sound level meter, unfortunately, does not in any way discriminate between noise from the particular machine of interest and the noise from adjacent or peripheral machines. In fact, instrumentation microphone characteristics are designed to be omnidirectional. For best results, measurements obtained with a quiet background are preferred. Unfortunately, this is not always possible; however, methods to account for background noise do exist, and presented here is a basic field technique that provides good first-order results.

To determine if background noise is a significant influence:

1. Turn off the machine, or eliminate the noise from the source of interest, and measure and record the overall ambient background noise level.
2. Turn on the machine or source of interest, and measure and record the overall noise level.
3. If the difference is 10 dB or more, the background is usually insignificant.

This conclusion follows from our previous consideration when combining sound levels. In Chapter 2, we noted that if the difference between sound levels is 10 dB or more, the contribution of the source of lower level is less than 0.5 dB or, in most noise control situations, insignificant.

Should the difference be less than 10 dB, the contribution of the background noise can be determined approximately by utilizing the chart in Fig. 5.9. Here the chart is entered at the abscissa (horizontal axis), which is a scale of the difference between total noise level and background level. From the intersection with the ordinate (vertical axis), one obtains the number of decibels to be subtracted from the total level. This difference is the overall noise level of the machine or source of interest. An example will clarify the method.

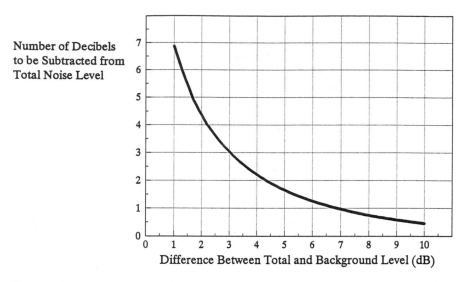

Number of Decibels
to be Subtracted from
Total Noise Level

Difference Between Total and Background Level (dB)

Figure 5.9   Corrections for background noise.

Example
   The noise from a machine tool located on a final assembly line
   is to be measured.   The background noise from adjacent equipment, measured on the sound level meter, is 92 dB overall.
   The total or combined level with the machine tool also running
   is 99 dB overall.   What is the overall sound level of the machine tool?
Solution
   The difference between the background noise level and the total
   noise level is 7 dB.   Entering the chart in Fig. 5.9 at 7 dB, we
   obtain at the intersection approximately 1 dB to be subtracted
   from the total noise.   The machine tool noise level corrected for
   background noise is 99 −1 = 98 dB.

It is easy to see that when the difference is small (i.e., 3 dB or
less), the influence of the background noise is substantial.   In
fact, when the difference is exactly 3 dB, the overall levels of
the background noise and the source are the same.   Measurements
taken under these conditions must be considered somewhat unreli-
able.   It is therefore good measurement practice to avoid measurements where the background levels are equal to or greater than
the source of interest.
   In summary, for reliable and accurate overall sound level measurements, every effort should be made to reduce the background

level to a minimum. Measurements taken when the difference in level between total and background is 3 dB or less should be strenuously avoided and considered, at best, first-order approximations.

5.1.8 Calibration

The fundamental accuracy of sound level meters depends foremost on microphone calibration. Although sound level meters are generally rather stable instruments, in time, sensitivity may vary. As such, it is a basic premise of sound measurement to acoustically calibrate the instrument immediately prior to and after field measurements.

There are numerous ways to calibrate a sound level meter. We shall limit the discussion to field calibration methods. Portable field calibrators usually consist of a vibratory piston or an oscillating diaphragm which is applied over the microphone capsule. The vibrating piston or diaphragm produces a precise, discrete sound level at the sensing element of the microphone. The sound level meter is then adjusted, usually by a small adjustment screw or digital offset, until the meter reading and sound level of the calibrator coincide. With this method, calibration to within ± 0.2 dB can be achieved. This is well within the accuracy required of most noise control measurements. Shown in Fig. 5.10 are examples of three different types of field calibrators. Figure 5.10a shows a loudspeaker diaphragm type, which produces a test tone that has a sound level of 94 dB and a frequency of 1000 Hz. Figure 5.10b shows a motor-driven piston type which produces a sound level of 114 dB at 250 Hz. This type of calibrator is called a pistonphone and since they are mechanically driven, the calibration frequency is limited to lower frequencies, usually 500 Hz or less. Figure 5.10c shows a multifunction acoustic calibrator which can produce calibration tones at many different frequencies and sound levels. This type of calibrator can also test RMS and peak detector circuits.

Most sound level meters also have built-in electrical calibration circuits for checking internal amplifier gain and related detector circuitry. This calibration is emphatically *not* a rigorous acoustical calibration since the sound transducer is not involved and assures only that the meter electrical circuits are functioning properly. The overall acoustical calibration could be in substantial error even though the electrical system may function satisfactorily. In short, there is no substitute for application of a portable vibrating piston or diaphragm calibrator to assure an accurate field calibration. It is also not good practice to interchange one manufacturer's sound level meter with another manufacturer's calibrator. The microphone cavity size and diaphragm position are critical for proper acoustical coupling and essential to accurate calibration.

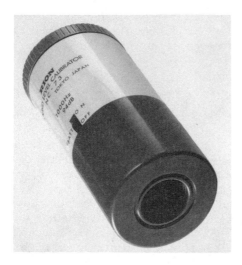

(a)

(b)

Figure 5.10   Field calibrators for sound level meters. [(a)—(b)
Courtesy of Scantek, Inc., Silver Spring, Md.; (c) courtesy of
Brüel and Kjær Instruments, Inc., Decatur, Ga.]

    Finally, the precise sound level produced by a calibrator and
the level measured by a meter are influenced to some degree by
barometric pressure. Normal daily variations in barometric pressure at a given location will not induce measurement errors greater
than 0.3 dB. However, pressure changes such as encountered in
aircraft flights or even on excursions up high mountains can induce
errors of 1 to 3 dB. It is therefore strongly recommended that the

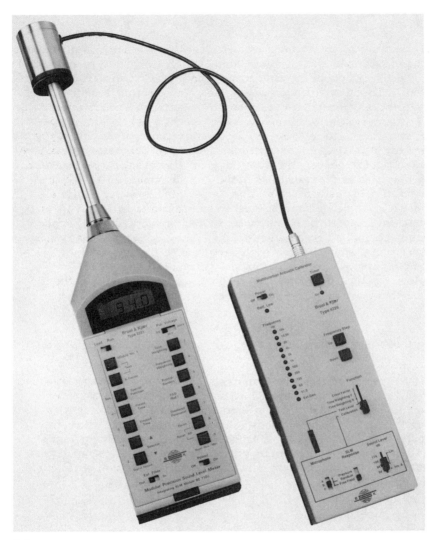

(c)

manufacturer's instruction manual be consulted for corrections when dynamic barometric excursions are anticipated.

Although field calibration is appropriate for most applications, manufacturers usually recommend annual laboratory calibrations. These methods are more accurate and produce results that are traceable to the National Institute of Standards and Technology (NIST) [formerly the National Bureau of Standards (NBS)].

5.1.9 Accuracy

The accuracy of sound level meters depends primarily on the quality of the microphone and the detector. Other associated components are usually far superior with respect to integrity and frequency response. To aid the acoustical engineer in selecting a sound level meter, the American National Standards Institute specification ANSI S1.4-1983 provides for three grades of sound level meters, Types 0, 1, 2, and for a special-purpose limited-function instrument, Type S. The Type 0 instrument or system, designated Laboratory Standard, is intended for use in the laboratory as a reference standard and accordingly is not required to satisfy the environmental requirements for field instruments. The Type 1 instrument, designated Precision, is intended for accurate measurements in the field and laboratory. The Type 2 instrument, designated General Purpose, is intended for general field use, that is, measurement of typical environmental sounds when high frequencies do not dominate. The Type S, designated Special Purpose, may be designed for any of the three grades but is not required to contain all of the functions required for a nonspecial-purpose sound level meter [2].

5.2 FREQUENCY ANALYSIS

The importance of frequency analysis as an acoustic diagnostic and design tool cannot be overemphasized. The ability to measure a spectrum of an acoustic signal can often enable one to identify and separate specific noise sources. This information can allow the acoustic engineer to rank-order noise sources in terms of signal level and frequency content. The advantage associated with this is that noise control efforts can become more effective, efficient, and systematic.

5.2.1 Bandpass Filter

The primary tool used to derive frequency information from a complex noise signal is the bandpass filter. An ideal bandpass filter is a frequency-selective device which transmits certain frequency components of the input signal and completely attenuates the remainder of the signal. The frequency components of the signal that are transmitted are determined by the passband of the filter. The frequency response characteristics of an ideal bandpass filter are illustrated in Fig. 5.11. The passband of a filter is the frequency band which falls between the upper and lower cutoff frequencies. The difference between the upper and lower cutoff

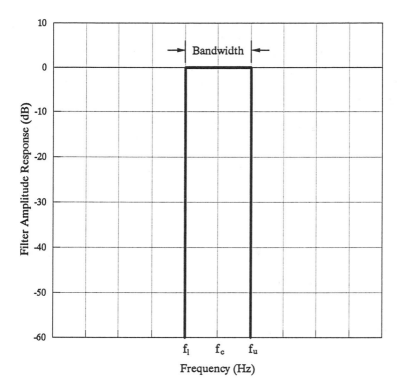

Figure 5.11   Frequency response characteristics of an ideal bandpass filter.

is called the filter bandwidth. Quite often, bandpass filters are
described in terms of their band center frequency and their
bandwidth. As will be seen later, detailed specifications control
the shape, bandwidth, and center frequencies for bandpass filters.

Typically, bandpass filters are separated into two different
classes: (1) constant percentage bandwidth filters (octave, 1/3
octave, etc.) and (2) constant bandwidth filters.

Constant percentage bandwidth filters are the most common
filters used in acoustic frequency analysis.   When measuring
an acoustic spectrum, one notes that the bandwidth of each filter used maintains the same percentage of its center frequency
throughout the spectrum.   For example, a 10% bandwidth filter
centered at 100 Hz has a bandwidth of 10 Hz.   Similarly, a 10%
bandwidth filter centered at 10 kHz has a bandwidth of 1 kHz.
Thus, for constant percentage bandwidth analysis, the filter

bandwidth at low frequencies is relatively small, whereas at high frequencies, the bandwidth may be quite large.

Constant bandwidth filters maintain a fixed bandwidth over the entire frequency spectrum being measured. These filters can have a very narrow bandwidth (often less than 1 Hz), and thus an analysis that utilizes constant bandwidth filters is sometime called a narrow band analysis.

In general, when plotting a spectrum which utilizes a constant percentage bandwidth analysis, the abscissa (horizontal axis) is normally scaled with a logarithmic frequency scale. For constant bandwidth analysis, the abscissa is usually scaled with a linear frequency scale.

5.2.2 Octave and 1/3 Octave Band Filters

The two most common filter types used in acoustical analysis are octave band and 1/3 octave band filters. The upper and lower cutoff frequencies for octave band filters can be calculated from the following equations:

$$f_1 = 2^{-1/2} f_c \tag{5.2}$$

$$f_u = 2^{1/2} f_c \tag{5.3}$$

where

f_c = the octave center band frequency   [Hz]
f_1 = the filter's lower cutoff frequency   [Hz]
f_u = the filter's upper cutoff frequency   [Hz]

The preferred center frequencies for an octave band analysis are: 31.5, 63, 125, 250, 500, 1000, 2000, 4000, 8000, and 16000 Hz. Using Eqs. (5.2) and (5.3), one can see that the upper cutoff frequency for each octave filter is the same as the lower cutoff frequency for next higher filter. In this way, the passbands of all 10 octave band filters completely cover the frequency range from 22 to 22,624 Hz. Therefore, if one were to measure the sound pressure level of the same signal consecutively passing through each of these filters, an octave band spectrum could be produced. Two different ways of displaying an octave band spectrum are presented in Fig. 5.12.

The upper and lower cutoff frequencies for 1/3 octave band filters can be calculated from the following equations:

$$f_1 = 2^{-1/6} f_c \tag{5.4}$$

$$f_u = 2^{1/6} f_c \tag{5.5}$$

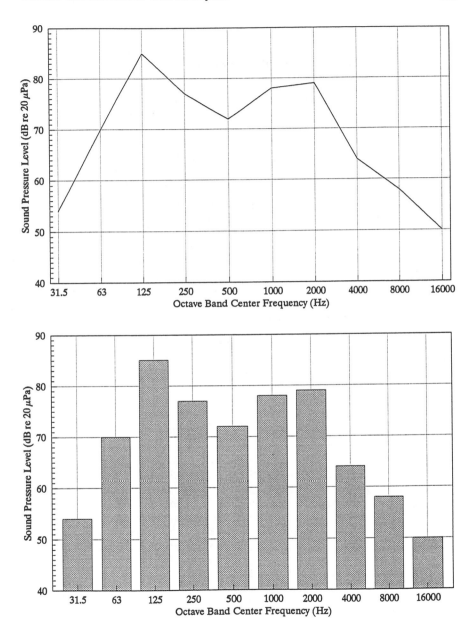

Figure 5.12 Two different ways of displaying an octave band spectrum.

where

f_c = the 1/3 octave band center frequency [Hz]
f_l = the filter's lower cutoff frequency [Hz]
f_u = the filter's upper cutoff frequency [Hz]

These filters obviously have a bandwidth which is narrower than the octave band filter. In fact, it should be no surprise that it takes three adjacent 1/3 octave band filters to cover the same frequency band as an octave band filter.

Although it would be nice if octave band and 1/3 octave band filters had the frequency response characteristics of the ideal filter, which is illustrated in Fig. 5.11, the practical limitations associated with electronic filters prevents this from occurring. Real octave band and 1/3 octave band filters differ from the ideal filter by having *ripple* in the passband and filter *skirts* outside of the passband. These characteristics are illustrated in Fig. 5.13. The effect of the ripple is to cause slight variations in signal level (usually less than 1 dB) for signals which fall in the passband. The effect of the filter skirts is to cause overlapping between adjacent filters. The filter cutoff frequency is usually defined as the frequency where the signal has been attenuated by 3 dB. This means that a signal which falls exactly at a cutoff frequency will be measured equally by two adjacent filters. However, both filters will attenuate the signal by 3 dB, and thus, if one were to sum the contribution from both filters, the total level would be correct. A complete set of typical characteristics for a commercially available octave band filter set is shown in Fig. 5.14. It should be emphasized that filter shape characteristics are rigidly controlled by the American National Standards Institute (ANSI) and by the International Electrotechnical Commission (IEC) [4,5].

Finally, it should be noted that, although this discussion has concentrated on octave and 1/3 octave band filters, other constant percentage bandwidth filters exist. These include 1/2, 1/12, 1/24, and other fractional bandwidths. Few measurement standards exist which require these bandwidths; however, the higher resolution measurements associated with the lower bandwidth filters (1/12 and 1/24) can provide detailed diagnostic information. The general equations for the upper and lower cutoff frequencies for an nth octave band filter are given by the following equations:

$$f_l = 2^{-1/2n} f_c \tag{5.6}$$

$$f_u = 2^{1/2n} f_c \tag{5.7}$$

where

f_c = the 1/nth octave band center frequency [Hz]
f_l = the filter's lower cutoff frequency [Hz]

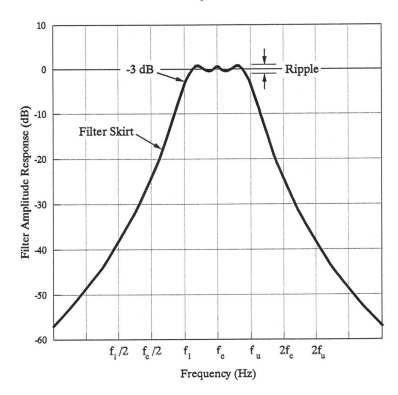

Figure 5.13   Frequency response characteristics of a typical octave bandpass filter.

f_u = the filter's upper cutoff frequency   [Hz]
n = 1, 2, 3, 12, 24, etc.

5.2.3   Spectrum Analyzers

Spectrum analyzers are instruments which combine the features of level measurement instrumentation with the frequency selectivity capability of bandpass filters. As with bandpass filters, there are two types of spectrum analyzers: (1) constant percentage bandwidth analyzers and (2) constant bandwidth analyzers.   Each type will be discussed separately.

Constant Percentage Bandwidth Analyzers

The most common constant percentage bandwidth analyzers utilize octave and 1/3 octave band filters to measure frequency spectra.

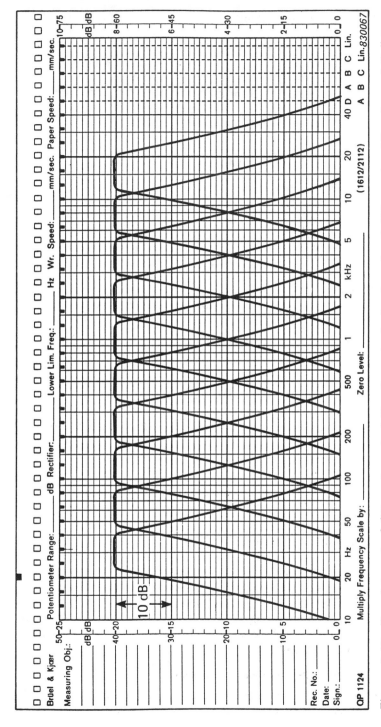

Figure 5.14 Complete set of filter characteristics for a typical octave band filter set. [Courtesy of Brüel and Kjær Instruments, Inc., Decatur, Ga.]

162

The simplest form of this type of instrument is the serial analyzer. As illustrated in Fig. 5.15 the octave band analyzer is a sound level meter which includes a bank of sequential octave band filters. This bank of filters can be built directly into the sound level meter or can be an optional package which can be attached to the sound level meter body. Similarly, a more extensive bank of 1/3 octave band filters can be incorporated into the sound level meter, yielding a 1/3 octave band analyzer. The principle of operation of these instruments is quite simple. In order to measure a spectrum, the user sequentially selects each filter, one at a time, and measures the sound level in each band. Normally, data is manually plotted on a scaled graph, with the octave or 1/3 octave band center frequency (Hz) on the abscissa (horizontal axis) and sound levels (dB) on the ordinate (vertical axis). Several commercially available octave band and 1/3 octave band analyzers are presented in Fig. 5.16.

The process of generating octave band and 1/3 octave band spectra can be a tedious operation. In particular, measuring 1/3 octave band spectra covering the audible frequency range (20 Hz to 20 kHz) requires 31 separate measurements. As a result, a second type of analyzer exists. The *real-time* or *parallel* analyzer (illustrated in Fig. 5.17) measures all frequency bands simultaneously. The obvious advantage of this type of analyzer is that

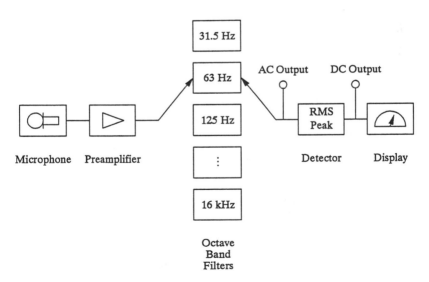

Figure 5.15 Block diagram for a serial octave band analyzer.

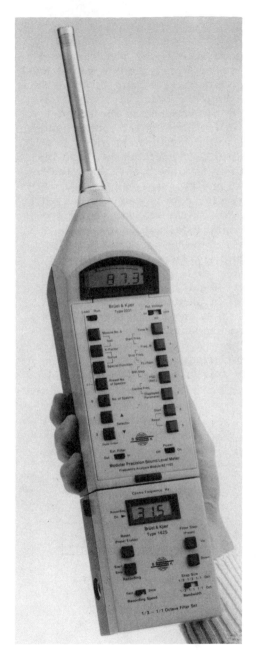

(a)

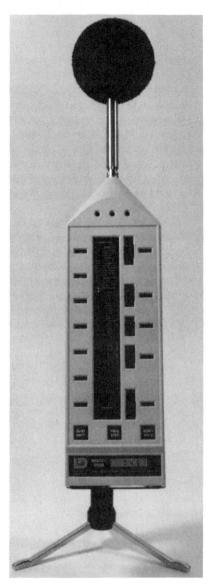

(b)

Figure 5.16 Portable serial octave and 1/3 octave band analyzers.
[(a) Courtesy of Brüel and Kjær Instruments, Inc., Decatur, Ga.;
(b) courtesy of Larson Davis Laboratories, Provo, Ut.]

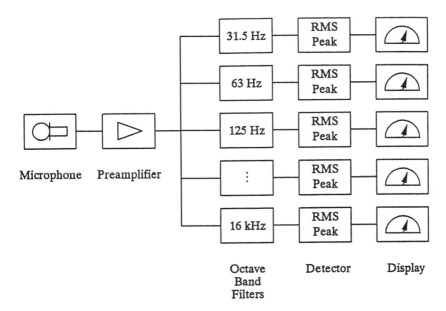

Figure 5.17 Block diagram for a parallel octave band analyzer.

it can greatly reduce the length of time it takes to measure a spectrum. This can be particularly important when an event is transi-
tory (e.g., a truck or train passby). Today, real-time analyzers
are generally implemented using digital signal processing techniques.
Several commercially available real-time octave and 1/3 octave band
analyzers are presented in Fig. 5.18.

Constant Bandwidth Analyzers

The most common type of constant bandwidth analyzer is the fast
Fourier transform (FFT) analyzer. FFT analyzers operate by digi-
tally capturing a short sample of a time signal and then calculating
a discrete approximation of the Fourier integral transform [6]. The
result of this calculation is a constant bandwidth spectrum. The
filtering, or resolution characteristics of these analyzers vary considerably. The number of constant bandwidth filters generated is
usually described as the number of lines in the analysis. It is not
uncommon for FFT analyzers to have the ability to produce 400,
800, 1600, or even 3200 lines of resolution. The line or filter

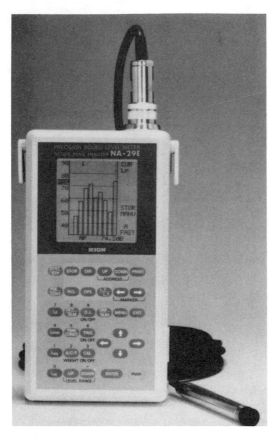

(a)

Figure 5.18   Portable parallel octave and 1/3 octave band analyzers.
[(a), (c) Courtesy of Scantek Inc., Silver Spring, Md.; (b) courtesy of Brüel and Kjær Instruments, Inc., Decatur, Ga.]

spacing in an FFT analysis can be calculated by using the following equation:

$$\text{Filter spacing} = \frac{F_{max} - F_{min}}{\text{no. of lines}} \qquad (5.8)$$

where

F_{max} = the highest frequency measured [Hz]
F_{min} = the lowest frequency measured [Hz]

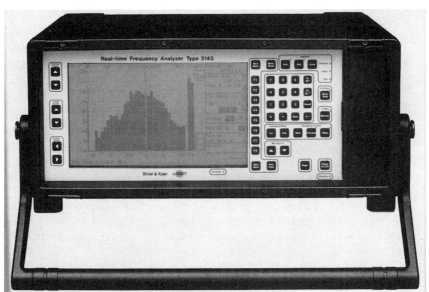

(b)

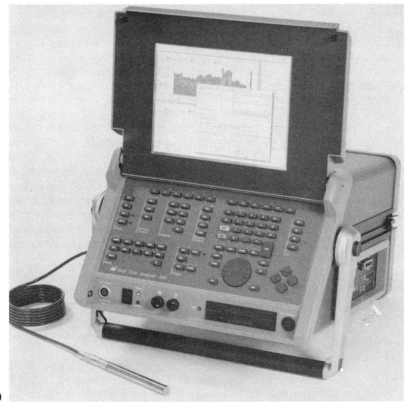

(c)

As an example, consider an analysis with 400 lines of resolution and a frequency range of 0—20,000 Hz. Here, the frequency spacing between each filter is 50 Hz (i.e., 20,000 divided by 400). As can be seen, by selecting the analysis range and number of lines, any degree of resolution is possible.

The need for high-resolution analysis is best illustrated in the example presented in Fig. 5.19. Here we see how the noise from a refrigerator compressor can be analyzed using an octave band analysis, a 1/3 octave band analysis and a 800-line FFT analysis. Note that as the bandwidth decreases, the resolution of the discrete tones clearly increases.

Space limitations do not permit further discussion of the powerful capabilities of these frequency analyzers. However, the reader is encouraged to consult References 3 and 6 at the end of this chapter.

5.3 GRAPHIC LEVEL RECORDER

The graphic level recorder (see Fig. 5.20a) plays a fundamental role in sound measurement. This instrument provides a permanent, reproducible, hard-copy trace of the noise level time history, as illustrated in Fig. 5.20b. As an accessory to sound measurement instrumentation, graphic level recorders can be used over long periods of time to record time-varying noise levels such as those found outdoors.

Most recorders are of a servomechanism type wherein the servo drives a marking pen over an accurately advancing paper strip chart. The resolution of the noise level versus time trace is controlled by selecting strip chart paper speed and pen writing speed. With respect to paper speed, typical recorders offer a range of 0.0003 to 100 mm/s. Pen or marker writing speeds will typically range from 2 to 2000 mm/s.

Few guidelines can be given for the operation of graphic level recorders since each application requires some judgment. For example, if one wants to record the time history of a cyclical noise pattern produced by an automatic bar machine, two primary considerations must be made:

1. If the duration of the cycle is, for example, 10 s, a paper speed must be selected that will record four or five machining cycles over, say, 500 mm, or approximately 18 in. As such, a speed of 50 mm/s or thereabouts would be selected. If the paper speed were too slow, machine cycles would be crowded into too short a length of paper, making the cycle difficult to define.
2. To assure accurate definition of the time-varying noise level, the pen writing speed must be selected accordingly. If too slow a writing speed is selected, the result is a heavily *damped*

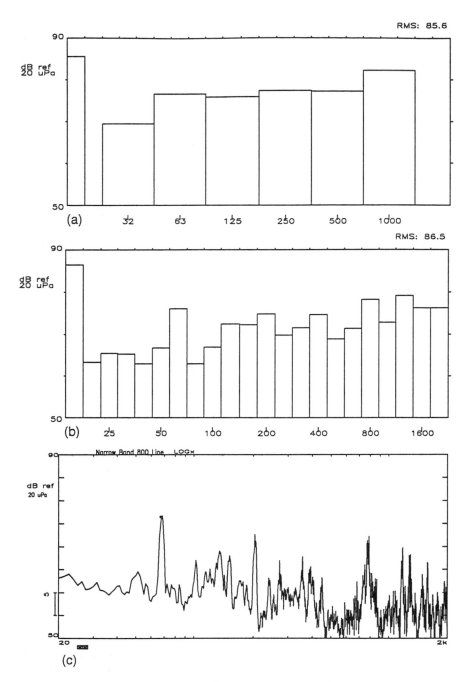

Figure 5.19 Examples illustrating the differences between various resolutions for spectrum analysis: (a) octave band, (b) 1/3 octave band, (c) 800-line FFT.

169

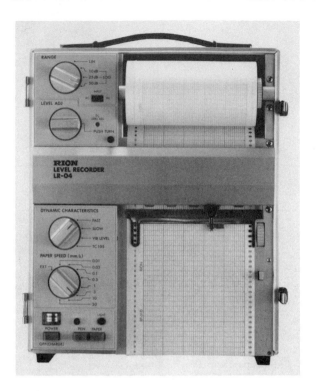

(a)

Figure 5.20   (a) Portable graphic level recorder. [Courtesy of Scantek Inc., Silver Spring, Md.] (b) Graphic level record showing A-weighted community levels versus time. [Courtesy of GenRad, Concord, Mass.]

average of the levels. If too fast, the dynamic excursions of the pen yield an ill-defined trace.

In short, for the graphic time trace to be accurate, careful attention must be given to selecting the paper and pen writing speeds that yield representative results. This attention is generally a matter of experience and judgment.

In addition to sound level versus time history measurements, the recorder can be used for recording spectral noise levels. Here, records of narrow-band or one-third octave or octave band data are recorded by synchronizing the paper speed of the recorder with the corresponding filter bands or sweep of the spectrum analyzer.

In summary, the graphic level recorder provides the acoustical engineer a permanent, easy-to-read record of noise data. When

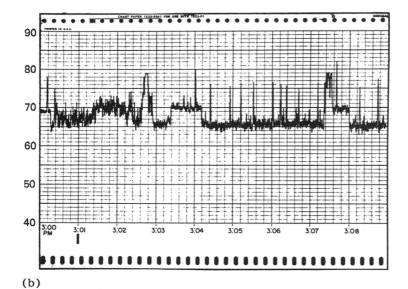

(b)

Figure 5.20   (Continued)

observations over a long time period are desired, the information on
the strip chart can be studied in a much more extensive manner
than can a few readings of a meter. In addition, by overlaying the
strip charts, level changes and duration can be clearly noted and
analyzed for diagnostic purposes.

5.4 TAPE RECORDERS

Tape recorders in noise control engineering are a valuable tool. By
recording sound signals on a tape recorder, the acoustic engineer
is able to bring the sound source into the laboratory for detailed
analysis. Here, the signals can be replayed and analyzed as many
times as necessary to obtain a full understanding of the character
of the signal. This can be particularly valuable when the sound
signal is an event of short duration such as a train or truck pass-
by. With the tape-recorded signal, it is possible to obtain a spectrum of the event with a serial analyzer. Finally, tape recorded
signals can be stored and archived as part of a permanent record.
    Field tape recorders vary from small, lightweight cassette re-
corders to slightly heavier reel-to-reel types. Today, tape recorders utilize either analog or digital methods for storing data on
magnetic tape. Two commercially available instrumentation recorders are presented in Fig. 5.21. As will be seen, from the standpoint

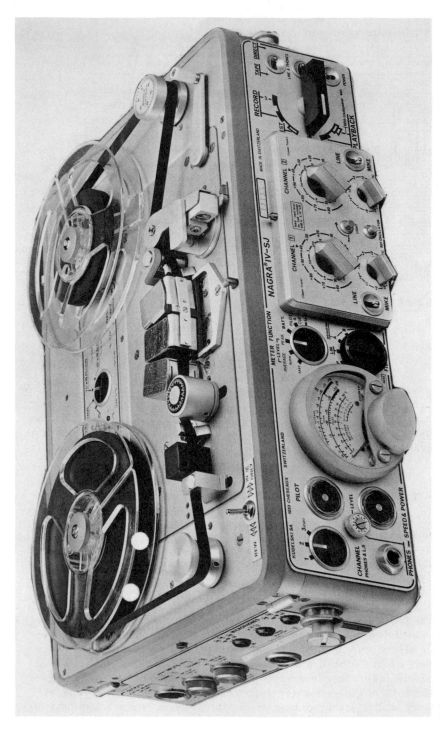

(a)

172

(b)

Figure 5.21 Portable tape recorders for field sound measurement.   (a) Analog reel-to-reel tape recorder. [Courtesy of Nagra Magnetic Recorders, Inc., New York, N.Y.] (b) Digital DAT recorder. [Courtesy of Sony Corporation of America, Montvale, N.J.]

of accuracy, the specifications of most digital audio tape (DAT) recorders far exceed the specifications of analog recorders.   Ideally,
during playback, the tape recorder should reproduce the recorded
signal exactly.   However, due to various limitations associated with
frequency response, internal noise, and tape speed control, there
are inherent limits to the accuracy of tape-recorded signals.

With respect to a tape recorder's frequency response, one would
like a flat response over the audible frequency range.   Most analog
tape recorders utilize an amplitude modulation technique for recording signals, and due to tape head size limitations, these recorders
have difficulty recording low-frequency signals.   Generally, this
type of tape recorder has a flat frequency response (±3 dB) over
the frequency range of 40—20,000 Hz.   Recording low-frequency phenomena such as infrasound (below 20 Hz) with this type of tape recorder is virtually impossible.   A second type of analog tape recorder
utilizes frequency modulation.   Although there is no lower limit to the
frequency response using this technique, there usually is an upper
frequency limit of approximately 10 kHz.   Finally, utilizing digital
technology, a flat frequency response from DC to 20 kHz (±1 dB)
is normally obtained in instrumentation DAT recorders.

The signal-to-noise ratio is a parameter used to describe the
presence of unwanted internal instrument and tape noise.   This
determines the useful range in terms of signal levels that the recorder can accurately reproduce.   With analog recorders, a signal-
to-noise ratio of 40—60 dB is common.   With DAT recorders, this
ratio is often greater than 80 dB.

Variations in the tape speed can produce undesirable effects
called *wow* and *flutter*.   These effects are measured as a percentage of signal frequency.   For analog records, a figure of 0.1% to
0.05% is normal.   For digital recorders, this quantity is so small
that it is unmeasurable.

For field recordings, a sound level meter is usually utilized in
conjunction with a tape recorder.   Here the AC output signal for
the sound level meter is connected directly to the signal input of
the tape recorder.   To assure use of the entire dynamic range of
the tape recorder, the sound level meter AC signal level must be
adjusted to be close to the tape recorder's maximum level.   To
reach this level, the range control amplifier of the sound level meter is adjusted.   For example, for relatively low sound levels, the
lower ranges of the sound level meter are selected to amplify the
signal to the recorder.   Correspondingly, signals from higher noise
levels are attenuated to avoid *overloading* the recorder input amplification stages.   In short, the microphone and the amplification
stages of the sound level meter are used to bring the analog
sound signal to the recorder at the proper level.

Calibration of the tape recording is accomplished by placing
a portable calibrator with precise known sound levels over the

microphone and recording the signal. By noting the sound level meter's full-scale level at calibration and announcing subsequent changes of the range control amplifier during recordings, the amplification changes can be accounted for at playback. For example, if at calibration, the sound level meter attenuator selected was the 90 dB range and later a recording is made on the 70 dB range, a correction of −20 dB is applied to the recorder playback output signal.

5.5 DOSIMETRY

Noise dosimeters are widely used for monitoring noise in workplace environments where sound levels may be hazardous to hearing. Basically, the dosimeter integrates a weighted function of sound pressure or sound pressure level over a time period to determine noise dose, which is a percentage of permissible exposure criteria. Shown in Fig. 5.22 are several dosimeters which are designed for measuring daily noise exposure in accordance with various regulations. Most often these instruments are worn on the person, with the microphone located on the shoulder. When not worn by a person, they may be used as an area monitor.

The dosimeter can be broken down into three basic components operating in series: (1) a sound level meter, (2) an integrator, and (3) a readout device. The overall response and quality of the sound level meter is controlled by standards, and minimum standards are usually specified by regulations. The integrator section electronically integrates a power function of the mean-square signal over time. The specific power function to be integrated depends on the applicable exposure criteria. A rather general expression which describes the operation mathematically is [7]:

$$D = \frac{100}{T_c} \int_0^T 10^{\frac{L-L_c}{q}} \, dt \quad [\%] \tag{5.9}$$

where

D = percentage exposure [%]
T_c = criterion sound duration   [usually 8 h]
T = measurement duration   [h]
t = time [h]
L = sound level (a function of time)   [usually dBA]
L_c = criterion level   [usually dBA]
q = criterion exchange rate parameter   [dB]

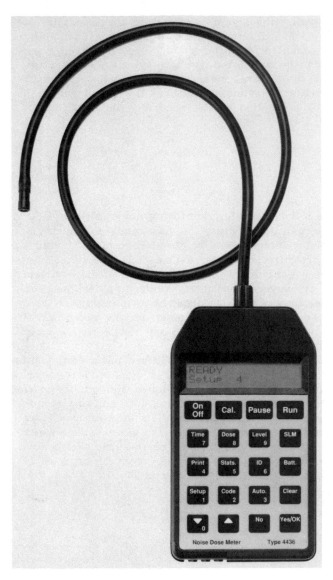

(a)

Figure 5.22   Personal noise dosimeters.   [(a) Courtesy of Brüel and Kjær Instruments, Inc., Decatur, Ga.; (b) courtesy of Larson Davis Laboratories, Provo, Ut.; (c) courtesy of Quest Electronics, Oconomowoc, Wis.]

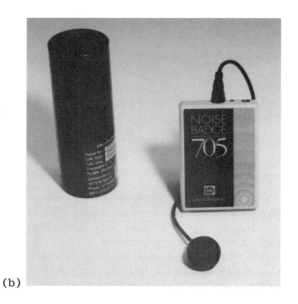

(b)

(c)

Figure 5.22 (Continued)

For discrete time intervals at a constant sound level, Eq. (5.9) can be rewritten as [8,9]:

$$D = \frac{100}{T_c} \sum_{i=1}^{n} t_i \, 10^{\frac{L_i - L_c}{q}} \tag{5.10}$$

where

L_i = weighted sound level in the ith time interval   [usually dBA]
t_i = time spent in the ith interval   [h]

Before moving on to an example, further discussion is necessary on the exchange rate parameter. The exchange rate parameter is generally related to the exposure accumulation factor of the criteria. For example, the Occupational Safety and Health Administration (OSHA) exposure criteria reduce permissible exposure by one-half when the noise level increases 5 dBA. Other international criteria reduce exposure by one-half when the noise level increases 3 dBA. Because of the biomedical implications, it is likely that the proper exchange rate and threshold level L_c will be the subject of some controversy for some time to come. As such, we shall conclude the discussion by noting that at this time:

$q = 10$ for an exchange rate of 3 dB
$q = \dfrac{5}{\log 2}$ for an exchange rate of 5 dB

With respect to the criterion level L_c, 90 and 85 dBA are most often used. Let us now consider an example.

Example
Calculate the daily (8-h) noise exposure of a machine operator where the noise levels are constant at (1) 90 dBA and (2) 100 dBA. Use an exchange rate of 5 dB and a criterion level of 90 dBA.
Solution
Since the sound level L is a constant throughout the day (i.e., 90 dBA), Eq. (5.10) reduces to:

$$D = \frac{100T}{T_c} 10^{\frac{L - L_c}{q}}$$

and, for condition (1)

T = 8 h
T_c = 8 h
L = 90 dBA
L_c = 90

Upon substitution, we get

$$D = 100 \left(10^{\frac{0}{q}} \right)$$

$$= 100\%$$

However, for condition 2,

$$L = 100 \text{ dBA}$$

and upon substitution, we get

$$D = 100 \left(10^{\frac{100-90}{5/\log 2}} \right)$$

$$= 100(10^{2\log 2})$$

$$= 400\%$$

The real significance of this example can be seen by referring to Appendix E, where the current OSHA criteria are presented and discussed in detail. Note in Fig. E.1 of Appendix E that for constand levels of 90 and 100 dBA, the permissible durations (T) are 8 and 2 h, respectively. Hence, a dosimeter worn by the operator for 8 h at 90 and 100 dBA would indeed register 100% and 400% exposure accordingly, as shown in the example.

With respect to readout, some dosimeters are self-contained in that the percentage exposure is given as a digital readout on the face of the instrument. Others require additional and separate readout-indicating equipment.

Finally, because most exposure criteria do not accumulate below a specified threshold level, dosimeters must respond accordingly. This simply means that the integrator is idle below a preset level, say 90 dBA or 80 dBA, and no exposure is accumulated.

With respect to calibration, dosimeters are calibrated in the same manner as sound level meters. In summary, for dosimeters there are numerous major variations in system flexibility (criterion level, threshold level, exchange rate, etc.). However, the response and operation are controlled by standards, the two most relevant being ANSI S1.25-1978 (American National Standards Specification for Personal Noise Dosimeters) and ISO R 1999 (Assessment of Occupational Noise Exposure for Hearing Conservation Purposes). Their proper operation and environmental limitations follow the basic guidelines one considers when using a sound level meter, as discussed in Section 5.1.

5.6 SOUND INTENSITY ANALYZERS

The quantities, sound intensity, and sound intensity level were introduced in Chapter 2. It was shown that for a freely propagating plane wave there is a direct relationship between sound pressure and sound intensity.

$$I = \frac{\bar{p}^2}{\rho c} \quad [W/m^2]$$ (5.11)

where

$\bar{p}^2$ = mean square sound pressure   (Pa)
ρc = characteristic impedance of air   (mks rayls)

This implies that to measure sound intensity in a free field, only sound pressure measurements are required. These can be performed with a sound level meter and appropriate calculations can be applied. However, in sound fields that are not free fields, this relationship between sound pressure and sound intensity does not apply. A more general equation for sound intensity is given by [3]:

$$\vec{I} = p \times \vec{u} \quad [W/m^2]$$ (5.12)

where

p = sound pressure   [Pa]
$\vec{u}$ = particle velocity in the direction that the sound intensity is
    being measured   [m/s]

This implies that simultaneous measurements of pressure and particle velocity are required to measure sound intensity in a complex sound field. The measurement of pressure entails the straightforward use of a microphone. However, the measurement of particle velocity is more complicated and beyond the scope of this text. Specialized sound intensity probes do exist and are commercially available to accomplish this purpose (see Fig. 5.23). The signals produced by these probes can be used to calculate both of these quantities simultaneously.

A sound intensity analyzer processes the electrical outputs of the probes to directly measure the vector quantity sound intensity. Often, sound intensity analyzers utilize octave and 1/3 octave band filters in order to produce sound intensity spectra. Two commercially available sound intensity analyzers are presented in Fig. 5.24.

5.6.1 Sound Intensity Applications

There are two main applications for sound intensity measurements. The first is for the determination of the sound power of a noise source. In this application, several sound intensity measurements are made over a conformal surface which completely encloses the noise source. The sound power is then calculated from the following equation [3]:

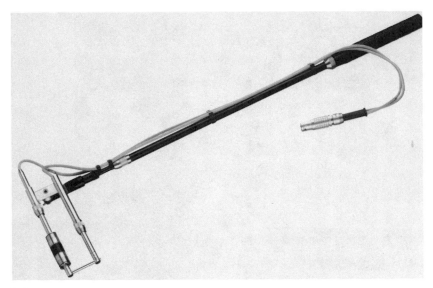

(a)

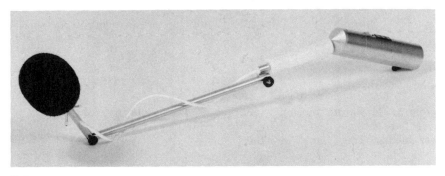

(b)

Figure 5.23 Sound intensity probes. [(a) Courtesy of Brüel and Kjær Instruments, Inc., Decatur, Ga.; (b) courtesy of Scantek Inc., Silver Spring, Md.]

$$W = \int_{S} I_n dS \quad [W] \tag{5.13}$$

where

I_n = sound intensity normal to the surface area $[W/m^2]$
dS = an element of area on the conformal surface $[m^2]$

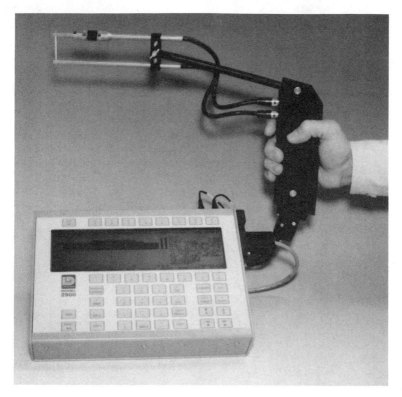

(a)

Figure 5.24   Sound intensity analyzers. [(a) Courtesy of Larson
Davis Laboratories, Provo, Ut.; (b) courtesy of Brüel and Kjaer
Instruments, Inc., Decatur, Ga.]

The unique aspect of measuring sound power in this manner is
that it can be performed in complex sound fields and in the presence of background noise. Once obtained, the sound power data
can be very useful. For example, the sound power emissions from
several machines can be compared to determine dominant sources.
With this information, noise control is applied in a systematic manner, treating those sources that contribute most to the overall
level.

A second application of sound intensity measurement is for
noise source localization. In this application, the directional characteristics of the sound intensity probe are utilized. Here, many
measurements are made at equal spaces on a grid which surrounds
the noise source. Regions of high sound intensity identify localized

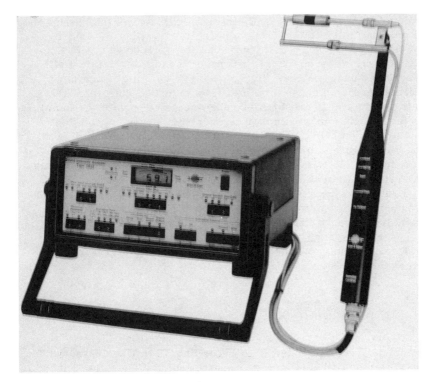

(b)

Figure 5.24   (Continued)

noise sources.   Therefore, to reduce noise emissions, treatments in these areas will be most effective.

The topic of sound intensity and its measurement is a broad field with many pertinent industrial noise control applications. For further discussion of these applications, the reader is encouraged to consult Reference 3 at the end of this chapter.

5.7   ENVIRONMENTAL OR COMMUNITY NOISE MONITORS

When performing environmental noise surveys, one must recognize that variations in noise levels will continually occur. Environmental noise monitors are sound level meters that have

been developed to measure and describe the temporal characteristics of community noise. The distinctive feature of environmental noise monitors is that they can be left unattended for
several hours, or even days, to gather sound level information.
This requires long battery life, rugged packaging (including
the microphone), and the capability to logically compress vast
amounts of data. In most environmental noise monitors, sound
levels are continuously sampled during the entire period that
the monitor is left in the field. Automatic start and stop times
can be programmed into the instrument, and an internal clock
is used to log all measurements. A digital processor is utilized
to calculate statistical descriptors on an interval basis (usually
hourly) and to store this information. Stored data is later
downloaded to a computer or directly to a printer. This data
might include hourly listings of a wide range of the community
noise descriptors that are discussed in detail in Chapter 16.
Several commercially available community noise monitors are presented in Fig. 5.25.

5.8 DATA RECORDING

Accurate and comprehensive data recording is also an essential factor in obtaining reliable sound measurements. Therefore, it is
strongly recommended that a data recording sheet be prepared
prior to gathering any field data. As a minimum, the following
items should be considered as vital information when applicable
[1]:

Sound source description
1. Photograph of equipment, manufacturer, size, model num-
    ber
2. Operating conditions, speed, power, product
3. Isolation mounts
4. Description of secondary sound sources

Environment
1. Sketch or photograph showing location of sound sources
2. Physical description of walls, ceilings, floors, building,
    trees, ground, and relevant reflecting surfaces
3. Dimensions of room, if indoors
4. Meterological conditions, wind velocity, temperature, hu-
    midity

Instrumentation
1. List of instruments including type, model, and serial numbers
2. Type of calibrator, model and serial number
3. Length of cables

(a)

(b)

Figure 5.25 Commercially available environmental noise analyzers.
[(a) Courtesy of Larson Davis Laboratories, Provo, Ut.; (b) courtesy of Brüel and Kjær Instruments, Inc., Decatur, Ga.]

Measurement data
1. Date, project location
2. Engineer and observers
3. Calibration level and method
4. Test time
5. Transducer locations and orientation
6. Weighting networks
7. Ballistic modes, i.e., fast or slow
8. Background or ambient noise level
9. Relevant measurement standards or procedures followed

Other items which should be recorded but cannot be anticipated:
1. Unusual operating conditions
2. Failures or malfunctions
3. Equipment replacement
4. Pretest and posttest calibration differences
5. Meterological changes during a test

In summary, attention to data recording detail will always be appreciated later when processing data or writing a measurement summary report.

5.9 SUMMARY

It should not be inferred that all or most sound measuring instruments or applications thereof have been considered in this chapter. In fact, only the most basic instruments, systems, and measurement methods have been presented. We shall, however, discuss other more specialized instruments and measurement methods as they apply in later chapters. In Appendix C there is also a comprehensive list of national and international standards, test codes, and procedures that apply to sound measurements and instrumentation.

Finally, if there is one point that must be emphasized, it is that for reliable and accurate sound measurements and analyses, there is no substitute for a thorough knowledge of the instrumentation and extensive *hands-on* experience with the field and laboratory equipment.

REFERENCES

1. A. P. Peterson and E. E. Gross, Jr. *Handbook of Noise Measurement.* GenRad, Concord, Massachusetts, 1972.
2. Specification for Sound Level Meters. *American National Standards Institute S1.4-1983.* American National Standards Institute, New York, 1983.

3. Cyril M. Harris. *Handbook of Acoustical Noise Measurements and Noise Control.* McGraw-Hill, New York, 1991.
4. Specification for Octave-Band and Fractional-Octave Band Analog and Digital Filters. *American National Standards Institute S1.11-1986.* American National Standards Institute, New York, 1986.
5. Octave-Band and Fractional-Octave Band Filters. *International Electrotechnical Commission IEC 225-1966.* International Electrotechnical Commission, Geneva, Switzerland, 1986.
6. R. B. Randall. *Frequency Analysis.* Brüel and Kjær, Naerum, Denmark, 1987.
7. Specification for Personal Noise Dosimeters. *American National Standards Institute S1.25-1978.* American National Standards Institute, New York, 1978.
8. *Instructions and Applications: Noise Dose Meters, No. 2.* Brüel and Kjær, Naerum, Denmark, 1972.
9. T. A. Dear. *Applications of the Dosimeter to Noise Control Measures.* Noise-Con Proceedings, Poughkeepsie, New York, 1973.

BIBLIOGRAPHY

Beranek, L. L. *Noise and Vibration Control.* McGraw-Hill, New York, 1971.
Faulkner, L. *Handbook of Industrial Noise Control.* Industrial Press, New York, 1975.
Harris, Cyril M. Handbook of Acoustical Measurements and Noise Control. McGraw-Hill, New York, 1991.
Methods for Measurements of Sound Pressure Levels. *American National Standards Institute S1.13-1971.* American National Standards Institute, New York, 1971.
Randall, R. B. *Frequency Analysis.* Brüel and Kjær, Naerum, Denmark, 1987.

EXERCISES

5.1 What is the sensitivity change of a microphone using a widescreen as shown in Fig. 5.8 at
a. 2 kHz?
b. 10 kHz?

Answer: (a) −1 dB
(b) −2 dB

5.2 A milling machine operating in a machine shop produces a noise level of 90 dBA. With the machine off, the sound level is 85 dBA. What is the approximate noise level of the milling machine alone?

Answer: 88.5 dBA

5.3 The background noise level in a food packaging area is 93 dBA. A new filling machine is installed, and the level increases to 96 dBA. What is the approximate noise level of the new filler?

Answer: 93 dBA

5.4 What is the hourly L_{eq} at a site where the sound level is 70 dBA for the first half hour and 80 dBA for the second half hour?

Answer: 77.4 dBA

5.5 Determine the approximate lower and upper cutoff frequencies for an octave band analyzer for the octave bands whose center frequencies are
a.   63 Hz
b.   4000 Hz

Answer: (a) 44.5 and 89 Hz;
(b) 2828 and 5657 Hz

5.6 Determine the bandwidth of a one-third octave band filter whose center frequency is
a.   180 Hz
b.   3150 Hz

Answer: (a) 41.8 Hz;
(b) 732 Hz

5.7 Constant percentage analyzers derive their name from the fact that the bandwidth is a constant percentage of the center frequency. Calculate the percentage for
a.   A one octave band analyzer
b.   A one-third octave band analyzer

Answer: (a) 71%
(b) 23%

5.8 Calculate the frequency resolution line spacing for an FFT analysis which covers from 0—1 kHz with 1600 lines.

Answer: 0.625 Hz

5.9 Determine the daily (8-h) noise exposure of a machine operator where the noise levels are constant at 95 dBA. Use an exchange rage of 5 dB and a threshold of 90 dBA.

Answer: 200%

5.10   Calculate the daily noise exposure of a pump room mechanic where noise levels are constant at 88 dBA.   Use an exchange rate of 5 dB and a threshold of 90 dBA.

Answer: 200%

5.11   Determine the daily exposure of a power plant engineer whose daily exposure is 90 dBA for 2 h and 95 dBA for 6 h.   Use an exchange rate of 5 dB and a threshold of 90 dBA.

Answer: 175%

II
NOISE CONTROL METHODS

At this point, the reader has a basic knowledge of the physics of sound and an introduction to those techniques for measurement and diagnostic analysis. These are the fundamental tools required to (1) identify and characterize the major sources of noise and (2) assess the magnitude of the required treatment to meet noise control design goals. In short, we are in a position to treat the problem.

One systematic approach which can be applied to most noise control problems is to break the problem down into a source-path-receiver diagram. Very simply, one recognizes that sound arises from a source or sources, travels over a path or paths, and affects a receiver or listener. Now it is a basic premise in noise control engineering that to reduce noise sensed by a receiver one must (1) lower the source noise through redesign or replacement; (2) modify the propagation path through enclosures, barriers, or vibration isolation; (3) protect or isolate the receiver; or (4) use some combination of each. Generally, source noise reduction is the most desirable in that cumbersome path modification and/or uncomfortable personal protection are avoided.

With source noise reduction, then, as top priority, we shall now focus attention on noise control methods.

6
ACOUSTICAL MATERIALS

A prominent acoustical consultant once said, "All materials are acoustical materials, but some are better than others." In this chapter we shall consider the various types of acoustical materials, their acoustical properties, and those parameters used to describe these properties.

Acoustical materials can be divided into three basic categories: (1) absorbing materials, (2) barrier materials, and (3) damping materials. Absorbing materials are generally resistive in nature, either fibrous, porous, or, in rather special cases, reactive resonators. Classic examples of resistive material are fibrous glass, mineral wools, felt, and polyurethane-type foams. Resonators include hollow core masonry blocks, sintered metal (honeycomb backed), etc. Effective barrier materials have one basic common property: dense mass. The most effective barrier materials also have a high degree of internal damping which qualitatively is described as *limpness*. Sheet lead is the best example of a dense massive limp barrier material. Damping materials are usually relatively thin coatings of plastic polymers, metal, epoxy, or glue which can be adhered to sheet metal panels, gears, machine parts, etc. With these coatings applied, the response of an impact blow to a sheet metal panel is a *dull thud* rather than a *ring*. Let us now consider each type in detail.

6.1 ABSORBING MATERIALS

The actual absorption of acoustical energy in porous and fibrous materials occurs as a transfer between aerodynamic and thermodynamic energies. In the case of porous materials, there is adiabatic and isothermal heat transfer due to the gaseous expansions and rarefactions along the *pore* edges and in the voids themselves. Which process is actually occurring depends to a large extent on the frequency of the incident energy. For fibrous materials, the fibers are forced to move, and a temperature rise occurs due to fiber bending and fiber-to-fiber friction. In short, the acoustic energy is transferred into heat in the material itself.

Now the parameter that best describes the absorption of materials is the absorption coefficient α. The absorption coefficient is essentially a measure of the acoustical energy absorbed by the material upon incidence and is usually expressed as a decimal varying between 0 and 1.0. As we shall soon see, the absorption coefficient depends dynamically on the angle of sound wave front incidence, and frequency.

To better understand the concept of acoustical absorption, let us consider the basic tube method for measuring the normal incidence absorption coefficient. The method outlined follows closely the American Society of Testing and Materials (ASTM) Standard C384 [1].

In the tube method, a sample of the material is placed at the end of the tube, as illustrated in Fig. 6.1. Discrete frequency sound waves generated by the loudspeaker propagate down the tube, impinge upon the sample, and are reflected. A standing wave interference pattern results due to the superposition of the incident and reflected wave. The following parameters are then measured with the movable microphone or probe:

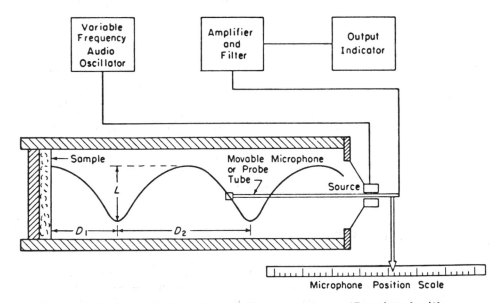

Figure 6.1 Acoustic impedance tube apparatus. (Reprinted with permission from the *Annual Book of ASTM Standards*, Copyright American Society of Testing and Materials, Philadelphia.

L = difference in decibels between the maximum and minimum sound
    pressure levels in the standing wave pattern in the tube
D_1 = distance from the face of the specimen to the nearest minimum
    in standing wave pattern, measured in any convenient units
D_2 = distance from the first to the second minimum in standing wave
    pattern, measured in the same unit as D_1

Now it can be shown [1−3] that the normal incidence sound absorption
coefficient (α_n) is given by

$$\alpha_n = 1 - \left(\frac{z/\rho c - 1}{z/\rho c + 1}\right)^2 \quad \text{[unitless]} \tag{6.1}$$

where

   z = specific normal acoustic impedance = r + jx (rayls)

   r = specific normal acoustic resistance

   $j = \sqrt{-1}$

   x = specific normal acoustic reactance

   ρc = characteristic acoustic impedance of free air (z, r, and x are
      customarily expressed in terms of their ratio to ρc)

Now Eq. (6.1) can be rewritten in terms of L [4,5], the difference between the maximum and minimum sound levels of the standing wave
measured in the tube:

$$\alpha_n = 1 - \left(\frac{\log_{10}^{-1}(L/20) - 1}{\log_{10}^{-1}(L/20) + 1}\right)^2 \tag{6.2}$$

Consider an example.

Example

The measured difference between adjacent maximum and minimum
levels in a standing wave tube was 8 dB at 1000 Hz. What is
the normal incidence absorption coefficient for the sample at
1000 Hz?

Solution

From Eq. (6.2), with the measured sound level difference L = 8
dB, we get

$$\alpha_n = 1 - \left(\frac{\log_{10}^{-1}(8/20) - 1}{\log_{10}^{-1}(8/20) + 1}\right)^2$$

$$= 0.815$$

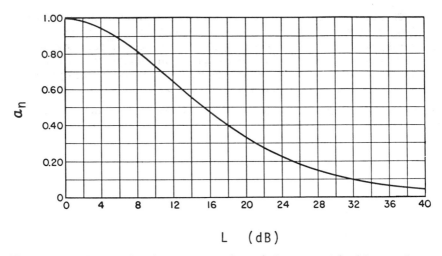

L   (dB)

Figure 6.2 Chart showing the relation of the normal incidence absorption coefficient α_n to the difference in decibels L between the measured maximum and minimum sound levels.

Hence, the normal incidence absorption coefficient is 0.815, or, on a percentage basis, about 82% of the incident energy was absorbed by the sample. Equation (6.2) is presented graphically in Fig. 6.2, which simplifies calculations and provides sufficient accuracy for most noise control applications.

Although this method yields directly the normal incidence absorption coefficient, unfortunately, in most practical situations, the noise is not normally incident. Therefore, to account for a wide range of incidence angles, a more applicable coefficient commonly called the statistical absorption coefficient α_{stat} can be determined when the resistive and reactive components of the impedance are known. These impedance components follow also from measurements determined by the *tube* method, which, as a matter of interest, is commonly called the impedance tube.

To determine the acoustic impedance $z = r + jx$ from tube measurements, it is again necessary to determine L, the standing wave ratio in decibels; D_1, the distance from the face of the specimen to the first minimum (referring to Fig. 6.1); and D_2, the distance between two successive minimums. The measured values are then substituted into the following equation [6] to obtain the specific acoustic impedance ratio:

$$\frac{Z}{\rho c} = \frac{r}{\rho c} + \frac{jx}{\rho c} = \coth(A + jB) \qquad \text{[unitless]} \qquad (6.3)$$

where

$$A = \coth^{-1}[\log_{10}^{-1}(L/20)] \text{ (unitless)}$$

$$B = \pi(1/2 - D_1/D_2) \text{ (unitless)}$$

Computational charts of Eq. (6.3) are shown in Figs. 6.3 and 6.4, from which $r/\rho c$ and $x/\rho c$ may be taken directly from the measured values of L and D_1/D_2. Consider an example.

Example

In an impedance tube the measured parameters at 500 Hz were L = 8, D_1 = 5 in., and D_2 = 11 in. What are the values of the resistive and reactive components of the impedance?

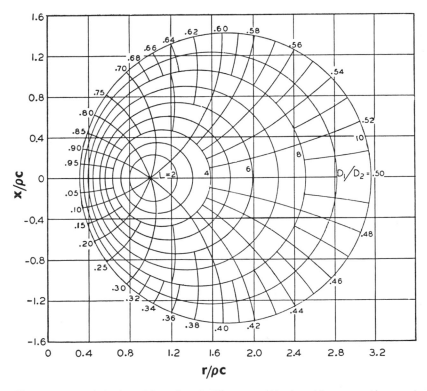

Figure 6.3 Relationship of specific acoustic impedance ratios and measured parameters L and D_1/D_2, L = 0 to 10. (Reprinted with permission from the *Annual Book of ASTM Standards*, copyright American Society of Testing and Materials, Philadelphia.)

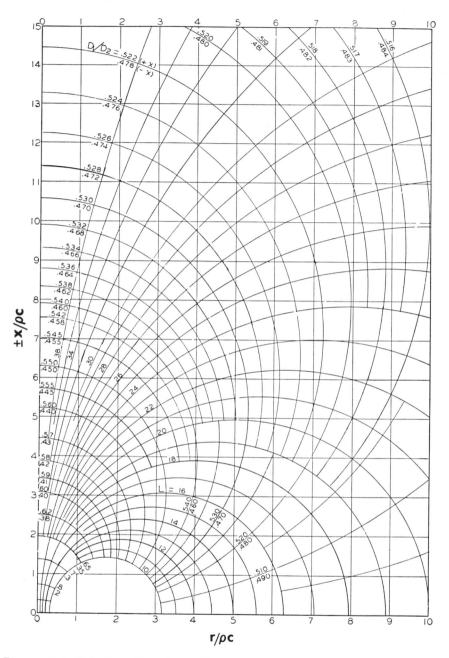

Figure 6.4 Relationship of specific acoustic impedance ratios and measured parameters L and D_1/D_2, L = 10 to 40. (Reprinted with permission from the *Annual Book of ASTM Standards*, copyright American Society of Testing and Materials, Philadelphia.)

Solution

$$\frac{D_1}{D_2} = \frac{5}{11} = 0.455$$

From Fig. 6.3, with $L = 8$ and $D_1/D_2 = 0.455$, the resistive component is $r/\rho c = 2.3$, and the reactive component is $x/\rho c = -0.7$ (approximately).

We now have the elements to calculate the statistical absorption coefficient, but a few more definitions are required. The ratio given in Eq. (6.3) is the specific acoustic impedance ratio,

$$\xi = \frac{Z}{\rho c}$$

and for convenience of calculation, it is desirable to define also the reciprocal of the impedance ratio η, which is called the specific acoustic admittance ratio:

$$\frac{1}{\xi} = \eta = \mu + j\kappa \qquad \text{[unitless]} \tag{6.4}$$

where

μ = specific acoustic conductance ratio

κ = specific acoustic susceptance ratio

In terms of these parameters, it is possible to compute the difference in intensity of the incident and reflected waves and obtain the absorption coefficient, which is the fractional loss of sound intensity, from the following [6]:

$$\alpha(\theta) = \frac{4\mu \cos \theta}{\kappa^2 + (\mu + \cos \theta)^2} \qquad \text{[unitless]} \tag{6.5}$$

Finally, statistically averaging over all incident angles θ, the statistical absorption coefficient is obtained:

$$\alpha_{stat} = 8\mu \left[1 - \mu \ln \left(1 + \frac{2\mu + 1}{|\eta|^2} \right) + \frac{\mu^2 - \kappa^2}{\kappa} \tan^{-1} \left(\frac{\kappa}{|\eta|^2 + \mu} \right) \right]$$

[unitless] $\hspace{5cm}$ (6.6)

Equation (6.6) has been computed for a wide range of absorption coefficients and is illustrated graphically in Fig. 6.5 [7]. The statistical absorption coefficient is given in terms of the specific resistance ratio $r/\rho c$ and specific reactance ratio $x/\rho c$, which are obtained directly from the impedance tube measurements. Consider an example.

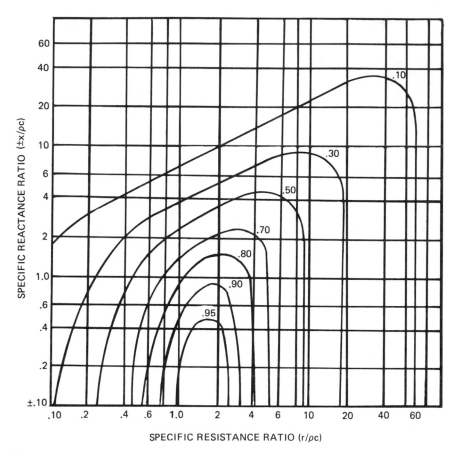

Figure 6.5 The statistical absorption coefficient in terms of the specific acoustic resistance and reactance ratios. (From Ref. 7.)

Example

From impedance tube measurements, the specific reactive and resistive ratios of a sample at 1000 Hz were found to be −2.0 and 4.0, respectively. What is the statistical absorption coefficient at 1000 Hz?

Solution

From Fig. 6.5, for $x/\rho c = -2.0$ and $r/\rho c = 4.0$, the statistical absorption coefficient α_{stat} is 0.70 (approximately).

The justification for the extent of this discussion can be seen by noting that for effective absorption, say values larger than 0.90, the following conditions must be met, referring again to Fig. 6.5:

1. The specific resistance ratio must be in the range $0.7 < r/\rho c < 3$.
2. The specific reactance ratio must be in the range $-1 < x/\rho c < 1$.

These ranges are rather narrow and limit drastically the number and types of materials which are effective absorbers.

It should be emphasized that the specific reactance ratio is naturally low for many fiberous or porous materials; thus condition 2 is easily met. However, condition 1 is satisfied in fibrous and porous materials only by careful control of the density, porosity, and thickness. For this reason, many *soft* and *fuzzy* materials used in packing crates and applied as a quick "fix" in noise reduction programs are disappointingly poor absorbers. For design purposes, the value usually selected to assure good absorption is

$$\frac{r}{\rho c} = 1.5 \quad \text{(approximately)} \tag{6.7}$$

Therefore, to assure effective absorption, from Eq. (6.7) the specific acoustic resistance r of fibrous or porous materials must be

$$r = 1.5\rho c$$

and for air, given $\rho c = 415$ mks rayls,

$$r = 1.5 \times 415$$

$$= 622 \text{ mks rayls}$$

In summary, for effective sound absorption, the acoustic impedance of the material and medium must be nearly equal or, as commonly expressed, *matched*.

For fibrous and porous materials, the real or resistive component of the acoustic impedance is usually determined experimentally. Here the flow resistance r of the material is calculated from measurements of the flow velocity and pressure drop across the sample [8]:

$$r = \frac{SP}{U} \quad \text{[mks rayls]} \tag{6.8}$$

where

P = air pressure difference across the test specimen (Pa)
U = volume velocity of the airflow through the specimen (m^3/s)
S = area of the specimen (m^2)

More often for homogenous materials, the resistivity r_0, which is the resistance per unit thickness, is the preferred parameter. The resistivity is then just a natural extension of Eq. (6.8) and is given by

$$r_0 = \frac{SP}{TU} \quad \text{(mks rayls/m)} \tag{6.9}$$

where T is the thickness of the material (m). In this way, the total resistance of a piece of absorbing material can be determined simply from its dimensions.

Example

The published resistivity of a certain polyurethane foam is 5000 rayls/m. What is the resistance and specific acoustic resistance ratio for a sample 10 cm thick? Use 415 mks rayls for ρc.

Solution

The resistance of the sample is just the resistivity multiplied by the thickness,

$$r = r_0 T$$

and for r_0 = 5000 rayls/m and T = 0.1 m, we get

$$r = 500 \text{ rayls}$$

The specific acoustic resistance ratio is then

$$\frac{r}{\rho c} = \frac{500}{415}$$

$$= 1.20 \qquad [\text{unitless}]$$

Comparing this value to the absorption criteria mentioned a few paragraphs earlier, i.e., $r/\rho c \cong 1.5$, one might conclude that a 10-cm-thick section of the material in the example is probably a reasonably good sound absorber. It must be emphasized again, however, that absorption depends strongly on the frequency and incidence angle of the sound. As such, for most noise control purposes, the statistical absorption coefficients are usually experimentally obtained by a more modern technique utilizing a diffuse sound field. In this method, controlled by commercial standard (ANSI/ASTM C423) [9], the decay rate of the sound is measured in a reverberant room containing a large sample of the material under study. By comparing the decay rate of the bare room to the rate including the sample, the statistical absorption coefficients at the preferred octave band center frequencies are determined. Because of the diffuse incidence of the sound energy, many acoustical engineers feel that absorption data obtained by the room method will probably be more representative in field applications. This subject is controversial, and we shall not take a position other than to say that differences of 10% to 20% are not uncommon between the two methods. However, as a practical matter, a difference of this magnitude is often not critical to a noise control design effort. In addition, because of the preponderance of data from the room method on commercially available absorbing materials, we shall hereafter make that inference. As a matter of interest, the coefficients obtained by the room method are called the *sabine absorption coefficients* after the noted acoustician Wallace Clement Sabine, who was a pioneer in the area of sound and noise control.

Shown in Fig. 6.6 is the typical sound absorption performance of commercially available acoustical-quality polyurethane foam. Note the increase in absorption at low frequencies as the foam thickness in-

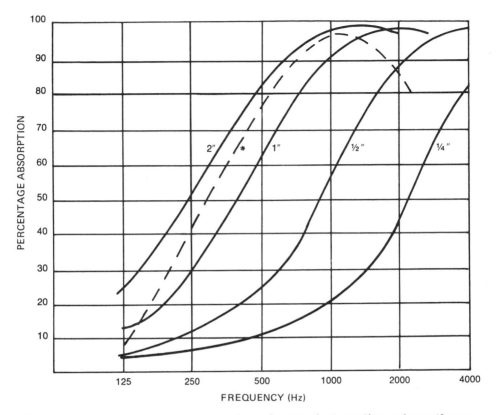

Figure 6.6 Typical sound absorption of acoustical-quality polyurethane foam at various thicknesses. *Represents 1 in. foam with protective film. (Courtesy of the Soundcoat Company, Inc., Deer Park, N.Y.)

creases. From this example, it is clear that significant absorption at low frequencies (below 500 Hz) is available only for resistive materials whose thickness is 1 in. or greater. Table 6.1 lists the absorption coefficients of some other commercially available materials commonly used in noise control.

Frequently, the performance of absorbing materials is given as a single number rating called the *noise reduction coefficient* (NRC). The NRC is just the mean or arithmetic average of the absorption coefficients at 250, 500, 1000, and 2000 Hz. For example, if we take the absorption coefficients of 1-in.-thick fibrous glass at 250, 500, 1000, and 2000 Hz (from Table 6.1) and compute the mean, we get

$$\frac{0.23 + 0.48 + 0.83 + 0.88}{4} = 0.60$$

Table 6.1 Statistical Absorption Coefficients for Common Commercially
Available Absorbing Materials

Material	Frequency (Hz)					
	125	250	500	1000	2000	4000
Fibrous glass (4 lb/ft^3)						
1 in. thick	0.07	0.23	0.48	0.83	0.88	0.80
2 in. thick	0.20	0.55	0.89	0.97	0.83	0.79
4 in. thick	0.39	0.91	0.99	0.97	0.94	0.89
Polyurethane foam (open cell)						
1/4 in. thick	0.05	0.07	0.10	0.20	0.45	0.81
1/2 in. thick	0.05	0.12	0.25	0.57	0.89	0.98
1 in. thick	0.14	0.30	0.63	0.91	0.98	0.91
2 in. thick	0.35	0.51	0.82	0.98	0.97	0.95
Hairfelt						
1/2 in. thick	0.05	0.07	0.29	0.63	0.83	0.87
1 in. thick	0.06	0.31	0.80	0.88	0.87	0.87

Thus, the NRC of the 1-in.-thick fibrous glass is 0.60. This parameter is most often used in architectural acoustics and in particular
with designing or selecting suspended ceilings for control of reverberation. We shall consider more applications of this parameter in a later
chapter.

Shown in Fig. 6.7 are some commercially available absorbing materials of the porous type. One of the most common applications of
these materials is to reduce reverberant sound buildup in small machinery enclosures. In this application they are usually installed on or
in the walls of the enclosure. To facilitate installation, a pressure-
sensitive adhesive is applied to one side, which simplifies attachment.
In addition, to avoid the absorption of oil, solvents, etc., a thin 1- or
2-mil (0.001- or 0.002-in.) plastic film facing is applied to one surface
or totally encapsulates the material. This lightweight material reduces
slightly the absorption of higher frequencies but usually improves the
performance in the lower range, as can be seen by again referring
to Fig. 6.6.

It should be noted from Table 6.1 that the absorption performances
of the foam and fibrous glass are nearly equal. However, because of
the ease of handling, forming, and cutting, the foams are often preferred. On the other hand the foams do not possess the resistance to

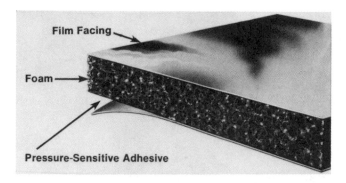

(a)

(b)

Figure 6.7 Examples of commercially available foam-absorbing material showing (a) applied film and adhesive and (b) anechoic wedge shape with improved low-frequency performance. [Courtesy of (a) Specialty Composites, and (b) Illbruck, Inc., Minneapolis, Minn.]

oils, chemical solvents, heat, and other harsh industrial environments of fibrous glass or mineral wools.

For example, shown in Fig. 6.8 is a saw blade guard where the interior was lined with acoustical-quality foam. With this treatment, a noise level reduction of 6 to 8 dB was measured in the 1000-Hz octave band at idle (not cutting). For saw blades at idle,

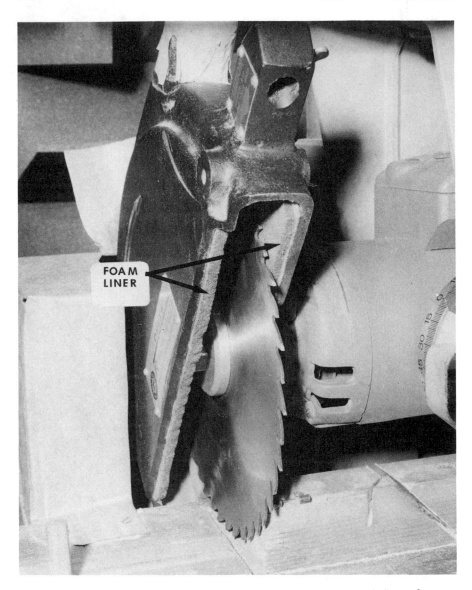

FOAM
LINER

Figure 6.8 Saw blade guard treated with an absorptive lining of acoustical-quality foam. (Courtesy of the Soundcoat Co., Deer Park, N.Y.)

1. The noise is aerodynamic in origin, as the blade tips shear through the air.
2. There may also be resonant blade vibration present.

It should be emphasized that aerodynamic noise or blade-radiated noise generated outside of the guard would not be affected by the absorbing materials inside the guard.

6.1.1 Fibrous Metal

Sintered metal-absorbing materials made by diffusion bonding of metal fibers are also commercially available. These materials, generally produced in sheet form, provide many advantages over foam or fibrous glass materials. In particular, they have high temperature resistance and structural features such as stiffness, weldability, machinability, and so forth. Further, these materials can be cleaned, perform well when wet, and are nonflammable.

Shown in Fig. 6.9 are examples of the surface character of these materials. Generally, thin sheets of these materials (0.020" to 0.080") are used in conjunction with a solid metal backing plate. That is, the fibrous metal sheet is spaced ¼" to 1.0" away from the backing plate. The spacer is often a metallic honeycomb-like material which provides rigid mechanical support. If the flow resistance of the sheet and the spacing between sheet and plate are carefully selected, effective absorption is achieved.

In short, if a fiber metal sheet having the correct resistance is placed ¼ wavelength from the plate, peak absorption occurs near the frequency corresponding to the selected wavelength. In a sense, the configuration can be "tuned" to provide optimal performance at a given frequency. Consider the following example.

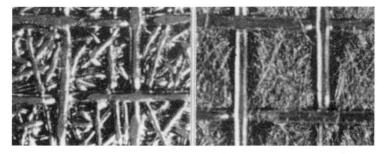

Figure 6.9 Surface characteristics of fiber metal absorbing materials (× 15). [Courtesy of Technetics Corporation, De Land, Fla.]

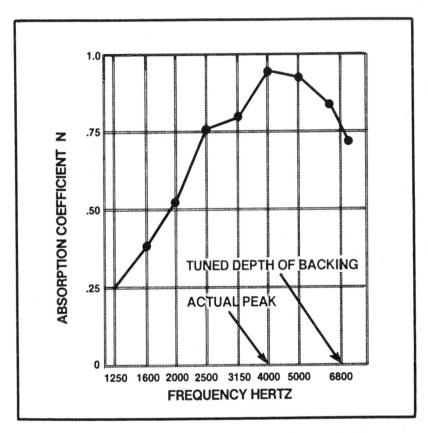

Figure 6.10 Normal incidence absorption coefficient vs. frequency for ½" cavity depth. [Courtesy of Technetics Corporation, De Land, Fla.]

Shown in Fig. 6.10 is the acoustical performance of a fiber metal sheet separated from the backing plate by a $\frac{1}{2}$"-thick metallic honeycomb spacer. Note that the peak absorption (4000 Hz) occurs a little below the tuned depth of the spacing cavity (6700 Hz). It must be emphasized that the absorption mechanism for these composites is quite complex, given that the diameter of the honeycomb cell is reactive in nature.

In summary, fibrous metal materials have been used extensively in the gas turbine, aerospace, and petrochemical industries. In these applications, other lightweight absorbing materials often do not have the integrity to withstand the harsh thermal environments or heavy mechanical stress.

6.2 BARRIER MATERIALS

The parameter that describes the isolation or *sound-stopping* capability of a wall or panel is the transmission coefficient τ. Now the transmission coefficient is defined as the ratio of the sound power transmitted across a partition to the power incident. In mathematical form, we have

$$\tau = \frac{W_t}{W_i} \qquad (6.10)$$

where

W_t = sound power transmitted (W)
W_i = sound power incident (W)

Recognizing the relationship among sound power, sound intensity, and sound pressure, it can be shown that for a homogeneous wall and plane waves the transmission coefficient is [10]

$$\tau(\theta) = \frac{I_t}{I_i} = \frac{|p_t^2|}{|p_i^2|}$$

$$= \left\{ \left[1 + \eta \left(\frac{\omega \rho_s}{2\rho c} \cos \theta \right) \left(\frac{\omega^2 B}{c^4 \rho_s} \sin^4 \theta \right) \right]^2 \right.$$

$$\left. + \left[\left(\frac{\omega \rho_s}{2\rho c} \cos \theta \right) \left(1 - \frac{\omega^2 B}{c^4 \rho_s} \sin^4 \theta \right) \right]^2 \right\}^{-1} \qquad \text{[unitless]}$$

$$(6.11)$$

where

I_t = sound intensity transmitted (W/m^2)

I_i = sound intensity incident (W/m^2)

p_t = sound pressure transmitted (Pa)

p_i = sound pressure incident (Pa)

θ = angle of incidence (rad)

η = composite plate loss factor (unitless)

ρ_s = plate surface density (kg/m^2)

B = plate bending stiffness per unit width (N/m)

ρ = density of medium (kg/m^3)

c = speed of sound in medium (m/s)

Equation (6.11) is not very useful in this form due to the strong dependence on incidence angle. As such, to bring the sound transmission concept into a more consistent and practical form, a logarithmic term called the transmission loss TL was defined and is given as

$$TL = 10 \log \frac{1}{\tau} \qquad [dB] \qquad (6.12)$$

where τ is the transmission coefficient averaged over all angles of incidence.

Now by combining Eqs. (6.11) and (6.12) and under the assumption that the term $\omega^2 B/c^4 \rho_s$ is small, i.e., much less than 1 (which will have more significance later), the transmission loss of a panel can be approximated over a wide range of usefulness as

$$TL(\theta) \approx 10 \log \left[1 + \left(\frac{\omega \rho_s}{2\rho c} \cos\theta \right)^2 \right] \qquad [dB] \qquad (6.13)$$

Equation (6.13) is referred to as the *limp-wall mass law transmission loss* because the only characteristic of the wall involved is its surface mass. It is worth noting that for sound impinging at normal incidence to the wall ($\theta = 0°$) the transmission loss is greatest. Correspondingly, the TL vanishes at grazing incidence, i.e., at $\theta = 90°$.

In most practical applications, the sound is incident over a wide range of angles, and Eq. (6.13), after averaging, further reduces to the more common form

$$TL = 20 \log (f) + 20 \log (W) - C \qquad [dB] \qquad (6.14)$$

where

f = frequency (Hz)

W = surface density (lb/ft^2/in. or kg/m^2/cm)

$C = 33$ if W is in lb/ft^2/in.

$C = 47$ if W is in kg/m^2/cm

Equation (6.14) is the limp-wall mass law for random incidence, which agrees well with experimental field data with some important limitations. We shall discuss these later, but first an example.

Example

Calculate the transmission loss at 1000 Hz of sheet lead 1/64 in. thick.

Solution

From Table 6.2, we see that lead 1/64 in. thick has a surface density of approximately 1 lb/ft^2. Hence, upon substitution in Eq. (6.14),

$$TL = 20 \log(1000) - 20 \log(1) - 33$$

$$= 60 - 33$$

$$= 27 \text{ dB}$$

From this example, we would expect a transmission loss of 27 dB across a 1/64-in. sheet lead barrier at 1000 Hz, and this agrees well with measured field data. Further, sheet lead is the classic example of a *limp* barrier material which the assumptions leading to Eq. (6.14) required. This limpness or lack of stiffness is due to the molecular structure of the material. Unfortunately for acoustical engineers, most construction materials are not limp and respond to incident sound, as illustrated in Fig. 6.11

Table 6.2 Surface Densities of Certain Common Building Materials

Material	Surface Density	
	lb/ft^2/in. of thickness	kg/m^2/cm of thickness
Brick	10—12	19—23
Concrete blocks	8	15
Dense concrete	12	23
Wood	2—4	4—8
Common glass	15	29
Lead sheets	65	125
Gypsum board	5	10
Steel	54—56	108—112

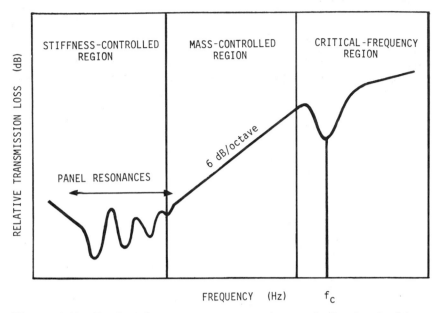

Figure 6.11 Typical frequency response (transmission loss) of homo-genous panels.

At the lower frequencies, the transmission loss fluctuates due to the presence of natural vibratory resonances of the wall. To understand this phenomenon, consider the result of impacting a steel plate or piece of sheet metal clamped at the edges. A ringing sound will be heard and, if analyzed, would contain discrete tones at the natural frequencies corresponding to the natural modes of vibration of the plate. There are an infinite number of these modes. At each of these natural frequencies, the barrier is nearly transparent to sound, or, better, the transmission loss approaches zero. In fact, the transmission loss would be zero if there were not some internal molecular *damping* present in the material.

As the name implies, damping has a strong effect on the transmission in the stiffness region. Materials such as steel have little internal damping and vibrate for a long time when struck, whereas materials such as rubber or lead have high internal damping and the sound decays very rapidly. With little internal damping, the resonant *dips* are steep and well defined; as the damping increases, the dips "smooth" out.

It follows, then, that the effectiveness of a barrier can be enhanced by addition of damping. We shall elaborate on this aspect when we discuss damping materials later in the chapter.

Above the stiffness-controlled region, we enter the mass-controlled region where the transmission is determined only by the mass. In this region, Eq. (6.14) applies, and on examination, the following conclusions can be drawn:

1. The transmission loss increases at a rate 6 dB per doubling the frequency.
2. Doubling the surface density also effects a 6-dB increase in transmission loss.

With respect to the first conclusion, we see that the acoustical performance of barrier materials generally increases with frequency just as was the case for absorbing materials. It is from this observation that the often-heard comment "low frequency noise is generally harder to control that high" has more meaning.

The second conclusion provides insight into the disappointing fact that doubling the thickness of a wall will yield a mere 6-dB increase in transmission loss. As mentioned earlier, only under rather controlled conditions can the human ear detect the difference. Fortunately, as we shall see later in the chapter, the mass law performance can be exceeded by the use of multiple panels with air cavities or in combination with absorbing materials.

At higher frequencies, in the critical frequency region, one further resonantlike effect appears which reduces transmission loss of barrier materials below mass law predictions. In this instance, a flexural *bending* wave is excited and propagates in the material, as illustrated in Fig. 6.12 Here the sound waves impinge upon the material at an incident angle θ_i such that the projected sound wave and the flexural wave are coincident or *in phase*. There is, in a sense, a continuous reinforcement of the flexural wave as the projected sound wave propagates along the panel. This condition is met when, referring again to Fig. 6.12 [10],

$$\lambda_f = \frac{\lambda}{\sin \theta_i} \qquad [m] \tag{6.15}$$

or

$$\sin \theta_i = \frac{\lambda}{\lambda_f} \qquad [m] \tag{6.16}$$

where

λ_f = wavelength of flexural wave in material (m)

λ = wavelength of incident sound (m)

θ_i = angle of incidence (rad)

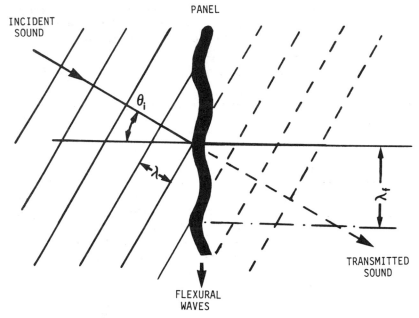

Figure 6.12 Coincidence of incident wave and flexural wave in a panel.

It should be noted that the coincidence condition can be met over a wide range of frequencies and incidence angles. However, in most common construction materials, the coincidence frequencies can be calculated to the first order by the following equation [11]:

$$f = \frac{c^2}{1.8Ut \, \sin^2\theta_i} \quad [Hz] \tag{6.17}$$

where

U = velocity of flexural wave in material (m/s)

c = speed of sound in air (m/s)

θ_i = angle of incidence

t = thickness of material (m)

Examining Eq. (6.17), one sees the following:

1. The coincidence frequencies vary inversely with panel thickness; i.e., the thinner the material, the higher the frequencies.
2. Coincidence occurs at different frequencies for different materials since the speed of sound depends on the kind of material.

3. The lowest frequency at which the coincidence occurs is at grazing
   incidence or $\theta_i = 90°$. This condition is called the critical frequency.

Thus, for this special case, $\theta_i = 90°$, one can compute the critical frequency f_c from Eq. (6.17).

Example
   What is the critical frequency for a plate of steel 1/4 in. (0.25
   in.) thick? For the speed of sound in steel, use 16,000 ft/s,
   and for air, 1100 ft/s.
   Solution
   The lowest coincidence frequency occurs when sin $\theta_i = 1.0$, and
   upon substitution in Eq. (6.17), we get

$$f_c = \frac{1100^2}{1.8 \times 1.6 \times 10^4 \times (0.25/12) \times 1.0}$$

$$= 2016 \text{ Hz}$$

   For reference and convenience, plotted in Fig. 6.13 [11] are the
critical frequencies as a function of plate thickness for some common
construction barrier materials.
   Because this coincidence phenomenon depends on flexural bending
of the material, high internal damping or the application of external
damping will smooth out or minimize the coincidence dips. Thus, from
this discussion, it is now easy to see why dense limp materials such
as sheet lead or lead and barium-loaded vinyls are preferred by
acoustical engineers.
   It should also be emphasized that in most noise control design
situations the acoustical engineer does not need to calculate the transmission loss of the barrier materials under consideration. The reason
for this is that for most popular barrier materials the transmission
loss has been measured experimentally in commercial laboratories
under strict standard control. The results of these measurements have
been published by the manufacturers or agencies and associations
sponsoring research and development. The experimental procedure
most often used by far follows the ASTM Standard E90 (Standard
Recommended Practice for Laboratory Measurement of Airborne Sound
Transmission Loss of Building Partitions). The basic procedure is to
mount the test specimen as a partition between two reverberation rooms.
Sound is then introduced into one of the rooms (the source room), and
measurements are made of the noise reduction (NR) between the source
room and the receiving room. The rooms are so arranged and constructed that the only significant sound transmission between them is
through the test specimen. Then the transmission loss is given by
[12]

$$TL = NR + 10 \log \left(\frac{S}{A} \right) \qquad [\text{dB}] \qquad (6.18)$$

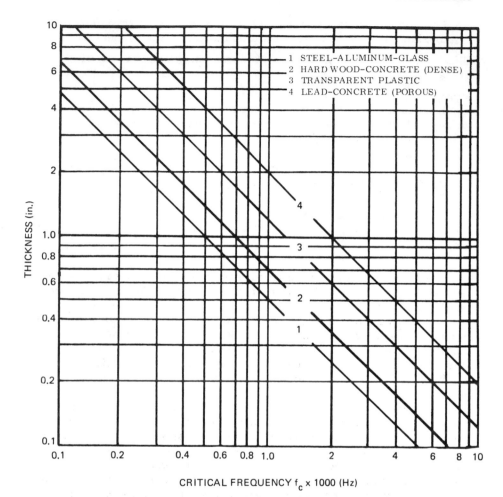

CRITICAL FREQUENCY f_c x 1000 (Hz)

Figure 6.13 Critical frequency f_c plotted as a function of plate thickness. This is the lowest frequency at which the coincidence effect is possible. At this frequency, the TL is sharply reduced.

where

    S = total area of the sound-transmitting surface of the test specimen (m^2)

    A = total absorption in the receiving room, expressed in units consistent with S

    NR = $L_s - L_r$, the noise reduction between the two reverberation rooms; i.e., the sound pressure level in the source room is L_s, and L_r is the sound pressure level in the receiving room

The noise reduction NR is measured at each of the preferred one-third octave bands from 125 to 4000 Hz. Shown in Fig. 6.14 is the actual transmission loss of a door as measured in a commercial laboratory under the ASTM E90 procedure. This door was not a homogeneous panel but rather a complex composite of sheet metal, mineral wool, gypsum, etc., arranged to enhance the transmission loss beyond the mass law. The point being made here is this: The acoustical performance of good-quality panels, walls, doors, windows, etc., is available to the engineer from laboratory measurements.

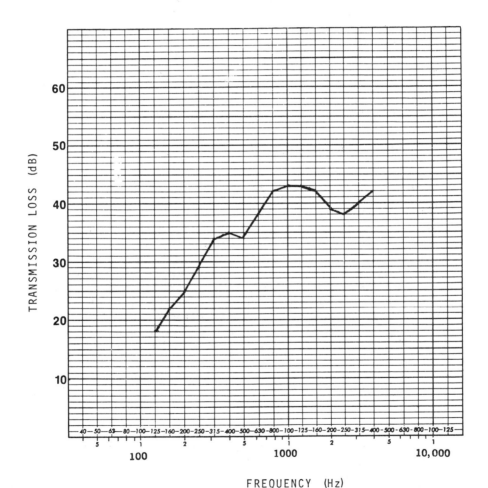

Figure 6.14 Actual transmission loss for a door of 9.5-lb/ft^2 surface density as measured in a commercial laboratory (ASTM E90).

Table 6.3 lists the spectral transmission loss of some popular materials, panels, and walls used in the design of acoustical enclosures
or as isolation barriers.

As one notes, transmission loss is highly dependent on frequency.
Numerous attempts have been made to condense this into a single number. By far the most popular is called the sound transmission class
(STC). The STC is determined by comparing the set of transmission
losses at all 16 frequencies to a set of standard curves as described
in ASTM Standard E413 (Classification for Determination of Sound
Transmission Class) [9].

The STC value is obtained by fitting a specified transmission loss
curve to the test data. When this curve is fit to the test data, its
value at 500 Hz is the STC value. These STC values provide rather
good measures of comparison for architectural structures such as
walls, doors, windows, etc. In this way, an architect can select
from an assortment of construction configurations well defined in terms
of laboratory measurements. We shall not pursue the matter further
at this time, and the interested reader is encouraged to consult the
References and Chap. 15, in which we deal with building acoustics.

6.2.1 Multiple Layer Panels

Because of the analytical complexity, we shall not discuss in detail the
physical behavior of multiple layer barriers. We shall, however, point
out that two 4-in.-thick masonry walls spaced 4 in. apart will perform
better as a barrier than a single 8-in. wall. Likewise, two 18-gauge
sheets of steel separated by 1 in. of absorbing foam will provide more
transmission loss than if mounted *back to back*. In general, as the
spacing between the barrier layers increases, the transmission loss increases. The magnitude of the increase also depends to the first
order on the acoustical coupling between the layers, which is related
to the impedance match or mismatch of the materials at each interface.
Fortunately, for most design purposes, measured values are generally
available, and we shall illustrate their application along with design
guidelines in Chap. 7.

6.2.2 Composite Barriers

Frequently a barrier wall will be composed of two or more elements,
i.e., doors, windows, etc., and the average transmission loss of the
composite wall TL_c can be calculated to the first order from [13]

$$TL_c = 10 \log\left(\frac{1}{\bar{\tau}}\right) \qquad [dB] \qquad\qquad (6.19)$$

Table 6.3 Transmission Loss of Some Common Materials Used for Acoustical Enclosures and Isolation Barriers

Material	lb/ft^2	Frequency (Hz)						
		125	250	500	1000	2000	4000	8000
Lead								
1/32 in. thick	2	22	24	29	33	40	43	49
1/64 in. thick	1	19	20	24	27	33	39	43
Plywood								
3/4 in. thick	2	24	22	27	28	25	27	35
1/4 in. thick	0.7	17	15	20	24	28	27	25
Lead vinyl	0.5	11	12	15	20	26	32	37
Lead vinyl	1.0	15	17	21	28	33	37	43
Steel								
18 gauge	2.0	15	19	31	32	35	48	53
16 gauge	2.5	21	30	34	37	40	47	52
Sheet metal (viscoelastic laminate core)	2	15	25	28	32	39	42	47
Plexiglas								
1/4 in. thick	1.45	16	17	22	28	33	35	35
1/2 in. thick	2.9	21	23	26	32	32	37	37
1 in. thick	5.8	25	28	32	32	34	46	46
Glass								
1/8 in. thick	1.5	11	17	23	25	26	27	28
1/4 in. thick	3	17	23	25	27	28	29	30
Concrete, 4 in. thick	48	29	35	37	43	44	50	55
Concrete block, 6 in.	36	33	34	35	38	46	52	55

where

$$\bar{\tau} = \frac{S_1\tau_1 + S_2\tau_2 + \cdots + S_n\tau_n}{S_1 + S_2 + \cdots + S_n} \qquad \text{[unitless]} \qquad (6.20)$$

Now, since the denominator of Eq. (6.20) is just the total surface area S of the barrier; we get, finally,

$$\bar{\tau} = \frac{1}{S} \sum_{i=1}^{n} \tau_i S_i \qquad \text{[unitless]} \qquad (6.21)$$

where

τ_i = transmission coefficient of the ith element (unitless)

S_i = surface area of the ith element (m^2)

Consider an example.

Example

A wall with a transmission loss of 40 dB at 500 Hz separates two food packaging areas. A window with a transmission loss of 20 dB at 500 Hz is to be installed which will occupy 10% of the wall. Calculate the resultant composite transmission loss of the wall at 500 Hz.

Solution

From Eq. (6.20),

$$\bar{\tau} = \frac{S_1\tau_1 + S_2\tau_2}{S_1 + S_2}$$

$$= \frac{S_1\tau_1}{S} + \frac{S_2\tau_2}{S}$$

where

S_1 = surface area of the window

S_2 = surface area of the wall

S = total surface area, i.e., $S_1 + S_2$

τ_1, τ_2 = transmission coefficients of the window and wall, respectively

Now since $S_1/S = 10\%$ and $S_2/S = 90\%$,

$$\bar{\tau} = 0.1\tau_1 + 0.9\tau_2$$

From the definition or TL, referring to Eq. (6.12), we obtain

$$\frac{1}{\tau_1} = \text{antilog}\left(\frac{20}{10}\right)$$

$$= \text{antilog } 2$$

$$= 100$$

Therefore, $\tau_1 = 0.01$, and, correspondingly, $\tau_2 = 0.0001$. Upon substitution,

$$\bar{\tau} = 0.1 \times 0.01 + 0.9 \times 0.0001$$

$$= 0.00109$$

Finally, from Eq. (6.19),

$$TL_c = 10 \log\left(\frac{1}{0.00109}\right)$$

$$= 29.6 \text{ dB}$$

Hence, the transmission loss TL_c of the composite wall at 500 Hz is 29.6 dB.

It is obvious from this example that a small area of relatively low transmission loss can severely compromise the acoustical performance of a high-quality barrier. Further, as we shall see later, small sound *leaks* can virtually scuttle a well-designed acoustical enclosure.

6.3 DAMPING MATERIALS

In Sec. 6.2, we saw that the sound barrier properties of materials are governed primarily by mass, stiffness, and damping. Of these three, the term damping is the least understood and often the subject of confusion. In a strict sense, damping is used to describe the dissipation of mechanical energy associated with vibration, i.e., usually the conversion of this energy of motion to heat. A *damping material* is any material that is applied to a structural member to increase its damping. The noise reduction capability of applying damping materials derives from the fact that once the mechanical energy is dissipated, it is not reradiated in the form of airborne sound or conducted along structurally. For example, if one touches a *ringing* cymbal, the noise level decays sharply due to damping provided by the fingers. In this example, the mechanical energy associated with the vibrating cymbal is converted directly to heat. Because so little acoustical energy is involved, the temperature rise at the point of touch will not be noticed. In short summary, damping materials are an effective way of reducing the amplitude of mechanical vibration.

All materials have some inherent damping, and this damping mechanism is derived from complex dissipative internal molecular friction.

Specifically, during deformation, there is a rubbing or sliding action between molecules or molecular layers, which explains why the vibration amplitude of a fixed-free beam decays in a vacuum. Some materials, because of their molecular structure, can store a considerable amount of strain energy when deformed and dissipate it quickly because it "flows" like a viscous fluid. Most rubbers and plastics possess this property along with elasticity and are thus called viscoelastic materials. It is from this combination and from this generic type of material that the most effective damping treatments originate.

There are three parameters which are often used to describe the characteristics of damping materials, and they are to some extent interrelated: (1) loss factor, η; (2) decay rate, Δ; and (3) damping factor, ξ.

The loss factor η is defined in terms of the energy dissipated or the damping-to-stiffness ratio:

$$\eta = \frac{D}{2\pi W_0} \qquad \text{[unitless]} \tag{6.22}$$

where

D = energy dissipated per cycle of vibration

W_0 = average total (kinetic plus potential) energy of the vibrating system

The decay rate Δ is an experimental value obtained by measuring the decay of a freely vibrating sample. The amplitude decay generally varies exponentially with time and thus is linear if plotted logarithmically. Further, the loss factor η is related to the decay rate as follows:

$$\eta = \frac{\Delta}{27.3f} \qquad \text{[unitless]} \tag{6.23}$$

where

Δ = decay rate (dB/s)

f = decay frequency (Hz)

The damping factor ξ is defined as the ratio of material damping to critical damping:

$$\xi = \frac{C}{C_c} \tag{6.24}$$

where

C = viscous damping coefficient of material (N-s/m)

C_c = critical damping coefficient (N-s/m)

The significance of Eq. (6.24) can be seen when one realizes that the critical damping coefficient is the smallest value of C for which a sample does not oscillate when deflected and then released. In short, ξ is a measure of how close a material is to being critically damped or that fraction thereof.

The loss factor η is simply related to the damping factor as follows:

$$\eta = 2\xi \tag{6.25}$$

Now, when the damping factor is multiplied by 100, the percent of critical damping is obtained.

Equation (6.23), which relates loss factor to decay rate, is by far the most useful and practical of the relationships given. Specifically, the term used most often to describe the effects of applying damping material, the loss factor, is derived from measured experimental data.

With respect to the measurement procedure, there are numerous methods currently being used and some standards under study. However, from a noise control standpoint, the engineer is rarely interested in how much damping a material possesses but rather in what the noise reduction is after some damping treatment has been applied. As an effective noise control measure, the loss factor generally must exceed 0.1. Table 6.4 lists the range of loss factors for some common structural materials.

Referring again to the experimental determination of loss factor, the complex moduli apparatus and Geiger plate methods are the most commonly used for evaluating or comparing the dissipative performance of damping materials. There is some controversy over the range of applicability of these methods, and we shall not comment, but those readers desiring further knowledge of the details are referred to the References at the end of the chapter.

Table 6.4 Loss Factors for Some Common Materials

Material	Loss Factor, η
Metals	0.0001–0.002
Glass	0.0006–0.002
Concrete, brick	0.01–0.05
Sand, gravel	0.12–0.6
Wood, cork, plywood	0.01–0.03
Rubber, plastic	0.0001–10

Source: Ref. 14.

Let us now consider the practical applications of damping materials as a noise control measure. External damping can be applied in three basic ways:

1. By applying a coating of damping material (free layer) with a high loss factor directly to the surface
2. By applying a constraining thin sheet metal layer along with a coating of damping material
3. By designing and constructing the critical parts of a prefabricated constrained laminate composite, as illustrated in Fig. 6.15.

6.3.1 Free-Layer Damping

With respect to the free-layer single coat, these damping materials are supplied either in semifluid form for application with a spray gun or trowel or in precured sheets. One disadvantage with the spray gun or trowel application involves the often lengthy and often critical temperature-controlled curing process required. However, irregular and complex surfaces can be treated, and any thickness can be applied. The precured sheets are usually supplied with a pressure-sensitive adhesive for ease of application on reasonably flat surfaces. In addition they can be die-cut to cover or fit even the most complex surfaces shapes, as illustrated in Fig. 6.16.

One of the most dramatic noise reduction measures utilizing the free-layer-type material occurred in the area of mass transit railway systems. Measurements and analyses showed that the high-frequency *screech* commonly heard as a subway car moves slowly around a curve was due to the *stick-slip* excitation between the rail and the wheel. Further studies also showed that the principal radiation was the wheel or "truck" and not the rail. A thick application (3/8 in.) of a viscoelastic material was cured on the outer rim of the wheel, as illustrated

Figure 6.15 Prefabricated constrained laminate composite materials. (Courtesy of Antiphon Inc., Bloomfield Hills, MI.)

Figure 6.16 Examples of die-cut free-layer damping sheets. (Courtesy of the Soundcoat Company, Deer Park, N.Y.)

in Fig. 6.17. With this treatment, the noise level was reduced more than 30 dB, and the screech was virtually eliminated.

Design guidelines for applying free-layer damping materials can be summarized rather simply. With respect to thickness, (1) a thin coating on sheet metal, one-half metal thickness or 10% by weight, will eliminate the *ring* from shock excitation, and (2) an application of 2 to 3 times the part or plate thickness will bring the composite or combined loss factor of the treated plate to the range of 0.3 to 0.6 or to within 15% to 30% of critical damping.

Thickness

Shown in Fig. 6.18 is the typical effect of free-layer damping material thickness on the loss factor of sheet steel. Note that after relative

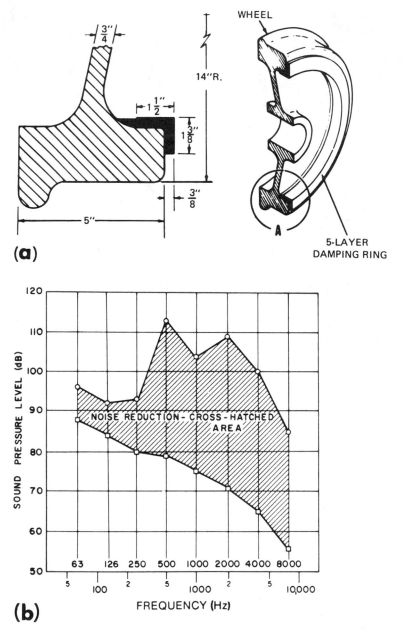

(a)

(b)

Figure 6.17 (a) Damping material applied to railway wheel and (b) octave band analysis showing noise reduction achieved. (Courtesy of the Soundcoat Company, Deer Park, N.Y.)

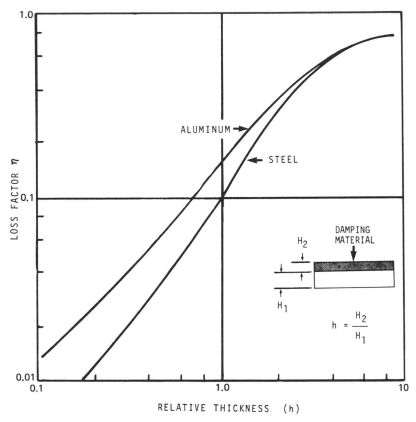

Figure 6.18 Chart showing the effect of damping layer thickness on the loss factor for aluminum and steel at 20°C.

thickness ratios above 3.0, little is gained in loss factor. This is not usually a problem since greater ratios are likely to be prohibitive from a practical standpoint. It must be emphasized that the effect of thickness as shown in Fig. 6.18 should be considered illustrative. It is far more complex than presented and depends to the first order on the ratio of the elastic moduli of the damping material and base material and secondarily on frequency. But more important is the effect of temperature.

Temperature

Shown in Fig. 6.19 is the typical effect of temperature on the loss factor for a commercially available damping material at a nominal relative thickness of 2.0. As one notes, there is a rather narrow range of

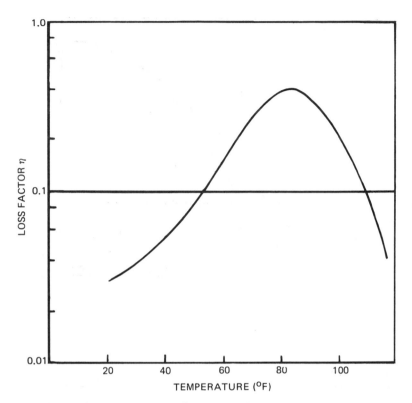

Figure 6.19 Dependence of the loss factor on temperature.

temperature over which the damping material is optimum, because at
lower temperatures, called the *glossy* region, the damping material is
hard and sometimes brittle. As such the viscous friction on which
the energy dissipation depends so strongly is reduced. Generally, at
higher temperatures in the *rubbery* region, the material is softer or
more fluidlike, with a corresponding decrease in dissipative capability.
In the *transition* region, however, the loss factor is at its peak, and
obviously, it is this temperature range to which one should design.
Now the transition region will vary from −150°F for some silicon
elastomers through 200°F for some epoxy and plastic materials and to
2000°F for some enamels. In short, one can, from commercially available materials, usually select a damping material whose damping will
be nearly optimum for the temperature range of interest. Consult
manufacturer's specifications, and follow closely the recommendations
regarding the effect of temperature on performance in all applications.
It is also important to remember that even though the composite loss
factor of the treated part is not optimum, it probably will be 1 or 2

orders of magnitude better than untreated parts. As such the noise reduction obtained is often surprisingly high.

Bonding

For effective damping, the viscoelastic material must be well bonded to the structural member. If the material selected is not self-adhering, a good-quality stiff adhesive such as epoxy should be considered.

Location

Finally, since energy is dissipated only when the damping material is distorted, application in the area of vibration antinodes (peak displacement amplitudes) is strongly recommended. In short, damping treatment applied only at nodal points (points of no strain) may as well be left in the can.

6.3.2 Constrained Layer Damping

When a rigid constraining layer is attached firmly to a viscoelastic layer, the effective damping is generally increased. The sharp difference is due to the addition of shear strain in both the viscoelastic and the constraining layer, as illustrated in Fig. 6.20. In the shearing deformation, the rate of energy conversion is much greater than with the tensile strain. In fact, it tends to dominate the damping.

With respect to the constraining layers for constrained-layer damping, they must possess high extensional stiffness. Metal sheets or foils are used because of their inherent high moduli, and steel is the first choice on the basis of modulus. Galvanized or stainless steels are preferred for their resistance to corrosion. Aluminum is the first choice on the basis of weight and ease of handling. Other good choices for special applications are brass, magnesium, or even titanium. High-performance composite materials, such as reinforced epoxy materials, can also be used with reasonable success in some applications.

Often these commercially available materials resemble metallic tape. That is, the constraining layer may be a *foil* as thin as 0.005 in. and ranging upward to 0.080 in., with a pressure-sensitive bonding adhesive on one side. The adhesive then becomes the viscoelastic layer. With respect to bonding, the criterion is to select an adhesive possessing good *peel* resistance and a high shear strength. Naturally, the adhesive must maintain a strong tack or adhesion to the substrate over the entire range of temperature exposure. Thin adhesive layers tend to have better adhesion with the added feature of higher shear stiffness.

The advantages of constrained-layer over free-layer damping systems can be summarized as follows:

1. Space saving: thinner treatments required
2. Weight saving: lighter treatments required

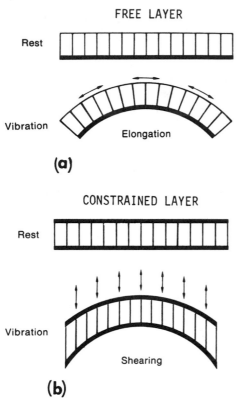

Figure 6.20 Free-layer and constrained-layer damping treatment illustrating (a) tensile strain and (b) shearing strain.

3. Durability: higher resistance to environmental attack
4. Installation: no drying or curing; no special equipment needed
5. Time saving: can be used immediately

The damping performance of constrained layers and problems relative to temperature and frequency follow closely those of free-layer material and will not be repeated. It is strongly recommended again that manufacturers' recommendations be closely followed in all applications.

Constrained-layer damping products are gaining in popularity and are useful in nearly any application in which damping compounds can be used. The only possible exceptions might be applications involving rough surfaces to which the product cannot readily be adhered or irregularly shaped objects to which sections of the product cannot readily be fitted. General applications include flat surfaces, cylinders, and shells. For flat surfaces we have panels such as machine enclosures, conveying chutes, cabinet walls, square ducts, doors, and partitions. For cylinders there are pipes, conduits, round ducts, curved panels, and shafts. Shells would include a large number of curved or spherically shaped structures such as tubs, bowls, and tanks.

Some configurations in which constrained-layer damping products can be especially useful because of their economy of weight and space are readily apparent. For instance, aircraft structures such as cabin partitions, walls, and floors; aerodynamically stressed structures such as wing surfaces and beams, and critical structures exposed to airframe vibration such as fuel tanks or electronic equipment have been treated successfully with constrained-layer damping products.

The heating, ventilating, and air-conditioning industries have also put constrained-layer damping materials to good use. These materials have been used to damp the panel vibrations in air-moving ducts which cause *booming*. They also damp vibrations from fans and heating units which might otherwise be transmitted along ductwork and cause the ducts to radiate noise. Significant noise reduction has also been reported where masonry and stone-cutting tools such as drills, chippers, etc., have been treated with damping materials.

Perhaps the most classic application of constrained-layer damping material is in the treatment of saw blades. Here a thin 0.010-in. steel

Figure 6.21 Thin constrained-layer damping material applied to a saw blade. The arrow shows the edge of treatment.

annulus is applied to one side of the blade, generally covering 70% to 80% of the blade, as shown in Fig. 6.21. With this treatment, field tests have shown that the high-frequency *whine* due to blade vibration in the idle (not cutting) mode of operation is eliminated (H. R. Mull, Bell and Associates, Norwalk, Conn., unpublished data, 1978). In addition, noise reduction of up to 10 dB is often achieved in the cutting mode of operation, especially in the stone masonry and ceramic industry.

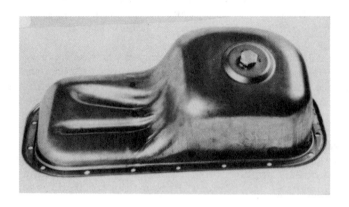

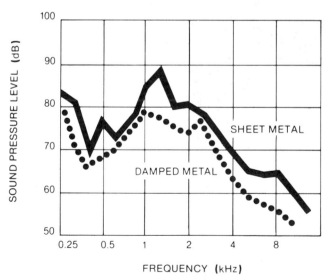

Figure 6.22 Oil pan cover constructed from constrained-layer laminate and spectral analysis showing the noise reduction obtained. (Courtesy of Antiphon Inc., Bloomfield Hills, Mich.)

6.3.3 Constrained-Layer Laminates

Finally, we come to the damping provided from constrained metal laminates. Here layers of sheet metal are bonded together with a high-
shear viscoelastic material. These materials possess slightly better
damping performance than the single constrained-layer material, but
this is not their most popular feature. With these laminates, critical
vibrating elements can be completely fabricated from the damping
material. Examples include machine enclosures, diesel engine oil
pans (see Fig. 6.22), rocker arm covers, conveyor chutes, cyclone
hoppers, etc. Another positive aspect of this approach is that the
risk of delamination of the constrained layer is minimal. Probably
the only disadvantage lies in the rather involved bending, forming,
cutting, and welding techniques which must be added to conventional
methods. It must be emphasized, however, that material suppliers
do provide detailed guidelines in these fabrication areas.

Along the same concept as constrained-layer metal, constrained-
layer wood panels are also commercially available. The transmission
loss of plywood, particle board, harboard, and so on can be dramatically increased by introducing a damping mastic (pressure-sensitive
adhesive) between the wood elements. See Fig. 6.23 for an illustration of the "sandwich-like" composite and Fig. 6.24 for a typical acoustical performance. These wooden panels have found extensive use in
boats, buses, stair and floor constructions, and so forth.

Figure 6.23 Commercially available constrained layer wood panels.
[Courtesy of Greenwood Forest Products, Inc., Lake Oswego, Ore.]

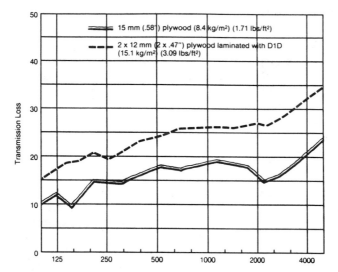

Figure 6.24 Acoustical performance of laminated plywood. [Courtesy of Greenwood Forest Products, Inc., Lake Oswego, Ore.]

6.4 ENERGY-ABSORBING CUSHIONS

Another related application of damping materials is in the treatment of part- or product-conveying chutes. For example, small metal parts are ejected from a punch press into a chute where they slide into a collection or *tote* bin. The impact of the parts on the chute is an impulsive *clatter*, and the subsequent slide is often a piercing *screech*. With the application of wear-resistant rubber or constrained-layer damping material to the chute in the form of an energy-absorbing cushion, considerable noise reduction can be achieved (see Fig. 6.25). It should be noted also that not only is there significant noise reduction, but the life of the chute is usually extended.

From these examples, it is easy to see that damping materials play a vital role in noise control. Often the role is *prime* in that the noise level at the source is significantly reduced. Used in combination with barrier materials, significant additional transmission loss can be acheived in those frequency ranges where resonant or coincident effects are present. It must also be emphasized that "a little damping goes a long way" in noise reduction. This is obvious from the classic example of placing a finger on a bell. The point is that often the entire vibrating surface need not be treated for effective results. This point is not being emphasized from a material-saving standpoint but rather to encourage the use of these materials in critical and strategic areas of part or panel deformation. Further, with the ease of application of most damping materials, quick tests can be conducted and performance evaluations made with little expended time or expense.

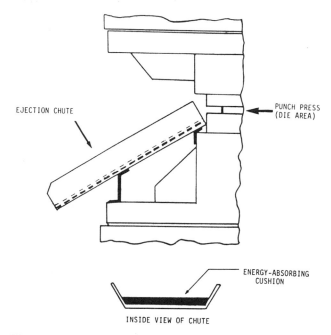

Figure 6.25 Illustration of punch press ejection chute and energy-absorbing cushion treatment.

6.5 SUMMARY

In final summary, it must be emphasized that for most noise control applications, no single material type is optimum, but rather a combination is usually preferred. Probably the best example of this is the composite of materials usually seen in a typical machine enclosure panel. Here, absorbing materials are used to reduce internal reverberant sound buildup, dense barrier materials provide acoustical isolation, and damping materials increase the transmission loss of the barriers. We shall see further examples of these composite materials and their use in later chapters.

REFERENCES

1. E. T. Paris. On the reflection of sound from a porous surface. *Proc. R. Soc.* 115(1927), 407.
2. Lord Rayleigh. *The Theory of Sound*, vol. 2. Macmillan, London, 1896, p. 161.
3. Lord Rayleigh. *The Theory of Sound*, vol. 2. Macmillan, London, 1896, p. 323.

4. H. O. Taylor. A direct method of finding the value of materials as sound absorbers. *Phys. Rev. 2* (1913), 270.

5. H. J. Sabine. Notes on acoustic impedance measurement. *J. Acoust. Soc. Am. 14*(2) (1942), 143.

6. L. L. Beranek. *Acoustic Measurements.* Wiley, New York, 1949, pp. 309–325.

7. *WADC Technical Report 52-204,* Handbook of Acoustic Noise Control, Physical Acoustics. vol. 1, supplement 1. April 1955.

8. Air Flow Resistance of Acoustical Materials. *American Society of Testing and Materials Designation C 522-73.*

9. Sound Absorption Coefficients by the Reverberation Room Method. *American Society of Testing and Materials Designation C 423.*

10. L. L. Beranek. *Noise and Vibration Control.* McGraw-Hill, New York, 1971, p. 280.

11. W. E. Purcell. Sound and Vibration, July 1977.

12. Standard Method for Laboratory Measurement. *American Society of Testing and Materials Designation E90-75.*

13. R. B. Tatge. *Noise Reduction of Barrier Walls.* Arden House, 1972.

14. L. L. Beranek, *Noise and Vibration Control.* McGraw-Hill, New York, 1971, pp. 453–456.

BIBLIOGRAPHY

Bell, L. H. *Fundamentals of Industrial Noise Control.* Harmony Publications, 1973.

Beranek, L. L. (Ed.). *Noise Reduction.* McGraw-Hill, New York, 1960.

Cremer, L., and M. Heckl. *Körperschall.* Springer, Berlin, 1967.

Faulkner, L. *Handbook of Industrial Noise Control.* Industrial Press, New York, 1975.

Graf, E. R., and J. D. Irwin. *Industrial Noise and Vibration Control.* Prentice-Hall, Englewood Cliffs, New Jersey, 1979.

Harris, C. M. *Handbook of Acoustical Measurements and Noise Control.* McGraw-Hill, New York, 1991.

London, A. The determination of reverberant sound absorption coefficients from acoustic impedance measurements. *J. Acoust. Soc. Am. 22* (March 1959), 263.

Morse, P. M. *Vibration and Sound,* 1st ed. 1936, p. 304.

Morse, P. M., and K. U. Ingard. *Theoretical Acoustics.* McGraw-Hill, New York, 1968.

Paris, E. T. On the coefficient of sound absorption measured by the reverberation method. *Philos. Mag. 5* (1928), 489.

Sabine, P. E. Specific normal impedances and sound absorption coefficients of materials. *J. Acoust. Soc. Am.* *12* (Jan. 1941), 317.

Standard Method of Test for Impedance and Absorption of Acoustical Materials by the Tube Method. *American Society of Testing and Materials Designation C 384-58.*

Willig, F. J. Comparison of Sound absorption coefficients obtained by different methods. *J. Acoust. Soc. Am.* *10* (April 1939), 293.

EXERCISES

6.1 The standing wave ratio measured in an impedance tube was 25 dB at 125 Hz, 20 dB at 250 Hz, 14 dB at 500 Hz, 12 dB at 1000 Hz, and 8 dB at 2000 and 4000 Hz. Calculate and plot the octave band normal incidence absorption coefficient for the test material.

Answer: 0.20, 0.33, 0.56
0.64, and 0.81

6.2 In an impedance tube at 2000 Hz, measured parameters on a sample were $D_1/D_2 = 0.478$ and $L = 25.5$ dB. What are the
a. Specific resistance?
b. Reactance ratios?
(Hint: Use Fig. 6.4.)

Answer: (a) 7.0
(b) −9.0

6.3 From measurements in an impedance tube at 1000 Hz, the specific resistive and reactive components of an absorbing material were found to be 4.0 and −0.4, respectively. What is the statistical absorption coefficient for the test material?
(Hint: Use Fig. 6.5.)

Answer: 0.80

6.4 From measurements, it is established that the absorption coefficient of 1/2 in. of acoustical-quality foam is 0.58 at 1000 Hz. For a particular installation using this material, it is desirable to have an absorption coefficient of at least 0.90. Based on typical performance of commercially available foams, how thick must the foam be to meet the required performance?
(Hint: Use Fig. 6.6.)

Answer: 1.0 in.

6.5 The air pressure difference across a test specimen 5 cm thick and 100 cm^2 in area was 1000 Pa. The volume velocity of the airflow was 5.0 m^3/s. What is the
a. Resistance of the specimen?
b. Resistivity of the specimen?

Answer: (a) 200 mks rayls;
(b) 4000 rayls/m

6.6 From Exercise 6.5, what is the specific acoustic resistance of the specimen? Use 415 mks rayls for the free-field impedance.
Answer: 0.48

6.7 The absorption coefficients of a ceiling tile at 250, 500, 1000, and 2000 Hz are, respectively, 0.40, 0.77, 0.98, and 0.83. What is the NRC value for the tile?
Answer: 0.745.

6.8 Show that as $\omega^2 B/c^4 \rho_s$ approaches 0,

$$TL(\theta) \approx 10 \log \left[1 + \left(\frac{\omega \rho_s}{2 \rho c} \cos \theta \right)^2 \right] \quad [dB]$$

6.9 Calculate the TL for sheet lead 1/32 in. thick which weighs 2 lb/ft^2 at
a.   1000 Hz
b.   500 Hz

Answer: (a) 33 dB;
(b) 27 dB;

6.10 Estimate the TL of dense concrete 8 in. thick at
a.   250 Hz
b.   1000 Hz
c.   4000 Hz

Answer: (a) 54.6 dB;
(b) 66.6 dB;
(c) 78.7 dB

6.11 Estimate the critical frequency for
a.   A plate of aluminum 1.4 in. thick
b.   Oak 1 in. thick

Answer: (a) 350 Hz;
(b) 700 Hz

6.12 A brick wall 35 × 10 ft has a TL of 48 dB at 1000 Hz. Two doors each 2.5 × 7 ft having a TL of 25 dB are to be installed in the wall. Estimate the TL of the composite wall.
Answer: 34.8 dB

6.13 The decay rate of a specimen mounted in an apparatus to measure loss factors was 10 dB/s at 120 Hz. What is the loss factor?
Answer: 0.003

6.14 From the results of Exercise 6.13, what is the damping factor?
Answer: 0.0015

7
ACOUSTICAL ENCLOSURES

Although strides of progress have been made in source noise reduction, some types of industrial and manufacturing equipment such as hammer mills, power presses, plastic grinders, diesel engines, etc., are still quite noisy. In addition, there are many equipment installations where the cost to develop noise reduction measures for old vintage equipment is prohibitive. In these situations, the acoustical engineer must consider enclosing the equipment in either a partial or total enclosure. The thought of an enclosure is initially repulsive to the plant engineers, the maintenance staff, and the operator because of the anticipated nuisance associated with the loss of visibility, accessibility, and added maintenance. These penalties are not inherent, and a well-designed enclosure can provide bonus features such as mist and smoke control.

It must be emphasized that noise reduction at the source always enjoys top priority, but with a systematic approach and careful attention to design detail, enclosures can be one of the most powerful noise reduction measures available to the acoustical engineer.

7.1 DESIGN METHODS

All too often the engineer designing an enclosure begins by considering what materials or combinations or materials will be selected for the enclosure. To be sure, this decision must be made, but several steps should be taken first which will simplify the selection process and assure a balanced design. These recommended steps are outlined below.

Step 1. As in any good design discipline, the initial step should be to establish design criteria (design goals) and to determine the corresponding acoustical performance required of the enclosure to meet the criteria. The design goals may be hearing damage risk criteria and/or annoyance criteria in noise-sensitive areas. Regardless, most criteria can be expressed in terms of octave band sound levels,

239

and for most cases, these will suffice. Therefore, the first step is
to establish a realistic octave band criterion at some location or perhaps several locations from the noise source. For industrial equipment, machines, motors, etc., the operator station and at ear level
3 ft from major machine surfaces are preferred locations. These locations are also consistent with the many recommended measurement
standards or procedures such as developed by industrial associations,
such as ASHRAE, NMTBA, and GAGI.

Step 2. The second step is to determine or predict the octave band
noise levels of the equipment at the locations selected in step 1. The
octave band noise level data can be measured either as installed or
from data as supplied by the equipment manufacturer. However,
actual measurements at the locations selected in step 1 are obviously
preferable. With the criteria established and measurements obtained,
the required acoustical performance of the enclosure can be calculated. To illustrate this method, consider the following example.

Example
    It is desirable to enclose three plastic grinders in separate en-
closures. The resultant noise levels must not exceed 90 dBA at
3 ft from the complex. What is the required acoustical performance
(noise reduction) of the enclosures? The measured noise levels
at 3 ft for one grinder are shown in line 2 of Table 7.1.

Table 7.1 Computation Table [1]

	Octave Band Center Frequency (Hz)						
	125	250	500	1000	2000	4000	8000
1. Octave band equivalent 90 dBA[a] (dB)	98	90	84	81	81	81	81
2. Measured noise levels at 3 ft (dB)	94	91	98	102	103	97	96
3. Accumulation due to three-grinder installation (dB)	5	5	5	5	5	5	5
4. Line 2 plus line 3; net noise level of three grinders (dB)	99	96	103	107	108	102	101
5. Line 4 minus line 1; noise reduction required (dB)	−1	−6	−19	−26	−27	−21	−20

[a]Not unique.

Solution

In this example, the criterion or design goal is 90 dBA; hence an octave band equivalent of 90 dBA is selected, as shown in line 1 of the computation table. It must be emphasized that this criterion is not unique but provides uniform levels for conservative design purposes. Hence, if all resultant levels are below these criteria levels, the combined overall level will also be below 90 dBA. Next the measured or manufacturer-supplied noise levels at the desired location are listed, as shown in line 2. Since there are three grinders, one must account for the sound accumulation. Using the chart in Fig. 2.12, we can see that three equal sources increase the level by 5 dB, which is included in line 3. The net noise level for the three grinders at 3 ft would then be the sum of line 2 and line 3, which is shown in line 4. The required spectral noise reduction of each enclosure is shown in line 5 and is just the difference between the criteria given in line 1 and the net noise levels of line 4. With the required acoustical performance (noise reduction) determined (line 5), the next step is to select the type of enclosure and materials which will meet these design goals.

7.2 ACOUSTICAL PERFORMANCE

There are two basic types of enclosures, total and partial. With a well-designed total enclosure, there is practically no lower limit to the resultant noise levels. Partial enclosures, on the other hand, have very serious acoustical performance limitations and must be dealt with accordingly.

First, let us define those parameters which establish and characterize the acoustical performance of an enclosure. If one measures the sound pressure level (1) near the inside wall of an acoustical enclosure $L_{p,1}$ and (2) outside $L_{p,2}$ at the location of interest, say the operator's station, the difference can then be considered the noise reduction (NR) of the enclosure:

$$NR = L_{p,1} - L_{p,2} \quad [dB] \tag{7.1}$$

Now the measured noise reduction as defined here does not necessarily provide direct quantitative information regarding the acoustical performance of the enclosure. To see this, note that if the outside measurement location is some distance away from the enclosure, there will be considerable attenuation due to divergence. As such, the difference in levels will include factors which are not related directly to the acoustical performance of the enclosure. Consider an example.

Example

Shown in line 1 of Table 7.2 are the octave band sound levels as measured near the inside walls of a compressor enclosure. Sound levels measured at a maintenance bench 15 ft from the enclosure are shown in line 2. What is the octave band noise reduction NR of the enclosure?

Solution

From Eq. (7.1), the noise reduction NR is just the difference between line 1 and line 2. The NR can also be seen graphically, as illustrated in the chart in Fig. 7.1.

Another popular measure of the acoustical performance of an enclosure is the insertion loss IL, which is defined as

$$IL = L_{p,1} - L'_{p,1} \qquad [dB] \tag{7.2}$$

where

$L_{p,1}$ = sound level measured at location of interest with enclosure on (dB)

$L'_{p,1}$ = sound level measured at same location with enclosure removed (dB)

The insertion loss IL as given in Eq. (7.2) is often a more useful measure of the acoustical performance, since the measured noise levels are at the same location of interest.

Table 7.2 Computation Table

	Octave Band Center Frequencies (Hz)							
	63	125	250	500	1000	2000	4000	8000
1. Sound levels inside compressor enclosure (dB)	82	86	90	92	90	84	75	68
2. Sound level at maintenance bench (dB)	77	74	72	68	62	52	41	33
3. Noise reduction NR; line 1 minus line 2 (dB)	5	12	18	24	28	32	34	35

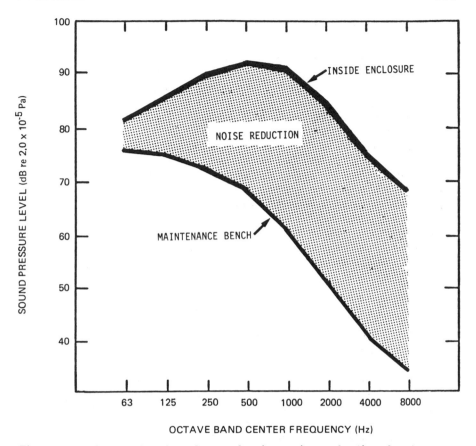

Figure 7.1 Octave band analyses showing noise reduction due to an enclosure.

    In summary, for performance evaluation, insertion loss IL, as defined by Eq. (7.2), is often more applicable, since one avoids an internal measurement as required by Eq. (7.1), noise reduction. However, it is not always easy to remove an enclosure, and there are possible errors associated with merely opening doors. Additional hazards are also encountered on a "before and after" installation measurement, since there are often 3 months or longer between measurements and many changes in source noise level can be anticipated in the interim. One thing is certain: If levels at the location of interest are below the criteria, design goals have been met regardless of the absolute acoustical performance of the enclosure. Let us now consider guidelines for enclosure design and construction.

7.2.1 Machine Enclosures

In large total enclosures, the most important design consideration is
the overall transmission loss of the walls. The transmission loss, as
we saw earlier, depends to the first order on surface density, and
hence there is a wide assortment available, depending on whether
thick masonry, modular panels, or relatively light acoustical curtains
are selected. For example, in large jet engine test cells, heavy concrete 12 to 18 in. thick is common, with TL values exceeding 70 dB.
We shall not spend much time in this area other than to emphasize
that when heavy-duty noise control enclosures are required, such as
test facilities for aircraft engines, large gas turbines, diesel engines,
etc., masonry wall construction should be the first consideration.
With masonry, only performance penalties associated with airflow and
penetrations such as windows, doors, etc., need further detail design effort, and we shall cover these areas later. Walls constructed
for use in buildings are also treated in detail in a later chapter.

7.2.2 Modular Panels

Often, large total machine enclosures are constructed from commercially available modular panels. Shown in Fig. 7.2 is an illustration
of the basic construction elements in these panels. Typically, these
panels are available in combinations of lengths of 4, 5, 6, 8, and 10
ft and widths of 2, 3, and 4 ft. With respect to thickness, 2 and 4 in.
are most common, with higher transmission loss for the thicker panels
as illustrated in Fig. 7.3. The outside sheet metal *skin* is typically
16-gauge galvanized steel, and the inside surface is typically 22-
gauge perforated steel, 10% to 30% open area. The interior sound-
absorbing *fill* is usually fibrous glass or mineral wool (4 to 6 lb/ft^3),
often protected from contaminants by a thin 1- or 2-mil polyethylene
film (bag).

   For added versatility, these panels are available with windows installed, usually double-glazed (two panes) and supported in rubber
isolator gaskets. Panels with high-performance acoustical doors and
seals are also available to the engineer and designer, to complete the
list of features.

   Because of the design flexibility of these panels, their applications
are limitless. However, shown in Fig. 7.4 are a few illustrations of
enclosures and barriers which can readily be constructed from the modular panels. In addition, shown in Fig. 7.5 are some photographs
showing sequentially the major steps in erecting a room enclosure.
Note the ease of assembly and the number of access doors.

   In summary, total and partial enclosures constructed of these
modular panels provide a high level of acoustical isolation along with
ease of assembly and flexibility.

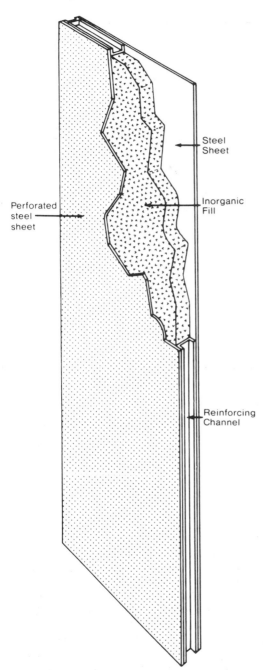

Figure 7.2 Conceptual sketch of a modular panel construction.
(Courtesy of Allforce Acoustics, a Division of Lord Corp., Erie, Pa.)

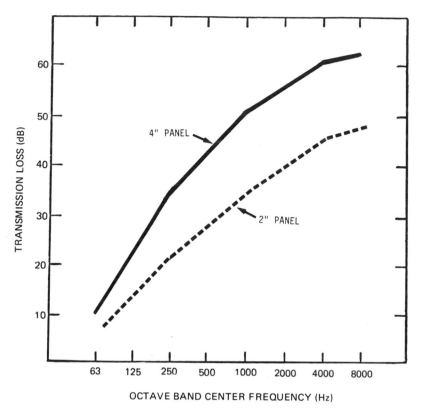

Figure 7.3 Typical transmission loss of modular panels 2 and 4 in.
thick.

7.3 ACOUSTICAL LEAKS

The factor limiting the acoustical performance of enclosures is often
the number and size of acoustical *leaks*. Shown in Fig. 7.6 is a chart
for calculating the effect of an acoustical leak. Note that an enclosure
with a transmission loss potential of 45 dB is reduced to 20 dB if an
opening of 1.0% is present. Further, the performance reduces to a
mere 10 dB if the opening is 10%. From this example, it is easy to
see that great care must be given to design and construction detail
in order to minimize acoustical leaks. This is not always easy, for
air, stock, materials, products, scrap, etc., usually must be moved
in and out of the enclosure at high speeds. We shall provide specific
design guidelines for treatment of penetrations later in this chapter.
    Finally, for machine enclosures, quick operator accessibility and
continuous visibility are almost always required. It is meeting these

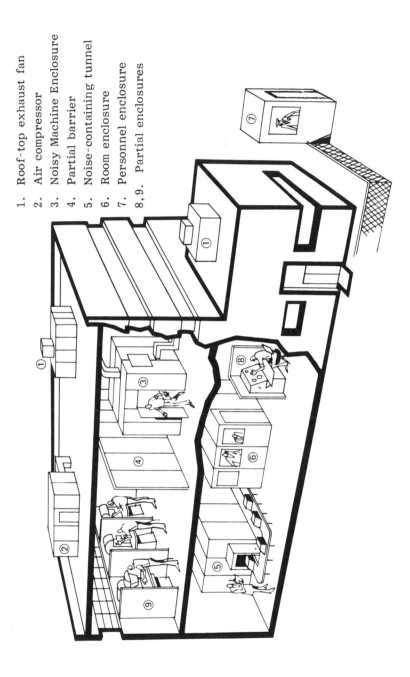

1. Roof-top exhaust fan
2. Air compressor
3. Noisy Machine Enclosure
4. Partial barrier
5. Noise-containing tunnel
6. Room enclosure
7. Personnel enclosure
8,9. Partial enclosures

Figure 7.4 Examples of total and partial enclosures constructed of modular panels. (Courtesy of Allforce Acoustics, a Division of Lord Corp., Erie, Pa.)

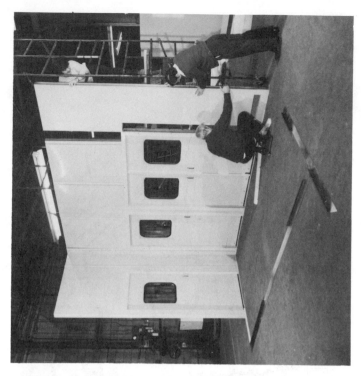

(b)

(a)

(c)

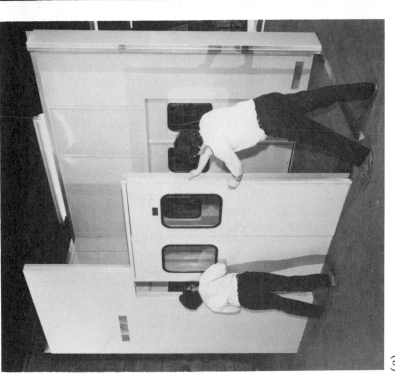

(d)

Figure 7.5 Sequentially, the major steps in erecting a room enclosure. (Courtesy of Eckel Industries., Inc., Cambridge, Mass.)

requirements while preserving the acoustical performance that challenges the design engineer. Shown in Fig. 7.7 is an example of one
of the most successful applications of totally enclosing a large machine.
Here automatic power presses have been acoustically isolated in mod-
ular-panel-type enclosures. Note the attention given to operator accessibility and visibility. With respect to acoustical performance, an
overall noise reduction of more than 20 dBA was achieved, and resultant levels in the immediate area were in the range of 82 to 85 dBA.
It should be mentioned that the presses and enclosures shown are
relatively small. In fact, enclosures for large *1000-ton* capacity
presses are the size of two-story houses. Other examples of large
industrial equipment that lends itself to total enclosures include cold
headers, swagers, bar machines, transfer presses, tumbling equipment, sand-blasting nozzles, large fans, compressors, etc.

7.4 PERSONNEL ENCLOSURES

For those areas such as boiler rooms, generator rooms, pump rooms,
in-plant offices, etc., where machine enclosures are not feasible, a
personnel room enclosure for the watch engineers or operators will
significantly reduce daily noise exposure. These personnel rooms are
often installed at control panels or located centrally such that critical
gauges or meters can be observed and recorded. In this way, only
short periodic excursions into the noisy areas are required for monitoring remote gauges, meters, etc. Shown in Fig. 7.8 is a photograph
of a personnel room enclosure installed in an industrial environment.
An overall noise level reduction of more than 20 dBA was achieved in
this installation.

In summary, the disadvantages of total enclosures are obvious;
costly additional floor space is generally required, and some production
penalties associated with machine operation and maintenance must be
anticipated. In short, there are usually some cumbersome access
and visibility problems to be solved.

An often-overlooked positive feature of the total enclosure is the
ease with which air contaminants can be collected. Here noxious or
toxic fumes, dust, oil mist, etc., common in industry, are locally contained in the enclosures and exhausted or vented to a central collection system. The enclosure approach is usually much more effective
than collection by hoods or with ceiling fans after dispersion throughout the production areas.

7.5 CUSTOM ENCLOSURES

When modular panel enclosures are not practical, a custom panel enclosure design approach can be undertaken. Here the enclosure

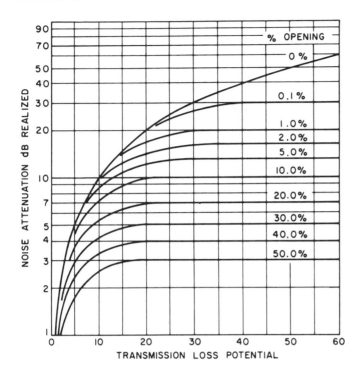

Figure 7.6 Effect of sound leaks on potential noise reduction for walls.

panels are often attached to machine surfaces and may only enclose the major sources of noise. Shown in Fig. 7.9 are examples of commercially available custom enclosures covering the tooling and stock reel area of automatic bar machines. Other examples of successful application of custom enclosures include small punch presses (eyelet machines), cold headers, cutoff saws, pumps, plastic molding equipment, grinders, hydraulic units, compressors, etc. (see Appendix I for extensive listing). It should be emphasized that because of the design detail involved in adapting these enclosures to complex surfaces and equipment configuration, the engineer must anticipate a rather extensive design and development program. In short, for these enclosures to meet both the acoustical and functional design goals, the following steps, as a minimum, are strongly recommended:

Phase I: preliminary design

1. Complete preliminary design drawings.
2. Review the preliminary design with cognizant production engineering, maintenance, and operating personnel.
3. Finalize the design.

Figure 7.7 Total enclosures for automatic power presses constructed of modular panels. (Courtesy of Neiss, Inc., Rockville, Conn.)

Phase II: prototype installation and evaluation

1. Fabricate and install the prototype enclosure.
2. Evaluate the prototype enclosure in terms of acoustical performance and production impact.
3. Review the evaluation, and finalize the design.

With respect to the design and construction of these custom enclosures, only guidelines can be given. The steps to establishing de-

Figure 7.8 All-purpose personnel enclosure installed in an industrial environment. (Courtesy of Industrial Acoustics Company, Bronx, N.Y.)

sign parameters, goals, or criteria which were given in the beginning of the chapter naturally apply here and will not be repeated.

With respect to panel construction, the most popular configuration is a combination of sheet metal and a commercially available composite of absorbing and barrier materials. Shown in Fig. 7.10 is an illustration of this laminate composite bonded to 16-gauge sheet metal, along with typical acoustical performance. Now, clearly, the performance can be improved by increasing either the surface density of the outer sheet metal layer or the inner septum or both. The inner 1-in. layer of foam provides absorption to reduce reverberant buildup, and the 1/4-in. layer decouples the barrier layers. To avoid *wicking* of water, oil, or industrial solvents, etc., the exposed surface of the foam is usually treated to close the pores or is covered with a thin plastic or vinyl film. In those cases where heavy wear can be expected, the foam can be protected with an additional layer of perforated or expanded sheet metal, 20% open area (minimum).

Additional barrier performance can be obtained by using the constrained-layer sheet metal laminates or applying a thin layer of damping material to the sheet metal layer. With more extensive design and development, the outer layer of these enclosure panels can be

Figure 7.9 Custom acoustical enclosures for automatic bar machines.
(Courtesy of Tamer Industries, Pawtucket, R.I.)

fabricated from molded fibrous glass, plastic, etc., which usually
adds some esthetic value.

Because of the individual aspects of custom enclosures, little
more can be said regarding design or material selection. As in the
case of modular panel enclosures, the overall performance of the enclosure will probably be determined by the unavoidable acoustical
leaks associated with conduit or stock penetrations, windows, doors,
access areas, etc. Let us now consider each of these areas and those
design guidelines which minimize acoustical leaks.

7.5.1 Penetrations

Rare is the enclosure installation where numerous penetrations into the
enclosure are not required. These penetrations include, for example,
electrical conduits, plumbing, stock feed, and discharge openings for

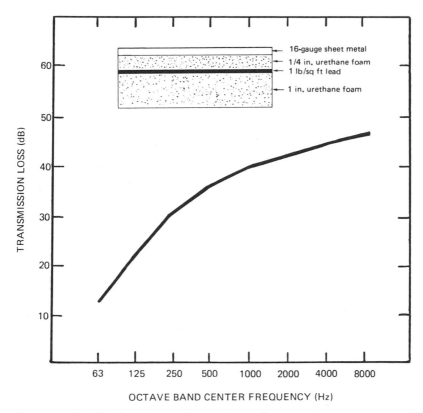

Figure 7.10 Typical transmission loss of composite material applied to sheet metal for use as an enclosure panel.

exchanging large volumes of air. Conduit or plumbing penetrations are rather easy to seal by working sheet lead (1/64 or 1/32 in.) or dense vinyl around the pipe or conduit, as illustrated in Fig. 7.11. Where the penetration openings are well oversized, the treatment illustrated should be applied to both the inner and outer sides of the panel.

With respect to moving air in and out of enclosures, there are two basic approaches. The first, and often the easiest, is to utilize either commercially available parallel baffle or circular flow-through silencers as a sound trap in the inlet and/or exit ducts. These types of si-lencers are mentioned specifically since they have very low pressure losses, which is often a major design requirement. Illustrated in Fig. 7.12 is a typical silencer installation for the intake of a fan enclosure. At a glance, one might think that the silencer is just a large hole, but as we shall see later, the noise reduction of the silencer is equiv-alent to the enclosure walls or better. We shall discuss silencer se-

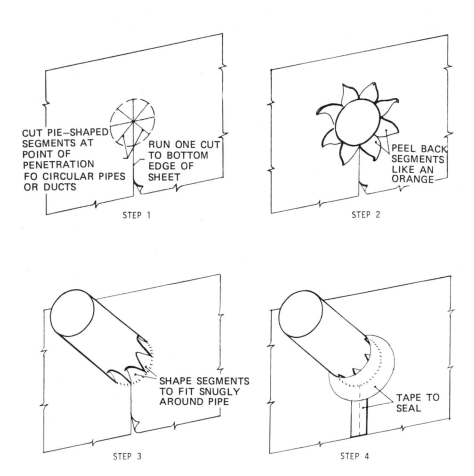

Figure 7.11 Technique for sealing penetrations with lead sheet. (Courtesy of Lead Industries Association, Inc., New York, N.Y.)

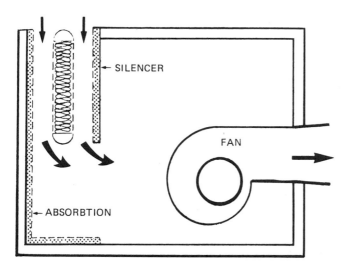

Figure 7.12 Absorptive silencer installed in a fan enclosure permitting airflow but containing fan noise.

lection criteria in detail in Chap. 8. This approach is especially applicable in enclosing combustion engines, forced draft fans, compressors, ventilation fans, rotary positive displacement blowers, burners, etc.

The second basic approach, which is especially applicable to small enclosures, is to include a series of lined bends or labyrinth sound traps, as illustrated in Fig. 7.13. In most installations, this approach will compromise the acoustical performance of the enclosure to some degree. However, if the lined bends are carefully constructed along the guidelines presented in Chap. 8, net overall attenuation of 15 to 20 dB can be readily achieved for dominantly high-frequency noise sources.

Illustrated in Fig. 7.14 is a total enclosure designed for a plastic scrap granulizer which includes a labyrinth sound trap for ventilation. This enclosure reduced noise levels by more than 17 dBA.

In summary, the use of commercially available silencers or carefully designed labyrinth sound traps will allow large volumes of air to be brought into an enclosure with little or no loss of acoustical isolation performance.

7.5.2 Access Panels and Doors

As mentioned, access panels and doors are often the culprits that sharply reduce the noise reduction capability of well-designed en-

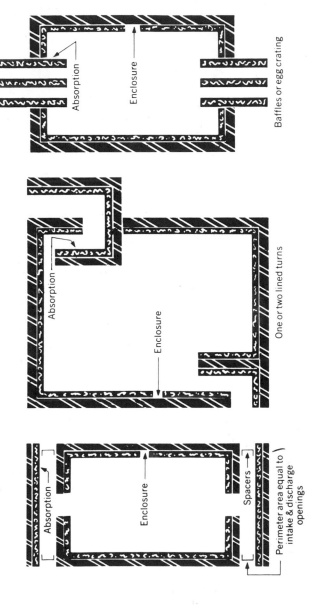

Figure 7.13 Examples of easy-to-construct sound traps for enclosures where low airflows for ventilation are required. (Courtesy of Soundcoat Co., Inc., Brooklyn, N.Y.)

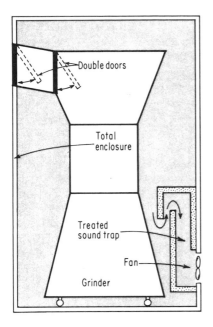

Figure 7.14 Schematic of scrap grinder totally enclosed in acoustic panels with double loading doors and ventilation through a sound trap.

closures. A few design guidelines for access doors will now be presented. A simple and very effective access panel door design is illustrated in Fig. 7.15. By far the most common error is to select a lightweight door material such as aluminum. As such the noise passes easily through the lightweight door, reducing the noise reduction capability of the enclosure. In addition, the door must obviously fit well and, if possible, be airtight. To assure an airtight fit, include labyrinth rubber seals and a positive pressure-type latch to compress the seals and resist vibration. Other equally good or better access door designs exist, but this configuration emphasizes the basic construction guidelines.

Where high acoustical performance doors are required for large room or test facility enclosures, commercially availabel units are strongly recommended. These doors are available as complete assemblies, that is, with jambs, sills, seals, and so forth. Shown in Fig. 7.16 are examples of commercially available acoustical quality doors. Shown in Fig. 17.16d is the actual acoustical performance of commercially available doors. To appreciate the scope of availability, refer to Figs. 7.17 and 7.18, which show actual heavy duty industrial door installations

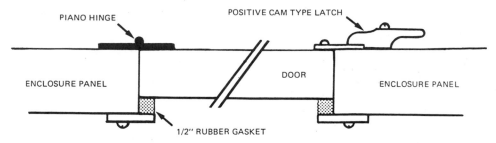

PIANO HINGE

POSITIVE CAM TYPE LATCH

ENCLOSURE PANEL

DOOR

ENCLOSURE PANEL

1/2" RUBBER GASKET

Figure 7.15 Schematic of enclosure access door illustrating major design features.

(a)

Figure 7.16 (a) A typical light duty door assembly. (b & c) Heavy duty acoustical quality swinging (b) and sliding (c) doors. (d) Acoustical performance of commercially available doors. [(a) Courtesy of Pioneer Industries, Carlstadt, N.J., (b-c) courtesy of Jamison Door Company, Hagerstown, Md.]

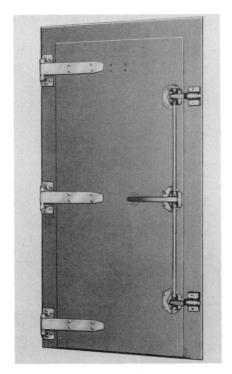

(b)

(c)

SOUND TRANSMISSION TEST RATINGS
Transmission Loss in db/Hertz — 1/3 Octave Bands

STC	100	125	160	200	250	315	400	500	630	800	1000	1250	1600	2000	2500	3150	4000	5000	Test Report Number	STC
48	34	35	40	42	44	45	44	46	46	44	45	47	49	50	52	53	56	56	RAL TL85-323	48
50	31	33	40	41	42	44	44	46	50	50	51	50	53	55	56	58	60	61	RAL TL85-321	50
52	34	36	39	43	44	46	46	48	50	51	50	51	54	55	57	58	59	56	RAL TL85-62	52
53	36	37	40	43	45	46	48	50	52	54	53	54	56	57	58	59	60	59	RAL TL85-64	53
54	39	40	41	44	46	48	49	51	53	54	54	54	56	57	58	60	61	61	RAL TL85-65	54
55	39	40	45	44	47	52	53	53	53	55	55	54	56	58	58	61	62	61	RAL TL85-182	55
57	39	42	45	46	48	53	52	55	57	58	57	56	58	60	60	62	63	62	RAL TL85-180	57
59	44*	52*	50*	50*	56*	56*	57*	59*	56	56	55	56	59	62	65	67	68	67	RAL TL86-317	59
59	44*	42*	48*	50*	55*	55*	58*	58*	57	56	56	56	59	62	65	67	68	66	RAL TL86-323	59
60	44*	52*	48*	51*	57*	56*	58*	59*	58*	57	57	59	60	63	66	68	69	68	RAL TL86-320	60
62	44*	50*	49*	51*	56*	57*	61*	60*	59	60	60	61	61	65	67	69	69	69	RAL TL86-322	62
51SL	38	41	37	44	46	44	46	49	49	49	50	52	53	52	53	56	54	50	RAL TL85-320	51SL
52SL	38	39	38	43	43	42	46	50	51	52	52	53	53	54	57	57	54	48	RAL TL85-319	52SL
53SL	43	42	41	46	47	45	48	51	51	54	53	53	53	55	58	60	58	53	RAL TL85-317	53SL

*Door performance equaled or exceeded transmission loss capabilities of laboratory

"SL" indicates sliding door ratings

(d)

Figure 7.16 (Continued)

Figure 7.17 Actual installation of heavy duty acoustical quality doors (STC-50) at diesel engine test facility. (Courtesy of Overly Manufacturing Company, Greensburg, Pa.)

Figure 7.18 Actual installation of acoustical quality door at jet engine test facility. (Courtesy of Böet American Company, Palm Beach Gardens, Fla.)

7.5.3 Windows

Windows always present a problem in enclosure design. However, the major problems can be simplified if two basic design features are followed:

1. The thicker the pane (glazing), the higher the transmission loss. Panes typically selected are safety glass or transparent plastic 1/4 to 1 in. thick. Two panes (double glazing) spaced 2 in. or more apart should be considered for extremely high noise levels such as gas turbine or jet engine control room windows. In addition, they should be installed nonparallel, i.e., canted slightly relative to each other. Further, in two-pane installations, a dessicant of some kind or heating elements must be placed between the panes to absorb moisture and prevent condensation or *fogging.*
2. Panes should be gasketed or mounted in rubber or dense polyurethane foam, as illustrated in Fig. 7.19. Here again, where high

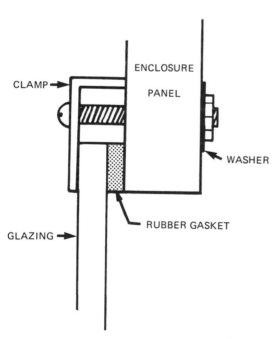

Figure 7.19 Window seal for enclosures.

Figure 7.20 Actual installation of acoustical quality window assembly in control room of jet engine test facility. (Courtesy of Böet American Company, Palm Beach Gardens, Fla.)

acoustical performance is required, commercially available window units are recommended. Shown in Fig. 7.20 is a window assembly used in a jet engine test facility. Note again the heavy construction features.

With respect to isolation performance, single-pane windows generally follow the mass law up to the coincidence region, where pronounced performance losses or dips of 10 dB or more can be anticipated. With commercially available window assemblies, performance equivalent to masonry block can be attained.

7.6 PARTIAL ENCLOSURES

When a total enclosure is not feasible or practical, a partial enclosure should be considered. Partial enclosures can be divided into two basic types:

Figure 7. 21 Power press, 75-ton capacity, with a die area enclosure; arrows show the stock feed and front partial enclosures. (Courtesy of Neiss, Inc., Rockville, Conn.)

1. Enclosures which totally enclose major noise sources but not the whole machine
2. Enclosures that only partially enclose a machine or noise source

With respect to the first type, shown in Fig. 7.21 is a photograph of a die area enclosure for a punch press (automatic mode of operation). In this particular installation, the die area was acoustically isolated by enclosing the frontal and rear entry portals and the stock in-feed and scrap exit penetrations. Specifically, in this photograph we see the stock in-feed and front portal fixtures. Note the high level of visibility and numerous access doors. It should also be noted that panel and access door construction details follow the basic guidelines for custom designs, as previously presented.

With respect to the second type, shown in Fig. 7.22 is a partial barrier enclosure which isolates the operator from adjacent noisy machines.

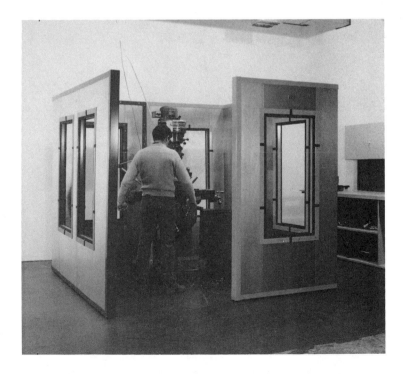

Figure 7.22 Example of partial barrier enclosure. (Courtesy of NEISS, Rockville, Conn.)

Shown in the photograph in Fig. 7.23 is an example of a partial enclosure which is quite effective and has wide application in food packaging areas. As mentioned previously, in a typical packaging line, empty bottles, jars, or cans are taken from a case and placed on a belt conveyor to be filled. Along the way to the *filler*, there are areas where the bottles or cans accumulate and impact each other.

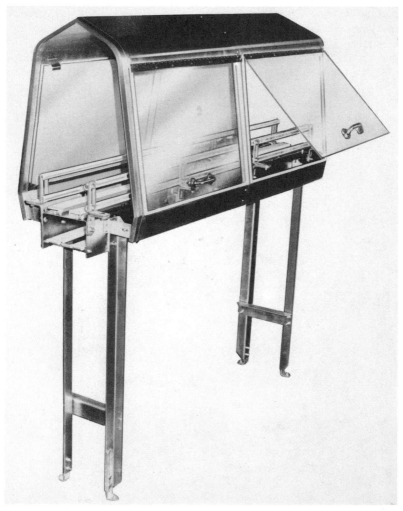

Figure 7.23 Partial enclosure for packaging line. (Courtesy of United McGill Corp., Groveport, Ohio.)

The result of these impacts is nearly continuous discrete noise at the natural frequencies or *bell tones* of the bottles or cans. An enclosure, as illustrated in Fig. 7.23, placed over the conveyor at critical accumulation points has been shown to reduce noise levels up to 12 dBA (H. R. Mull, Bell and Associates, Norwalk, Conn., unpublished data, 1979).

With respect to partial enclosure design, the following guidelines should be followed:

1. Enclose as many sides of the noise sources as possible.
2. Treat the enclosure walls or panels heavily with absorbing materials.

From these examples and design guidelines, it is clear that enclosing the major source of noise on a machine can play an important role in controlling noise. Unfortunately, often considerable noise or vibratory energy escapes or *flanks out* of the partial enclosure through support brackets in a structure-borne manner. The energy is subsequently radiated as sound, and the anticipated noise reduction of the enclosure is compromised. As such this partial enclosure approach often has serious limitations when it is the only noise reduction measure applied.

7.7 ACOUSTICAL CURTAINS

The use of acoustical curtains as enclosures or partial enclosures has grown sharply. Their popularity stems from their acoustical effectiveness, versatility, and ease of installation. Typically, the curtain materials are 0.5- or 1-lb/ft^2 lead- or barium-loaded (salt-loaded) vinyl. The smooth vinyl materials are *limp* and highly resistant to the industrial environment. Shown in Fig. 7.24 are some illustrations of typical curtain enclosure and installations. Here again the noise reduction is usually limited to the number of acoustical leaks and the amount of noise *flanking* over or under the curtain. As such, an overall noise level reduction of more than 10 dB is rarely achieved.

Another variation of the dense vinyl curtain is the transparent strip curtain, as shown in the photograph in Fig. 7.25. Here a can-filling machine was partially enclosed (no top) with transparent PVC strips (1/4 in. thick) with a 50% overlap. Overall noise reduction in the range of 5 to 8 dBA was achieved, with the major portion of the sound energy in the high-frequency range (above 500 Hz). These strip curtains can be obtained commercially in a variety of thicknesses, lengths, and overlaps. The acoustical performance of the strip curtains increase with both thickness and overlap. Shown in Fig. 7.26 is a combination of a quilted barrier system combined with curtains to provide a flexible enclosure.

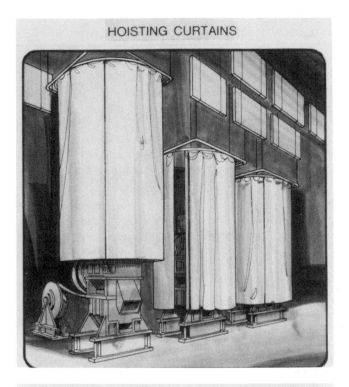

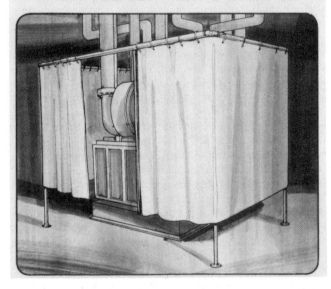

Figure 7.24 Examples of machinery enclosures utilizing dense vinyl curtains. (Courtesy of Industrial Noise Control, Inc., Addison, Ill.)

Figure 7.25 Overlapping strip curtain enclosure for a can-filling machine.

Figure 7.26 Partial barrier enclosure utilizing quilted barrier material and dense curtains. (Courtesy of Sound Seal Acoustical Products, Div. of United Process, Inc., Agawam, Mass.)

7.8 TELEPHONE ENCLOSURES

Often telephone conversation is severely limited in noisy industrial environments. Shown in Fig. 7.27 is an example of an acoustically treated
booth which greatly improves the intelligibility of telephone conversation. It should be noted that the interior of the booth is lined with
absorbing material and covered with perforated sheet metal for protection. Also shown in Fig. 7.27 is the octave band acoustical performance
for several models that are commercially available. Environments where
these treated booths offer excellent application are press rooms, weaving rooms, assembly areas, printing plants, packaging areas, cafeter-
ias, and so forth.

NOISE REDUCTION			
Frequency, Hz.	BD-602	BD-45	BD-45L
125	0	0	0.1
250	6.1	9.8	10.0
500	10.1	17.9	19.1
1000	7.8	15.4	16.7
2000	7.1	16.9	16.4
4000	7.8	17.0	16.8
8000	6.6	14.2	12.7
Overall	7.3	16.3	16.1

Tests were conducted with the
booth occupied by a 160 lb. man,
5'-10" tall.

(a)   (b)

Figure 7.27 (a) Commercially available treated telephone booth.
(Courtesy of Acoustics Development Corporation, St. Joseph, Mo.)
(b) Acoustical performance data.

7.9 SUMMARY

In final summary, it must be emphasized that successful enclosure
design and installation comprise one of the most challenging areas in
acoustical engineering. This discipline requires not only a thorough
knowledge of acoustical principles and materials but also extensive
experience and cleverness in mechanical design. It is hoped that the
basic design guidelines and examples presented here will provide
inspiration in bringing these elements together.

REFERENCE

1. L. H. Bell. *Fundamentals of Industrial Noise Control.* Harmony
   Publications, Trumbull, Connecticut, 1973.

BIBLIOGRAPHY

Beranek, L. L. *Noise and Vibration Control.* INCE, 1988.
Compendium of Materials for Noise Control, Department of Health,
   Education and Welfare, Washington, D.C., June 1975, prepared
   under Contract No. HSM-99-72-99.
Crocker, M. J. (Ed.). Noise control by the use of enclosures and
   barriers. In *Reduction of Machinery Noise.* Purdue University
   Press, Lafayette, Indiana, 1974.
Diehl, G. M. *Machinery Acoustics.* Wiley, New York, 1973.
Harris, C. M. *Handbook of Acoustical Measurements and Noise Control.*
   McGraw-Hill, New York, 1991.

EXERCISES

7.1 A hammer mill is to be enclosed to reduce the radiated noise.
Measurements show average noise levels 6 ft from the mill to be 90 dB
at 125 Hz, 97 dB at 250 Hz, 102 dB at 500 Hz, 106 dB at 1000 Hz,
99 dB at 2000 Hz, and 91 dB at 4000 Hz. What must be the spectral
transmission loss of the enclosure walls in order to meet on overall
90-dBA sound level criterion at 6 ft? Use the 90-dBA octave band
equivalent given in Table 7.1.

$$\quad Answer: \quad +8 \text{ at } 125 \text{ Hz}; \quad -25 \text{ at } 1000 \text{ Hz};$$
$$-7 \text{ at } 250 \text{ Hz}; \quad -18 \text{ at } 2000 \text{ Hz};$$
$$-18 \text{ at } 500 \text{ Hz}; \quad -10 \text{ at } 4000 \text{ Hz};$$

7.2 The noise level at 500 Hz inside a tumbling machine enclosure was 98 dB. The noise level at 500 Hz at an operator's station nearby was 81 dB. What was the noise reduction of the enclosure at 500 Hz?

Answer: 17 dB

7.3 The noise level at 1000 Hz at a quality control station near a punch press was 99 dB. With a dense vinyl curtain enclosing the press, the level at the station was 85 dB at 1000 Hz. What was the insertion loss of the curtain enclosure at 1000 Hz?

Answer: 14 dB

7.4 The transmission loss at 4000 Hz of an acoustical panel is 35 dB. A small penetration hole representing 2% of the total surface of the panel must be left open. Estimate the net transmission loss of the panel, including the 2% leak.

Answer: 17.5 dB

7.5 Noise levels at an operator's station near a shake-out screen in a foundry were 95 dB at 125 Hz, 98 dB at 250 Hz, 102 dB at 500 Hz, 108 dB at 1000 Hz, 105 dB at 2000 Hz, 100 dB at 4000 Hz, and 96 dB at 8000 Hz. Calculate the required insertion loss for an enclosure panel that would reduce the levels at the operator's station below 85 dB overall. (Hint: Adjust the octave band 90 dBA equivalent to 85 dBA on an energy basis.)

Answer: −2 at 125 Hz;
            −13 at 250 Hz;
            −23 at 500 Hz;
            −32 at 1000 Hz;
            −29 at 2000 Hz;
            −24 at 4000 Hz;
            −20 at 8000 Hz

7.6 Referring to Exercise 7.5, if another shake-out machine were installed on the other side of the operator's station, what increase in insertion loss would be required to meet the 85-dBA criterion?

Answer: 3 dB at each octave
            band center
            frequency

8

SILENCERS, MUFFLERS, AND ACTIVE NOISE CONTROL

Silencers and mufflers cover a wide range of noise reduction devices and must be considered one of the most powerful weapons available to the acoustical engineer. There is no technical distinction between a muffler or silencer, and the terms are frequently used interchangeably; i.e., one manufacturer will use a silencer and another a muffler for the same basic configuration. Despite the terms and myriad of configurations, the devices can be broken into three fundamental groups: absorptive/dissipative, reactive, and dispersive. Absorptive silencers contain either fibrous or porous materials and depend on absorptive dissipation of the acoustical energy. Reactive silencers contain no absorbing material but depend on the reflection or expansion of the sound waves with corresponding self-destruction as the basic noise reduction mechanism. The noise reduction of dispersive silencers usually comes from diffusing a high-velocity gas flow into smaller lower-velocity streams. Some silencers combine the elements of two or more types for extended performance. In addition, other functions such as water separation, filtering, spark arresting, heat recovery or exchange, etc., may also be present. In any case, they are usually installed in pipes or ducts to reduce sound transmission from one section of a gas flow system to another.

With a basic understanding of the acoustical properties of each type, the noise control engineer can usually select a silencer or combination of silencers which will effectively provide noise reduction regardless of the character of the noise. Difficulty generally is found not with finding a silencer with adequate acoustical performance but with dealing with problems such as size, weight, aerodynamic pressure losses, etc.

Showing extreme promise as a noise control measure is a concept called *active noise control*. Here, noise reduction is achieved by generating an "anti-noise" field which is superimposed on the source field. With careful attention to matching phase and sound pressure amplitude, a cancellation process ensues (interference), with resultant lower noise levels.

It should be emphasized that the basic approach of active noise control is not new [7], but it is only recently that the control theory and microelectronic components have reached the state of the art to produce practical results. We shall see that many of the inherent penalties of passive devices are avoided with the active noise control approach.

8.1 ACOUSTICAL PERFORMANCE PARAMETERS

Before examining the acoustical properties of the major types of silencers and mufflers, let us consider the parameters used to describe the acoustical performance. The following are the terms most commonly used:

1. Insertion loss (IL). Insertion loss is the difference between two sound pressure levels measured at the same point in space before and after a muffler has been inserted. Mathematically, insertion loss IL can be expressed as

$$IL = L_{p,1} - L'_{p,1} \quad [dB] \tag{8.1}$$

    where

    $L_{p,1}$ = sound pressure level measured at some point in space with no silencer (dB)

    $L'_{p,1}$ = sound pressure level measured at the same point in space with the silencer inserted (dB)

2. Transmission loss (TL). Transmission loss is the ratio of the sound power incident on the muffler to the sound power transmitted by the muffler.

3. Noise reduction (NR). Noise reduction is the difference between the sound pressure level as measured at the source side of a muffler and the sound pressure level measured at the receiving side. Noise reduction NR can be expressed mathematically as

$$NR = L_{p,1} - L_{p,2} \quad [dB] \tag{8.2}$$

    where

    $L_{p,1}$ = sound pressure level at the source side of a silencer (dB)

    $L_{p,2}$ = sound pressure level at the receiving or exit side (dB)

    It should be noted that experimental determination of noise reduction usually requires measurements inside the duct or pipes.

Of the three, insertion loss is the easiest to measure and is used most often by muffler or silencer manufacturers.

Generally, the acoustical performance of a silencer is given spectrally, that is, for each of the eight octave bands. With the spectral performance given, the acoustical engineer can now quantitatively select a silencer which will assure sufficient noise reduction or better *attenuation* (a general term) over the entire spectrum. Examples of the selection methods will be given as each type of silencer is presented.

High-performance silencer designs generally combine both absorptive and reactive elements in their construction. However, before discussing the more complex combinations, each type will be broken down into its simplest form and the basic characteristics of each examined. It should be emphasized that even the basic forms cannot always be strictly divided as purely absorptive or purely reactive. However, the grouping herein will be on the basis of the major noise reduction mechanism.

8.2 ABSORPTIVE SILENCERS

Absorptive silencers are widely used in treating noise where large volumes of air or gas are moved at relatively low static pressures. We shall therefore begin our discussion of silencers with these types, their performance, and their methods of selection.

8.2.1 Parallel Baffles

The simplest form of absorptive or dissipative silencers is the parallel baffle, as illustrated in Fig. 8.1. Basically, these baffles consist of an aerodynamically streamlined entrance and exit with perforated walls backed by highly absorbent acoustical material. The absorbing material is usually fibrous in texture, either glass or mineral wools.

The acoustical performance of parallel baffles depends primarily on three parameters:

1.   Thickness of absorbing materials
2.   Spacing of baffles
3.   Length of baffles

The effect of baffle thickness and spacing can be seen in Fig. 8.2 [1]. Note that the attenuation increases sharply at high frequencies as the spacing is narrowed. Note also that better performance at lower frequencies is obtained as the thickness of the absorbing material is increased. With respect to the length of parallel baffles, the acoustical performance increases as length increases. Shown in Fig. 8.3 [1] is the performance of baffles 6 in. thick, 12 in. on center (50% open area) for silencer lengths of 4, 8, and 12 ft. Note that the performance is not a linear function of length; i.e., doubling the length does not double the attenuation. The nonlinearity is due principally to the rapid absorption of high-order transverse modes in the first few feet of the silencer, leaving only a plane-wave-type of sound propagation. The resultant plane wave motion presents essentially a grazing incidence to the absorbing treatment, and hence little sound is absorbed.

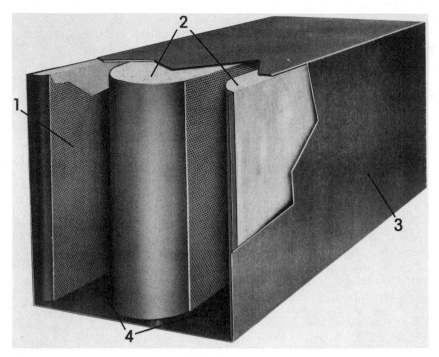

Figure 8.1 Typical parallel baffle-type duct silencer showing major construction elements. (1) perforated, galvanized steel; (2) bonded long fiber, acoustical fill; (3) galvanized sheet metal casing; (4) aerodynamically designed air passages. (Courtesy of United McGill Corporation, Groveport, Ohio.)

The performance of absorptive silencers can be sharply improved if the line of sight through the silencer is blocked or eliminated. Various curves and staggered patterns have been designed and are commerically available. The increased performance of a blocked line of sight baffle configuration is clearly evident from the performance data presented in Fig. 8.4 [1].

Parallel baffle-type absorptive silencers are available commercially from a large number of manufacturers. To meet a wide range of application, most manufacturers provide the units with face cross-sectional dimensions in increments of 6 in. and lengths of 3, 5, and 7 ft. Furthermore, the individual modules can be assembled as illustrated in Fig. 8.5 to fit almost any duct size or gas flow requirements. For design purposes, presented in Table 8.1 are typical attenuation values for parallel baffle-type absorptive silencers for common commercially available lengths of 3, 5, and 7 ft.

It must be emphasized that the acoustical performance of parallel baffles decreases sharply as gas flow velocity through the silencer reaches 3000 fpm. Acoustical performance corrections for flow velocity

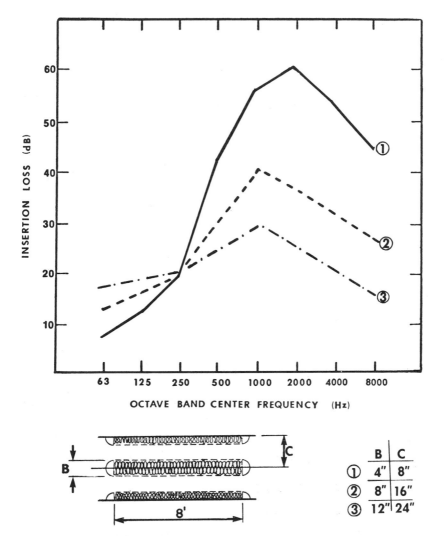

Figure 8.2 Effect of treatment thickness on insertion loss for parallel baffle-type silencers. (From Ref. 1).

are usually supplied by the manufacturers in their design specification sheets.

Because there is very little flow resistance in parallel baffles, they can be applied readily to those installations where pressure losses are critical such as in the air-conditioning or ventilating systems, forced draft fans, gas turbine facilities, etc. Shown in Fig. 8.6 are some examples of common parallel baffle silencer applications.

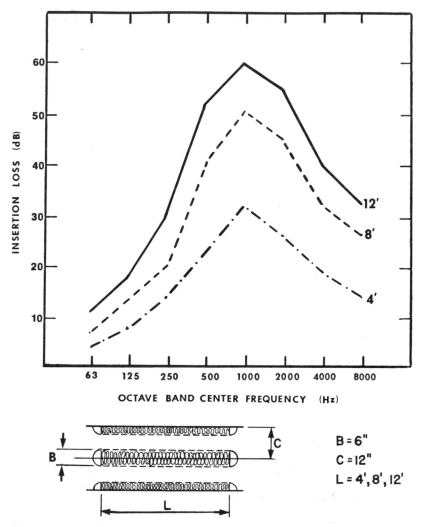

Figure 8.3 Effect of length on insertion loss for parallel baffle-type silencers. (From Ref. 1.)

8.2.2 Tubular Silencers

Another form of absorptive silencer is the tubular silencer. Illustrated in Fig. 8.7 is the typical construction of a tubular silencer. These silencers are really parallel absorbing baffles which have been *wrapped around* in order to interface simply with circular inlet or exhaust duct geometries. The spectral attenuation for tubular silencers

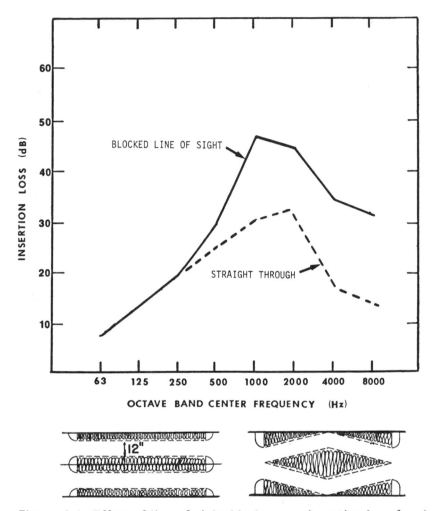

Figure 8.4 Effect of line of sight blockage on insertion loss for absorptive silencers. (From Ref. 1.)

is very similar to parallel baffles, i.e., good high-frequency performance but rather poor low-frequency performance. For example, shown in Fig. 8.8 is the typical performance of a tubular silencer with and without outer wall absorbing treatment. Note the higher performance with the outer wall treatment.

For heavy industrial application involving gas flow at high-pressure velocity and temperature, the outer walls of the silencer and the adapter flanges must be constructed of heavy-gauge metal, i.e., a minimum of 10 gauge.

Table 8.1 Typical Insertion Loss for Commercially Available Silencers

	Insertion Loss (dB) at Octave Band Center Frequencies						
Silencer Length (ft)	125 Hz	250 Hz	500 Hz	1000 Hz	2000 Hz	4000 Hz	8000 Hz
3	10	13	18	25	30	27	24
5	15	25	31	40	45	36	30
7	19	28	40	48	51	43	36

Consider now an example that illustrates the basic method for selecting absorptive silencers.

Example
  Measurements made at 3 ft from an unsilenced forced draft fan are shown on line 1 of Table 8.2. Select a silencer which will reduce the overall noise level below 90 dBA.
  Solution
  Presented in line 2 is a conservative design octave band equivalent to 90 dBA. The difference of line 1 and line 2 then represents the required attenuation of the silencer. Looking to Table 8.1 for

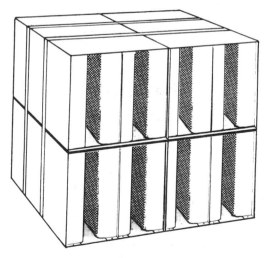

Figure 8.5 Four modular absorptive silencers assembled for use in a large duct. (Courtesy of Gale Corporation, North Brunswick, N.J.)

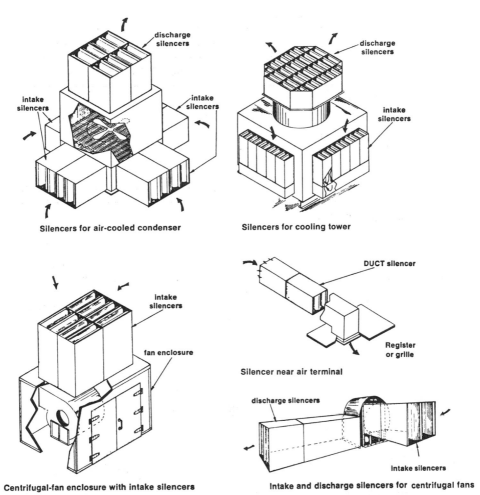

Silencers for air-cooled condenser

Silencers for cooling tower

Centrifugal-fan enclosure with intake silencers

Intake and discharge silencers for centrifugal fans

Figure 8.6 Examples of common absorptive-type parallel baffle silencer applications. (Courtesy of Industrial Acoustics Company, Bronx, N.Y.)

the typical attenuation of parallel baffle silencers, one can easily see that the 3-ft silencer is inadequate in the 250- and 500-Hz bands. Therefore, a minimum silencer length of 5 ft is required.

8.2.3 Effect of Flow

As mentioned earlier, the acoustical performance of absorptive silencers is reduced somewhat in the presence of airflow or gas flow.

Table 8.2 Computation Table

	\multicolumn{8}{c}{Octave Band Center Frequency (Hz)}
	63	125	250	500	1000	2000	4000	8000
1. L_p at 3 ft from forced draft fan (dB)	99	105	112	111	109	102	97	95
2. 90 dBA equivalent (not unique) (dB)	107	98	90	84	81	81	81	82
3. Line 1 minus line 2 (dB)	+8[a]	−7	−22	−27	−28	−21	−16	−13

[a]Meets design criteria.

Before going directly to this topic, two new terms, dynamic insertion loss and face velocity, need to be defined. Dynamic insertion loss is used to denote the net or actual insertion loss in the presence of airflow or gas flow. Face velocity v_f is the velocity of the air or gas at the face of the silencer and is given as

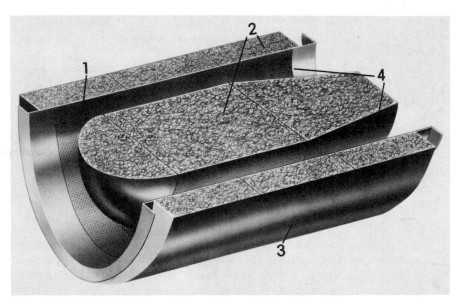

Figure 8.7 Cutaway view of absorptive tubular silencer showing major construction elements. (1) perforated, galvanized steel; (2) bonded long fiber, acoustical fill; (3) galvanized sheet metal casing; (4) aerodynamically designed air passages. (Courtesy of Industrial Acoustics Company, Bronx, N.Y.)

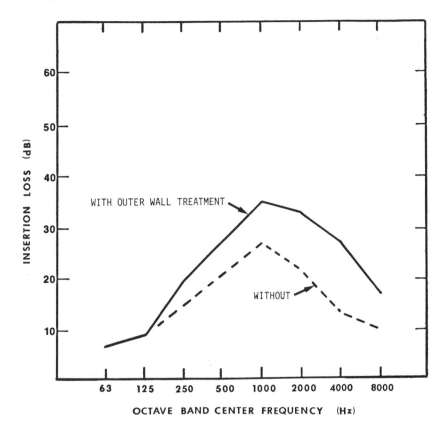

Figure 8.8 Insertion loss of tubular silencer with and without outer wall absorbing treatment.

$$v_f = \frac{V_f}{A_d} \qquad [\text{fpm}] \tag{8.3}$$

where

V_f = volume flow of air or gas (usually cfm)

A_d = cross-sectional area of silencer inset *face* (ft^2)

It should be emphasized that these terms and units are conventions used by most silencer manufacturers.

Shown in Table 8.3 are excerpts from a silencer catalog which shows the effect of flow on insertion loss. Note that up to 8 dB of dynamic insertion loss is lost at face velocities of +2000 vs −2000 fpm in the 250 Hz band for silencer model 7Es. This loss in performance can hardly

Table 8.3 Excerpt from Silencer Catalog Showing the Effect of Air
or Gas Flow on Insertion Loss[a]

Model no.	Octave band, Hz	1	2	3	4	5	6	7	8
	Silencer face velocity, fpm	63	125	250	500	1K	2K	4K	8K
		Dynamic insertion loss, dB							
3Es	−2000	6	10	19	29	36	34	23	15
5Es	−2000	11	19	24	40	52	49	27	17
7Es	−2000	11	21	40	53	54	54	38	24
10Es	−2000	14	30	40	57	58	56	47	29
3Es	−1000	5	10	17	28	36	34	23	16
5Es	−1000	11	17	23	39	50	49	32	20
7Es	−1000	11	22	38	50	55	54	43	27
10Es	−1000	12	31	44	54	55	56	50	31
3Es	+1000	4	8	15	25	34	34	24	17
5Es	+1000	8	13	20	35	50	49	36	24
7Es	+1000	8	18	35	48	54	53	46	32
10Es	+1000	10	26	42	55	54	57	51	38
3Es	+2000	3	7	13	23	32	34	24	18
5Es	+2000	7	12	19	33	48	49	36	25
7Es	+2000	7	16	32	46	54	51	48	34
10Es	+2000	8	23	39	55	54	57	52	41

[a]The number in the model denotes silencer length, i.e., 3 ft, 5 ft,
7 ft, 10 ft.
Note: Forward flow (+) occurs when noise and air travel in the
same direction, as in a typical supply or fan discharge system. Reverse flow (−) occurs when noise or air travel in opposite directions, as in a typical return or fan intake system.
Source: Courtesy of Industrial Acoustics Company, Bronx, N.Y.

be ignored and shows that the effect of flow velocity must be a
design consideration in all silencer installations.

8.2.4 Self-Generated Noise

Another effect which must be considered in selecting or specifying
silencers is the self-generated noise produced as the air or gas flows
through the silencer and exits. Obviously, the level of such noise is
the minimum noise level one can obtain. The noise level is a first-
order function of face velocity but can be substantially affected by
silencer geometry. In the latter case, self-noise can be minimized
by careful attention to good turbulence-free aerodynamic flow.

Self-generated noise is usually given in terms of sound power level
for various face velocities. Shown in Table 8.4 is an excerpt from a
silencer catalog showing the self-noise power levels for several silencers at given face velocities. Note that the self-noise increases
quite sharply with increasing face velocity. In fact, in some bands,
doubling the face velocity produced a level increase of 20 dB or more,
representing a power ratio increase of 100. This substantial increase
is not surprising, and we shall see in a later chapter that noise associated with airflow is indeed a dynamic function of velocity.

Table 8.4 Excerpt from Silencer Catalog Showing the Self-Generated
Sound Power Levels for a Range of Face Velocities[a]

	Octave band	1	2	3	4	5	6	7	8
	Hz	63	125	250	500	1K	2K	4K	8K
Model no.[b]	Silencer face velocity, fpm	Self-noise power levels, dB (10^{-12} W)							
	−2000	56	54	58	60	61	65	69	64
	−1500	47	47	52	55	57	63	64	54
	−1000	41	41	45	47	52	60	48	38
	+1000	42	35	33	32	34	33	27	22
	+1500	50	47	44	41	43	45	43	41
	+2000	60	57	54	50	49	53	53	50

(+) Forward flow; (−) reverse flow.
[a]The numeric in the model number denotes the silencer length.
[b]Es all sizes.
Source: Courtesy of Industrial Acoustics Company, Bronx, N.Y.

Now how does one use these data? These power levels simply represent the minimum acoustical power one can expect at the exit of a silencer for a given face velocity. For example, if a silencer were installed in a duct to reduce noise from a fan, and the fan noise were completely eliminated, a residual noise level would still remain due to the aerodynamic flow. If the silencer were model 7Es and with +2000 fpm, the octave band power levels would be as given in the last line of Table 8.4. Furthermore, the exit of the silencer can usually be considered a point noise source, and as such one can apply directly the methods given in Chap. 4 to evaluate the resultant noise level at any point in space. Consider an example.

Example

An absorptive silencer (model 7Es) 7 ft long was installed in the inlet duct of a fan. If the fan noise was completely eliminated, what is the sound pressure level 20 ft from the silencer at 500 Hz, given a face velocity of +2000 fpm?

Solution

The residual noise is due exclusively to airflow or self-generated noise. Hence, the sound power at 500 Hz and a face velocity of +2000 fpm is, from Table 8.4, 50 dB re 10^{-12} W for model 7Es. The sound pressure level at 20 ft is then given from Eq. (4.5):

$$L_p = L_W - 20 \log_{10}(r') - 0.5 \qquad [dB]$$

Upon substitution, we get

$$L_p = 50 - 20 \log_{10}(20) - 0.5$$
$$= 23.5 \text{ dB}$$

Thus the sound pressure level at 20 ft in the 500-Hz band would be 23.5 dB. It is obvious that the octave band sound pressure level for the entire spectrum could be calculated similarly.

One note of caution: There are usually rather strong directional factors to be considered, especially at the higher frequencies, i.e., above 1000 Hz. Here the noise tends to *beam* forward or be concentrated along the silencer axis, i.e., in the direction of flow.

8.2.5 Pressure Losses

Another factor, generally present, which provides design constraints on silencer selection is pressure loss. Dissipative-type silencers are most often used where high volumes of air or gas are being moved at relatively low *head* pressures, i.e., in the range of a few inches of water. As such static pressure losses in the system must be minimized.

In particular, in air-conditioning or ventilating systems, pressure
losses due to silencers must be a first-order design consideration.
Most silencer manufacturers provide the static pressure losses for
their equipment, again as a function of volume flow. Shown in Table
8.5 is a catalog excerpt where static pressure loss is given for a range
of volume flows. In accordance with basic aerodynamic principals,
pressure *drop* increases as the volume velocity increases. Therefore,
the only design guideline for minimizing pressure loss is to assure that
the flow velocity through the silencer is not increased. As a rule of
thumb, if the open flow cross-sectional area of the silencer is 1.25
to 1.5 times the cross section of the duct, pressure losses will be
minimal. Consider an example.

Example

What is the static pressure loss for a model 7Es (24 × 48) silencer
at a volume flow rate of 10,000 cfm?

Solution

Referring to Table 8.5, one sees that the static pressure loss is
in the range of 0.46 to 0.61 in. of water for a volume flow rate
of 10,000 cfm,

The parallel baffle silencers we have discussed thus far are usually
considered *duct silencers* in that their installation is dominantly in air
ducts for air-conditioning or ventilating applications. In other heavy
industrial applications such as jet engine test facilities, gas turbine
generators, etc., the same basic configuration is used, but the design is usually customized to reflect the more harsh environment. For
example, shown in Fig. 8.9 is the typical configuration of a gas
turbine installation. Note the dissipative acoustical treatment in the
inlet and exhaust, commonly called *splitters*. Now because of the
temperature and corrosive nature of the exhaust gases, the sheet
metal selected for these silencers must be of a heavier gauge (thicker)
than typical duct silencers and highly resistant to corrosion. In addition, because of high noise levels and flow velocities, the absorbing
materials require additional protection. For example, the absorbing
materials are usually encapsulated in glass cloth bags. In addition,
a wire screen or *scrubble* (steel wool 1 in. thick) is inserted between
the perforated panels and the encapsulated absorbing materials. It
must be emphasized that the specific acoustic resistance of the cloth,
screen, and/or scrubble must be no greater than the absorbing materials. If greater, these inserts would provide an acoustical barrier,
thus reducing the effectiveness of the absorbing material.

The effectiveness of splitter panels can be fully appreciated when
we consider that a large jet engine in a well-designed test cell is barely
audible at a distance of 100 ft. Generally all that can be heard is a
very low-frequency *rumble*. It should be noted that the length of the

Table 8.5 Excerpt from Silencer Catalog Showing State Pressure Loss of Es Silencer Modules for a Wide Range of Volume Airflow (cfm)

Nominal module size W × H, in.	Silencer face area, sq. ft	Model no. Weight, lb 3 Es	5 Es	7 Es	10 Es	Static pressure drop, inches H₂O										
						.05	.10	.15	.20	.25	.30	.40	.50	.60	.75	1.00
						.05	.11	.16	.22	.27	.33	.44	.55	.66	.82	1.10
						.08	.15	.23	.31	.38	.46	.61	.77	.92	1.15	1.54
						.10	.20	.30	.40	.50	.60	.70	.99	1.19	1.49	1.99
						Airflow in cfm										
6 × 12	0.50	22	40	55	77	237	336	411	475	531	581	671	751	822	919	1062
6 × 24	1.00	35	63	88	123	475	671	822	949	1062	1163	1343	1501	1644	1838	2123
6 × 36	1.50	49	87	122	171	712	1007	1233	1424	1592	1744	2014	2252	2467	2758	3184
12 × 12	1.00	33	56	78	111	475	671	822	949	1062	1163	1343	1501	1644	1838	2123
12 × 18	1.50	43	73	102	155	712	1007	1233	1424	1592	1744	2014	2252	2467	2758	3184
12 × 24	2.00	52	89	125	177	949	1343	1644	1899	2123	2326	2685	3002	3289	3677	4246
12 × 30	2.50	62	107	150	212	1187	1678	2056	2374	2654	2907	3357	3753	4111	4596	5307
12 × 36	3.00	74	125	176	250	1424	2014	2467	2848	3184	3488	4028	4504	4933	5516	6369
12 × 42	3.50	83	141	199	—	1662	2350	2878	3323	3715	4070	4699	5254	5756	6435	7430
12 × 48	4.00	93	158	226	—	1899	2685	3289	3798	4246	4651	5371	6005	6578	7354	8492
24 × 18	3.00	71	121	170	241	1424	2014	2467	2848	3184	3488	4028	4504	4933	5516	6369
24 × 24	4.00	86	147	207	293	1899	2685	3289	3798	4246	4651	5371	6005	6578	7354	8492
24 × 30	5.00	102	173	243	354	2374	3357	4112	4747	5308	5814	6714	7506	8222	9193	10615
24 × 36	6.00	117	204	288	405	2848	4028	4933	5697	6369	6977	8056	9007	9867	11031	12738
24 × 42	7.00	132	230	325	—¦	3323	4699	5756	6646	7430	8140	9399	10508	11511	12870	14861
24 × 48	8.00	147	256	362	—	3798	5371	6578	7595	8492	9303	10742	12010	13156	14708	16984
36 × 30	7.50	142	249	—	—	3560	5035	6167	7121	7961	8721	10070	11259	12333	13789	15922
36 × 36	9.00	162	284	—	—	4272	6042	7400	8545	9553	10465	12084	13511	14800	16547	19107
36 × 42	10.50	182	319	—	—	4985	7049	8633	9969	11145	12210	14098	15762	17267	19305	22291
36 × 48	12.00	204	355	—	—	5697	8056	9867	11393	12738	13954	16112	18014	19734	22063	25476

Source: Courtesy of Industrial Acoustics Company, Bronx, N.Y.

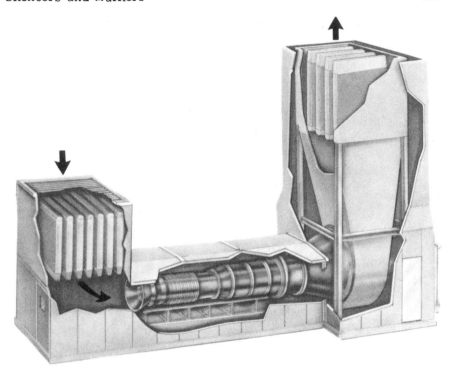

Figure 8.9 Typical gas turbine installation showing splitter silencers in the inlet and exhaust. (Courtesy of Burgess Industries, Dallas Tex.)

inlet and exhaust silencers for these applications ranges from 10 ft for a small gas turbine to 24 ft for a large jet engine test cell.

Due to the proprietary nature of this acoustical treatment, very little performance data have been published. However, shown in Fig. 8.10 is the octave band noise reduction measured across the inlet silencer of a small gas turbine test cell. In this installation the inlet splitter baffles were 6 in. thick on 12-in. centers (50% open) and 7 ft long. It should be noted that the chain saw used as a noise source was barely audible yet could be seen through the splitters less than 10 ft away.

Because of the immense scope of this subject and space limitation, we cannot add further to the design guidelines presented. However, from the actual examples and illustrations, it is clear that noise from

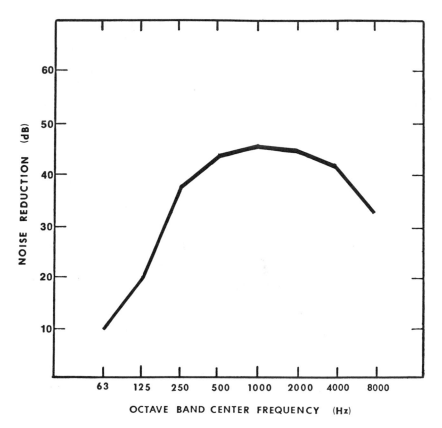

Figure 8.10 Noise reduction measured across inlet splitters of a gas turbine cell. (From H. R. Mull, Bell and Associates, Norwalk, Conn., unpublished data, 1972.)

the loudest industrial equipment can be controlled to the level required with sufficient absorptive silencer treatment.

Let us now consider other types of absorptive or dissipative silencer types and configurations.

8.3 LINED DUCTS

A very simple form of duct silencer can be fabricated by lining the
interior of the duct with absorbing materials. Typically, the lining
material chosen is acoustical-quality fibrous glass 1 to 3 in. thick.

The rigorous mathematical analysis for estimating the attenuation for lined ducts is quite involved and is beyond the scope of this text. However, an empirical formula for estimating the linear attenuation is [2]

$$\text{Attenuation} = 12.6 \left(\frac{P}{S}\right)\alpha^{1.4} \quad [dB/ft] \tag{8.4}$$

where

    P = perimeter of the duct (inside the lining) (in.)

    S = cross-sectional area of the duct (inside the lining) (in.^{2})

    α = Sabine absorption coefficient of the lining material obtained from laboratory measurements (unitless)

Equation (8.4) provides good agreement with field measurements for lower frequencies, i.e., 500 Hz and lower. However, at frequencies above 500 Hz, the attenuation rate falls off sharply after the first 3 to 5 ft. What happens here is the higher-ordered transverse pressure wave fronts are quickly attenuated in the first few feet due to impingement with the absorbent linings. The remaining lower-order nearly plane or plane wave fronts, propagating parallel to the absorbent lining, do not impinge on the lining and hence show little attenuation. The attenuation can be *restarted* by introducing a 90° or 180° elbow a few feet downstream, which disperses the plane wave front into higher transverse wave fronts which are again quickly attenuated.

In addition, for the best results in using Eq. (8.4),

1. The smallest duct dimension should lie in the interval of 18 to 6 in.
2. The ratio of duct height to duct length should not exceed 2:1.

Also, the acoustical performance will be affected if flow velocities exceed 4000 fpm. Here again the attenuation will be less if flow is in the direction of sound propagation and greater if opposed. Consider an example.

Example

    Calculate the attenuation for 10 ft of lined duct (12 × 12 in.) at 250 and 2000 Hz. The duct lining is fibrous glass 1 in. thick with measured Sabine absorption coefficients α of 0.40 and 0.80 in the 250- and 2000-Hz bands, respectively.

    Solution

    The perimeter P of the duct inside the lining is 4 × 10 in. = 40 in. The cross-sectional area of the duct inside the lining is 10 × 10 in. = 100 in.^{2} For α = 0.40 (250 Hz),

$$\alpha^{1.4} = 0.28$$

Upon substitution into Eq. (8.4), we get

$$\text{Attenuation} = \frac{12.6 \times 40 \times 0.28}{100}$$

$$= 1.41 \quad [\text{dB/ft}]$$

Hence for a lined duct 10 ft long, the net attenuation at 250 Hz is

$$1.41 \times 10 = 14.1 \text{ dB}$$

At 2000 Hz,

$$\alpha = 0.80$$
$$\alpha^{1.4} = 0.73$$

Therefore, upon substitution in Eq. (8.4), we have

$$\text{Attenuation} = \frac{12.6 \times 40 \times 0.73}{100}$$

$$= 3.68 \quad [\text{dB/ft}]$$

And for 10 ft of lined duct, the net attenuation at 2000 Hz is

$$3.68 \times 10 = 36.8 \text{ dB}$$

It should also be noted that the acoustical lining will serve as effective thermal insulation, thus often serving a dual purpose.

8.4 LINED BENDS

Duct elbows or *bends* lined with absorbing materials can also provide significant noise reduction. Shown in Fig. 8.11 is the typical acoustical performance for a lined bend whose principal dimensions are in the range of 1 to 2 ft. To achieve this level of performance, the absorptive lining must

1. Have an average absorption coefficient (NRC) in the range of 0.70 to 0.80
2. Extend 4 to 5 times the principal dimension on both sides of the bend

If the bend is 180° or *switched back*, an additional 5 dB can be expected at all frequencies above 250 Hz. Where pressure losses are extremely critical, the bend should be rounded with the possible inclusion of a treated guide vane.

Typical lining materials are essentially the same for bends as for lined ducts. In most cases, 1- or 2-in. acoustical-grade fibrous glass or mineral wool is used, being constrained and protected with perforated sheet metal or plastic film.

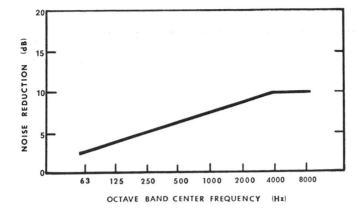

Figure 8.11 Typical noise reduction for a 90° lined bend.

Example
   What is the sound attenuation of a 90° lined bend and a 180°
   switchback in a duct at 2000 Hz?
   Solution
   From Fig. 8.11, we see that the attenuation for a 90° lined bend
   at 2000 Hz is approximately 9 dB. For a 180° switchback, the
   attenuation is typically 5 dB more than a 90° line bend, or 9 + 5 =
   14 dB.

8.5 LINED PLENUM

Illustrated in Fig. 8.12 is a plenum chamber lined with sound-absorbing
materials. Plenum chambers of this type are frequently used as silencers in air-conditioning and ventilating systems and in test facilities
to reduce flow velocity and turbulence. The attenuation of these devices comes from both dissipative and reactive effects, which renders
mathematical analysis quite difficult. However, an empirical relation
which provides good first-order results has been developed by Wells [3]:

$$\text{Attenuation} = 10 \log_{10} \left(\frac{1}{S_e [\cos(\theta)/2\pi d^2 + (1 - \alpha)/\alpha S_w]} \right) \quad [\text{dB}]$$

$$\tag{8.5}$$

$$= -10 \log \left[S_e \left(\frac{\cos \theta}{2\pi d^2} + \frac{1 - \alpha}{\alpha S_w} \right) \right] \quad [\text{dB}] \qquad (8.5a)$$

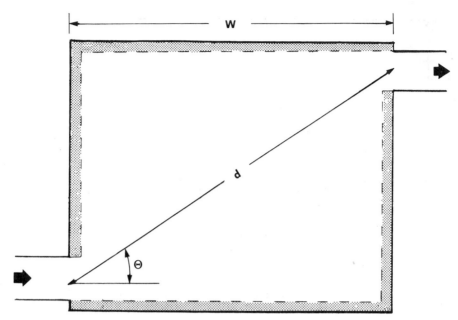

Figure 8.12 Acoustically lined plenum chamber silencer.

where

α = sabine absorption coefficient of the lining (unitless)

S_e = plenum exit area (ft^2 or m^2)

S_w = plenum wall area (ft^2 or m^2)

d = distance between entrance and exit (see Fig. 8.12) (ft or m)

θ = angle of incidence at the exit, i.e., the angle which the direction line d makes with the normal to the exit opening (rad)

$\cos \theta = w/d$ (unitless)

For frequencies sufficiently high so that the wavelength is less than the plenum dimensions, Eq. (8.5) is accurate within a few decibels. At lower frequencies, it is conservative, and the actual attenuation often exceeds the calculated value by up to 10 dB.

To optimize the performance of a plenum, one sees by examining Eq. (8.5) that

1. θ should be as close to $\pi/2$ or 90° as possible.
2. d should be as large as practical.

3. α should be as large as possible.
4. The exit duct cross section S_e should be minimal (often a problem from aerodynamic considerations).

Consider an example.

Example

A plenum chamber of the following dimensions is to be used in a high flow velocity refrigeration system:

$$S_e = 4 \text{ ft}^2$$

$$S_w = 400 \text{ ft}^2$$

$$w = 6 \text{ ft}$$

$$d = 12 \text{ ft}$$

The absorption coefficients for the acoustic lining are as follows:

Octave Band Center Frequency (Hz)

	125	250	500	1000	2000	4000
α (unitless)	0.25	0.50	0.80	0.90	0.80	0.75

Estimate the noise reduction for each octave band.
Solution
From the dimensions,

$$\cos \theta = \frac{w}{d}$$

$$= \frac{6}{12}$$

$$= 0.5$$

With the other dimensions, upon substitution in Eq. (8.5a), we obtain the attenuation in terms of the lining absorption coefficients:

$$\text{Attenuation} = -10 \log_{10} \left[4 \left(\frac{0.5}{2\pi \, 12^2} + \frac{1 - \alpha}{\alpha \times 400} \right) \right]$$

Now, for the 125-Hz band, $\alpha = 0.25$, and upon substitution we obtain

$$\text{Attenuation} = -10 \log_{10} \left[4 \left(0.0005 + \frac{1 - 0.25}{0.25 \times 400} \right) \right]$$

$$= 14.9 \text{ dB}$$

Continuing the calculations for the remaining octave bands, we obtain the attenuation, presented in tabular form:

Frequency (Hz)	125	250	500	1000	2000	4000
Attenuation (dB)	14.9	19.1	23.3	24.8	23.3	22.6

With the spectral attenuation determined, the adequacy of the plenum chamber in a particular application can be readily assessed in the same manner as illustrated for parallel baffles.

As mentioned earlier, plenum chambers are often used where high-velocity gas is being discharged such as for gas turbines, burners, etc. In these installations where pressure losses are not so critical, a heavy-duty *blast* panel can be installed opposite the inlet to protect the chamber walls and lining treatment. If this panel extends from the inlet to the exit, the *line of sight* from the inlet to the exit is blocked, and a sharp increase of up to 10 dB of additional attenuation is achieved above 1000 Hz.

For similar applications where a high degree of silencing is required, several chambers in series can be used. Here the resultant attenuation is approximately additive for noise above 500 Hz. At lower frequencies, the total attenuation can be approximated from

$$\text{Total attenuation} = (3N + A_0) \qquad [\text{dB}] \qquad (8.6)$$

where

N = number of chambers

A_0 = attenuation of original plenum chamber (dB)

Example

The attenuation of a plenum chamber is 12 dB at 250 Hz. Estimate the attenuation at 250 Hz if another identical chamber is added in series.

Solution

From Eq. (8.6),

$N = 2$

$A_0 = 12$ dB

Therefore, the total attenuation of the double chamber is

Total attenuation = $3 \times 2 + 12 = 18$ dB

8.6 ACOUSTICAL LOUVERS

Another form of parallel baffles is the acoustical louver, as illustrated in Fig. 8.13. In this configuration, each louvre *blade* is filled with

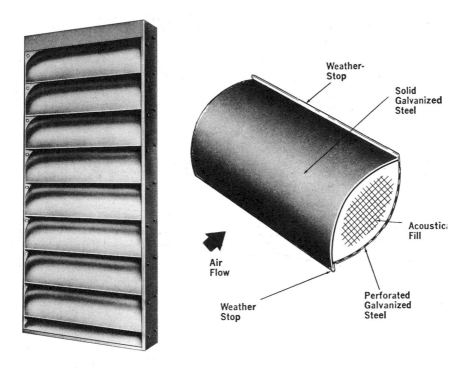

Acoustical Performance Data

	OCTAVE BANDS							
Octave Band Mid-Frequency, Hz.	1 63	2 125	3 250	4 500	5 1000	6 2000	7 4000	8 8000
TRANSMISSION LOSS IN DECIBELS								
Model R	5	7	11	12	13	14	12	9
Model LP	4	5	8	9	12	9	7	6

Figure 8.13 Basic construction features of acoustical louvers and typical acoustical performance. (Courtesy of Industrial Acoustics Company, Bronx, N.Y.)

absorbing material. With the underside of the blade constructed of perforated sheet metal, noise passing through the air passage is absorbed in a manner analogous to a parallel baffle. The acoustical performance TL of commercially available models is also shown in Fig. 8.13. Note that the performance is somewhat limited, especially in the low-frequency range.

There are many applications for these louver units such as cooling tower inlets, mechanical room inlets, fresh air intakes, external build-

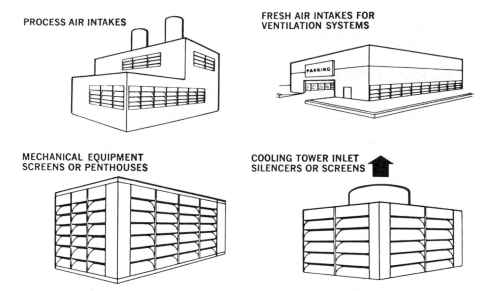

Figure 8.14 Examples of acoustical louver applications. (Courtesy of Industrial Acoustics Company, Bronx, N.Y.)

ing wall penetrations, etc. Some typical installations are illustrated in Fig. 8.14.

The two principal features which account for the popularity of louvers are (1) ease of installation and (2) relatively low pressure losses, i.e., less than 1.0 in. of water for face velocities less than 1000 ft/sec. These louver units are usually available in modular sizes of 12-in. increments on each dimension. A bird screen is usually available as an option from the manufacturer and is strongly recommended for external (outside) wall installations.

8.7 STACK-INSERT SILENCERS

Another form of absorptive silencer which has unique application is the stack-insert silencer. Often, noise-generating equipment such as fans, blowers, scrubbers, combustion chambers, etc., discharge both gas and noise into the atmosphere through a stack. The insert silencer is designated to be installed inside the stack, as illustrated in Fig. 8.15. If a noise problem is anticipated, the insert can be installed in the stack at the time of construction. If after construction a noise problem is determined (most often the case), the insert can then be lowered into the stack and fastened.

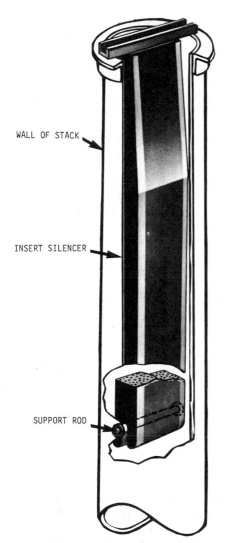

Model	Octave Band, Center Frequency, Hz.							
	63	125	250	500	1000	2000	4000	8000
	Net Insertion Loss, db							
24″ SSA3	3	6	12	22	27	20	12	8
48″ SSA3	6	12	22	27	20	12	8	4
96″ SSA3	12	22	27	20	12	8	4	2
24″ SSA5	4	8	16	29	36	24	15	10
48″ SSA5	8	16	29	36	24	15	10	6
96″ SSA5	16	29	36	24	15	10	6	4
24″ SSB5	5	11	21	30	40	30	17	10
48″ SSB5	11	21	30	40	30	19	12	7
96″ SSB5	21	30	40	30	19	12	7	5
24″ SSC5	6	13	24	33	41	22	14	12
48″ SSC5	13	24	33	41	22	14	12	8
96″ SSC5	24	33	41	22	14	12	8	4
24″ SSD5	9	18	30	51	41	36	23	14
48″ SSD5	18	30	51	41	36	23	14	7
96″ SSD5	30	51	41	36	23	14	7	6

Duct Velocity, ft /sec	50	75	100	125
FREE DISCHARGE (Mounted at Top of Stack)				
	ΔP inches of water			
SSA3	0.58	1.30	2.30	3.60
SSA5	0.80	1.80	3.20	
SSB5	1.30	3.00	5.30	
SSC5	2.30	5.20	9.10	
SSD5	4.40	10.00	17.40	
DUCT MOUNTED (Mounted at Least Five Stack Diameters Below Top of Stack)				
	ΔP inches of water			
SSA3	0.29	0.65	1.10	1.80
SSA5	0.48	1.10	1.90	3.00
SSB5	0.80	1.85	3.20	
SSC5	1.50	3.40	5.90	
SSD5	3.10	7.10	12.40	

Figure 8.15 Insert silencer installed in an exhaust stack. Also shown are typical acoustical performance and pressure loss (ΔP) data. (Courtesy of the Aeroacoustic Corporation, Jacksonville, Fla.)

The sound reduction is obtained through absorption. Here the basic construction of the insert is a sheet metal shell, enclosing acoustical-quality fibrous glass or mineral wools. The sides of the shell are perforated to permit sound wave impingement on the absorbing materials. Since the stack itself provides an outer containment shell, the configuration resembles the tubular silencer, as discussed earlier in this chapter. The acoustical performance of commercially available insert silencers are also shown in Fig. 8.15 along with aerodynamic pressure loss characteristics. The first two digits of the model number refer to the stack diameter in inches, i.e., 24 in. refers to a 24 in. diameter stack.

Note that the acoustical performance is appreciable, exceeding 20 dB in most models for the critical, hard-to-control 250-Hz octave band. Equipment discharged through a stack is generally rich in lower-frequency acoustic energy and in the 250-Hz band in particular.

Note also that pressure losses also are appreciable and cannot be ignored. However, these losses can be reduced considerably by installing the insert at least five stack diameters below the top of the stack.

Installation of these insert silencers is relatively simple. When the stack insert is to be installed at the top of the stack, it is supplied with a beam on the trailing edge that extends over the edges of the stack. After the stack insert is centered in the stack, this beam is either welded or bolted to the top of the stack. The lower edge of the insert is provided with a load-carrying circular support extending completely through the leading edge. This support is sized for a "snug" fit with a suitably sized rod or bar. After dropping the insert silencer into the stack, the rod or bar is passed through the support and welded.

8.8 REACTIVE SILENCERS

Reactive-type silencers generally consist of one or more expansion chambers wherein, as sound passes through, attenuation is achieved through reflective self-destruction. Probably the most common example of a reactive silencer is the automobile *muffler*. As we shall see, the acoustical performance of reactive silencers is rather selective spectrally, and hence in most applications the silencer must be designed or *tuned* to the discrete frequency character of the noise. In short, these silencers are most effective in applications associated with rotary equipment generating noise with dominantly discrete frequency character. Some common examples include internal combustion engines, compressors, rotary positive displacement blowers, vacuum pumps, etc.

Consider now some of the more basic types of reactive silencers and their applications.

8.8.1 Single Expansion Chamber

The simplest form of reactive silencer is the single expansion chamber, as illustrated in Fig. 8.16. By considering the reflection and expansion of sound waves as they propagate through the discontinuities at A and B, the transmission loss TL can be calculated and shown to be [4]

$$TL = 10 \log \left[1 + \frac{1}{4} \left(m - \frac{1}{m} \right)^2 \sin^2 kL \right] \quad \text{[dB]} \qquad (8.7)$$

where

m = silencer cross-sectional area/inlet and exit duct cross-sectional area = S_2/S_1

k = wave number = $2\pi / \lambda = 2\pi f/c$ (m^{-1} or ft^{-1})

L = chamber length (m or ft)

By examining Eq. (8.7), one sees that the magnitude of transmission loss depends on the value of m and also on the frequency from the argument of the sine function (kL).

More specifically, the character of the attenuation is periodic such that maximum transmission loss results when

$$kL = \frac{n\pi}{2} \ (n = 1, \ 3, \ 5, \ 7, \ . \ . \ ) \qquad \text{[unitless]} \qquad (8.8a)$$

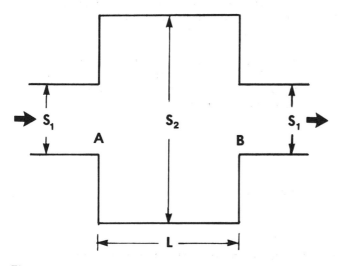

Figure 8.16 Single expansion chamber silencer of length L and cross-sectional areas S_1 and S_2.

and zero transmission loss occurs at

$kL = n\pi \quad (n = 1, 2, 3, 4, \ldots) \quad$ [unitless] $\quad$ (8.8b)

Now for design purposes, Eq. (8.7) can be solved for L, and using

$$k = \frac{2\pi}{\lambda}$$

we see that for a given m, maximum attenuation occurs when

$$L = \frac{n\lambda}{4} \quad \text{[m or ft]} \quad\quad\quad (8.9)$$

where

λ = wavelength of sound (m or ft)

$n = 1, 3, 5, \ldots$ (odd integers)

Since λ is related to frequency by the speed of sound, one can say that peak attenuation occurs at frequencies which correspond to a chamber length $L = \lambda/4, 3\lambda/4, 5\lambda/4, \ldots$ Because of this $\lambda/4$ relationship, these silencers are often called *quarter wave mufflers*.

In the same way, the attenuation vanishes at frequencies corresponding to chamber length $L = \lambda/2, \lambda, 3\lambda/2, \ldots$

A single cycle of Eq. (8.7) for various values of m is shown in Fig. 8.17. This is an especially practical presentation for design purposes. An example will illustrate the design selection methods.

Example

The fundamental discrete frequency tone for a diesel-engine-driven generator is 90 Hz. Design a single expansion chamber muffler with 20-dB attenuation at 90 Hz.

Solution

From Fig. 8.17, for 20-dB attenuation, the expansion or cross-section ratio m of the muffler must be at least 25 for a 20-dB attenuation including a little design margin. To size the chamber length for optimum performance, from Eq. (8.9),

$$kL = \frac{n\pi}{2}$$

or

$$L = \frac{\pi c}{2 \times 2\pi f} \quad \text{(for n = 1)}$$

$$= \frac{c}{4f}$$

Therefore, for f = 90 and taking c = 1100 ft/sec, we get

$$L = \frac{1100}{4 \times 90} = 3.06 \text{ ft}$$

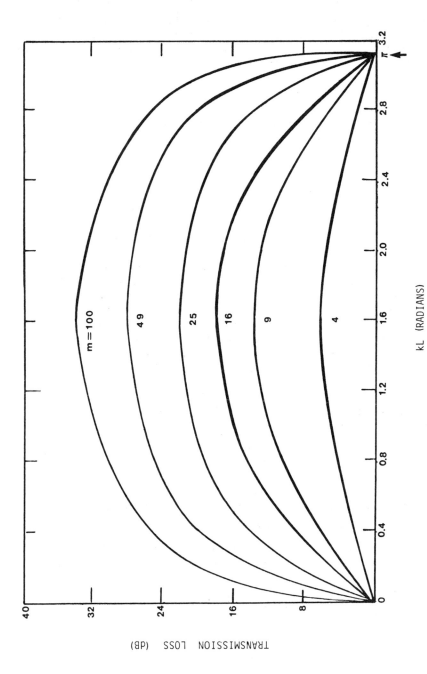

Figure 8.17 Transmission loss of a single expansion chamber of length L for a range of expansion ratios.

Hence, for 20-dB attenuation at 90 Hz, the chamber should be 3.06 ft long and have a cross-sectional ratio m of 25 or greater.

Returning to the design curves of Fig. 8.17, one notes that the effective attenuation actually extends over the range of $0.5 \leq kL \leq 2.6$. Therefore, by rewriting the inequalities, for the chamber length L, we obtain a range given by [5]

$$\frac{0.5c}{2\pi f} \leq L \leq \frac{2.6c}{2\pi f} \qquad [m] \qquad\qquad (8.10)$$

which provides additional design flexibility. To see this, let us return to the previous example. For f = 90 Hz, we get

$$\frac{0.5 \times 1100}{2\pi \times 90} \leq L \leq \frac{2.6 \times 1100}{2\pi \times 90}$$

or

$$0.97 \leq L \leq 5.06 \text{ ft}$$

Hence, a chamber whose length lies between 1 and 5 ft would provide more than 16 dB of attenuation at 90 Hz.

It must be noted that in these examples only the first *cycle*, i.e., $0 \leq kL \leq \pi$, of the design curves has been utilized. Additional chamber configurations with equivalent attenuation will be longer, usually less desirable from space and weight considerations.

Another factor which must be considered in expansion chamber design is the effect of high temperature on exhaust gases, common, for example, with internal combustion engines. This factor can easily be included in the design equations by noting that

$$c = 49.03\sqrt{°R} \qquad [ft/sec]$$

and upon substitution in the inequalities, we get the more general expression for chamber length L:

$$\frac{0.5(49.03\sqrt{°R})}{2\pi f} \leq L \leq \frac{2.6(49.03\sqrt{°R})}{2\pi f} \qquad\qquad (8.11)$$

where

°R = absolute temperature of exhaust gas (deg Rankine)

f = frequency of sound (Hz)

Returning again to the previous examples, if the exhaust gas is at 300°F = 459.7 + 300 = 759.7°R, the range of chamber length from Eq. (8.11) [5] is

$$\frac{0.5(49.03\sqrt{759.7})}{2\pi \times 90} \leq L \leq \frac{2.6(49.03\sqrt{759.7})}{2\pi \times 90}$$

$$1.2 \leq L \leq 6.2 \qquad [ft]$$

Hence, for effective attenuation, the chamber length must fall between 1.2 and 6.2 ft.

Comparing these results to the previous examples, one notes that for a temperature of 300°F, the chamber length must be increased slightly for effective attenuation. In this example, the effect of temperature appears secondary, but for optimum performance the effect is of the first order.

8.8.2 Multichamber Reactive Silencers

To increase and provide a wider range of attenuation, most manufacturers and suppliers of reactive silencers put several chambers in series. Generally, the critical length L of each chamber is varied such that for those areas in the spectrum where the attenuation vanishes for one chamber, another is at or near peak. In this way, those regions of little attenuation or *holes* are filled, and the muffler has wider application. Shown in Fig. 8.18 is an example of a typical multichamber silencer. Silencers such as these are commercially available in a wide range of sizes and with a variety of options.

Figure 8.19 is a page from a silencer catalog, where sizes, weights, typical attenuation, and pressure loss data are presented. Note that peak attenuation occurs in the lower-frequency ranges below 500 Hz for the reactive chamber silencers in contrast to the absorptive types (compare Table 8.1).

Note also that a variety of inlet and exit orientations is available. Consider now an example of the selection process.

Example

Select a silencer for a compressor inlet from the catalog page of Fig. 8.19 that will have more than 25-dB attenuation over the frequency range of 63 to 1000 Hz. The diameter of the compressor exhaust pipe is 8 in., and volume flow is 1000 cfm through the silencer. In addition, estimate the pressure loss.

Solution

From the attenuation curves of Fig. 8.19, we see that the UR series of silencers has the required attenuation of more than 25 dB over the range of 63 to 1000 Hz. From the dimension chart at right, the model UR-8 has an 8 in. diameter inlet which would interface the compressor inlet exactly. Thus the silencer selected would be the model UR-8, which is 97 in. long and 22 in. in diameter and weighs 410 lb. If space is a problem, a horizontal installation utilizing the model URY-8, with the inlet on the side, might be considered. With respect to aerodynamic considerations, the silencer inlet cross-sectional area is

$$A = \pi r^2$$
$$= \pi (0.33)^2$$
$$= 0.35 \text{ ft}^2$$

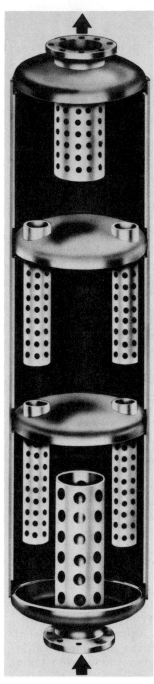

Figure 8.18 Cutaway view of multichamber reactive silencer. (Courtesy of Burgess Industries, Dallas, Tex.)

Hence, for a volume flow of 1000 cfm, the face velocity v is

$$v = \frac{1000}{0.35} = 2864 \text{ ft/s}$$

From the pressure drop curve, one sees that the pressure drop for a velocity of 2864 ft/s is less than 1.0 in. of water.

For even wider applicability, reactive chamber silencers with absorptive passages are also commercially available. With this combination, the attenuation is at a high level and nearly uniform over the entire audio spectrum.

8.8.3 Side Branch Resonator

Another form of reactive silencer is the side branch, cavity, or Helmholtz resonator, as illustrated in Fig. 8.20. Here we have a cavity of volume V connected to a duct through a connecting cylinder of diameter d. Now as a periodic sound wave propagates down the duct, the mass of gas in the cylinder is forced in and out of the cavity. The gas in the cavity acts like a spring as it alternately compresses and expands, due to the motion of gas in the connecting cylinder. With these elements, along with the resistance due to viscous forces, etc., the system is analogous to a simple mechanical spring mass system or a series electrical circuit containing a resistor, inductor, and capacitor. In each case, the system is classical, and these silencers are called Helmholtz resonators after the brilliant German physicist. More important, however, is the behavior or, better, the frequency response of these systems. Shown in Fig. 8.21 is the typical insertion loss characteristics of these devices. Note the single resonant frequency f_r.

Now it can be shown, and the reader is strongly urged to pursue the derivation from Ref. 6, that the resonant frequency f_r is given by

$$f_r = \frac{c}{2\pi} \sqrt{\frac{A}{lV}} \quad [\text{Hz}] \tag{8.12}$$

where

  c = velocity of sound (m/s or ft/s)

  V = volume of the cavity (m^3 or ft^3)

  L = cylinder length (m or ft)

  l = effective length of cylinder = $L + 0.8\sqrt{A}$ (m or ft)

  A = cylinder cross-sectional area (m^2 or ft^2)

Consider an example of the use of this design equation.

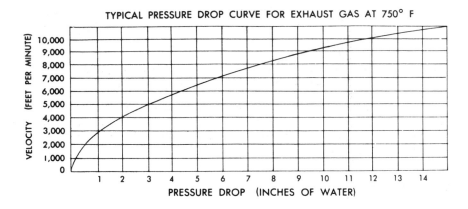

TYPICAL PRESSURE DROP CURVE FOR EXHAUST GAS AT 750° F

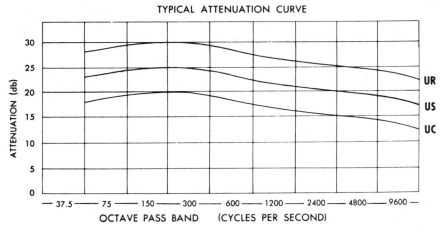

TYPICAL ATTENUATION CURVE

Figure 8.19 Page from reactive silencer catalog showing sizes, weights, and acoustical and aerodynamic performance. (Courtesy of Universal Silencer, Libertyville, Ill.)

Example

Design a side branch resonator which will attenuate a pure tone at 180 Hz in an air compressor exhaust duct 6 in. in diameter.

Solution

We want the resonant frequency f_r of the resonator to be 180 Hz. Therefore, utilizing Eq. (8.12), we have, using c = 13,200 in/s

$$180 = \frac{13,200}{2\pi} \sqrt{\frac{A}{lV}}$$

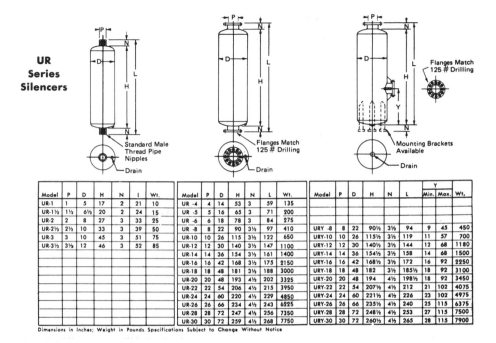

UR Series Silencers

Model	P	D	H	N	l	Wt.
UR-1	1	5	17	2	21	10
UR-1½	1½	6½	20	2	24	15
UR-2	2	8	27	3	33	25
UR-2½	2½	10	33	3	39	50
UR-3	3	10	45	3	51	75
UR-3½	3½	12	46	3	52	85

Model	P	D	H	N	L	Wt.
UR-4	4	14	53	3	59	135
UR-5	5	16	65	3	71	200
UR-6	6	18	78	3	84	275
UR-8	8	22	90	3½	97	410
UR-10	10	26	115	3½	122	650
UR-12	12	30	140	3½	147	1100
UR-14	14	36	154	3½	161	1400
UR-16	16	42	168	3½	175	2150
UR-18	18	48	181	3½	188	3000
UR-20	20	48	193	4½	202	3325
UR-22	22	54	206	4½	215	3950
UR-24	24	60	220	4½	229	4850
UR-26	26	66	234	4½	243	6225
UR-28	28	72	247	4½	256	7350
UR-30	30	72	259	4½	268	7750

Model	P	D	H	N	L	Y Min.	Y Max.	Wt.
URY-8	8	22	90½	3½	94	9	45	450
URY-10	10	26	115½	3½	119	11	57	700
URY-12	12	30	140½	3½	144	12	68	1180
URY-14	14	36	154½	3½	158	14	68	1500
URY-16	16	42	168½	3½	172	16	92	2250
URY-18	18	48	182	3½	185½	18	92	3100
URY-20	20	48	194	4½	198½	18	92	3450
URY-22	22	54	207½	4½	212	21	102	4075
URY-24	24	60	221½	4½	226	23	102	4975
URY-26	26	66	235½	4½	240	25	115	6375
URY-28	28	72	248½	4½	253	27	115	7500
URY-30	30	72	260½	4½	265	28	115	7900

Dimensions in Inches; Weight in Pounds. Specifications Subject to Change Without Notice

or

$$\frac{A}{lV} = 0.0073$$

At this point we have considerable design flexibility in selecting the actual configuration of the resonator. As a starting point and for ease of construction, let us select a round 1 in. diameter (ID) pipe for the connecting cylinder and let it be 2 in. long. Then solving for volume V and using the effective length l of the cavity, we have

$$V = \frac{\pi \times 0.5^2}{(2 + 0.8\sqrt{\pi\,0.5^2})0.0073}$$

$$= 39.7 \text{ in}^3.$$

Thus the volume of the cavity must be 39.7 in^3. Staying with cylindrical symmetry, if we pick a 4 in. diameter pipe from which to construct the cavity, we can solve for the cavity length as follows: Since the volume of a cylinder is $V = \pi r^2 h$, where r is the cylinder radius (inches) and h is the cylinder length (inches), then solving for h, we have

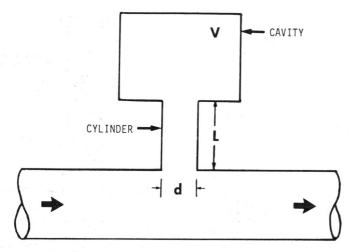

Figure 8.20 Major elements for a cavity or Helmholtz resonator.

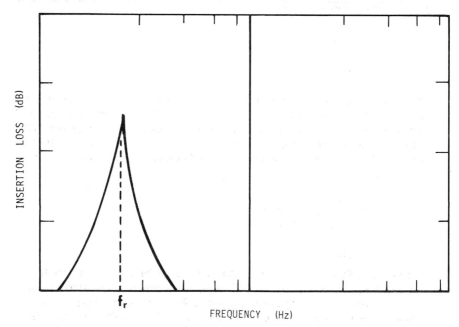

Figure 8.21 Typical insertion loss characteristics for cavity or Helmholtz resonator silencer.

$$h = \frac{39.7}{2^2 \pi}$$

$$= 3.16 \text{ in.}$$

Thus the cavity will be a cylinder 4 in. in diameter and 3.16 in. long.

The limitations of this silencer are obvious since the frequency range of effective attenuation is generally quite narrow. However, in special situations where one or two pure tones are dominant, especially at low frequencies, i.e., below 250 Hz, these resonators are highly effective.

8.9 SPECIALIZED SILENCERS

The number of specialized silencers and their applications are far too numerous to cover in detail. However, a few are worth mentioning in that their application and basic construction illustrates imaginative design approaches.

8.9.1 Tuned-Dissipative Silencers

A clever variation of the parallel baffle silencer is the tuned-dissipative silencer, as shown in Fig. 8.22 Here, each baffle contains cavities which can be designed or tuned for optimum acoustical performance at selected frequencies.

Each cavity also is treated with absorbing material along one of its sides. In this way, the width of spectral attenuation near the tuned frequencies can be increased. These silencers have been especially effective in reducing noise of large, high-volume fans. An added feature is that the acoustical performance of these silencers does not degrade by the buildup of particulates over the perforated sheet metal facing as is the case for conventional parallel baffle units. As such, they are popular in "dirty" applications such as coal-, wood-, and oil-fired power plants; sewage and solid waste treatment incinerators; coffee roasting exhaust fans; and so forth.

8.9.2 Vent or Exhaust Silencers

Wherever high-pressure steam or gas is vented or exhausted, extremely high-intensity noise levels result. Examples of such installations are steam boilers, relief and purge valves, gas process vents, switch valves, compressor blowoffs, autoclaves, etc. The sudden expansion of the gas produces high-velocity turbulent *jet* flow with

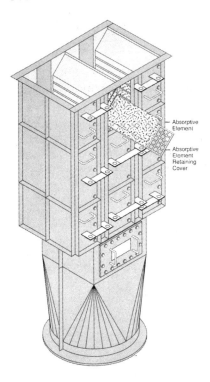

Figure 8.22 Tuned-dissipative silencer. (Courtesy of Transco Products, Inc., Chicago, Ill.)

corresponding broadband aerodynamic noise. The level and character of the noise from a steam vent, for example, can resemble a small jet engine, and the impact on a quiet neighborhood can be, to say the least, disturbing.

To control the noise from these vents, one basic design principle must be observed. The gas must be allowed to expand smoothly inside the silencer, and the exit flow velocity must be reduced. To achieve this design criterion, a series of 180° switchback expansion chambers in conjunction with absorbing materials is employed. Presented in Fig. 8.23 are the basic design features of a commercially available vent or exhaust silencer. Note that as the gas exits the first chamber it impinges on an impact plate constructed of heavy-gauge steel. To reduce the velocity of direct impingement, a perforated diffuser plate backed by scrubble is also included. In the next chamber, the gas flow velocity is reduced to approximately one-half its original velocity. In addition, the noise levels are reduced as the sound waves pass in

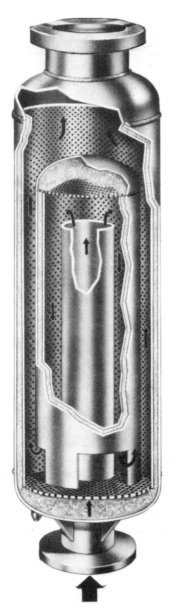

Figure 8.23 Cutaway view of a vent or exhaust silencer showing major design features. (Courtesy of Penn Separator Co., Fairfield, N.J.)

close proximity to the absorptive treatment contained behind the
perforated sheet metal. In the next chamber, the flow velocity is
again reduced, and the sound again is exposed to absorptive treatment. Here, however, the treatment is thicker than in the first
chamber, and hence better low-frequency noise attenuation is obtained.
The flow finally enters a large reflow or plenum chamber designed to
reduce turbulence and lower the final exit velocity. Because of the
relatively high pressures and unusually high temperatures typical
of vents, the containment shell of the silencer must be of heavy
gauge and the absorbing materials capable of withstanding temperatures of 1000°F or greater.

The acoustical performance of these silencers varies, but attenuation of more than 40 dB above 250 Hz is common. One point that cannot be overemphasized, however, is that the gas flow velocity at the
silencer exit must be sharply reduced, or very little attenuation will
be achieved. That is, the noise will be regenerated if the gas flow
velocity is returned to its original value. A good design criterion
for effective attenuation is to keep the exit velocity for those vents
under 100 ft/s or a Mach number of 0.1. To achieve this, a single
plenum chamber in series with the vent silencer will further reduce
the exit velocity and provide additional noise reduction.

8.9.3 Spark Arrest and Water Separator Silencers

In some applications, not only is noise a problem, but often water,
carbon particles, sparks, etc., are entrained in the gaseous discharge. Shown in Fig. 8.24 are the basic construction details of a
spark arrest and water separator silencer. In this configuration
there is a stationary element which imparts a rotational movement to
the gas flow. The heavier water, sparks, carbon particles, etc.,
are centrifugally thrust outward and collected, while the gas proceeds
on through the inner chamber. Silencing is accomplished in the usual
manner by employing reactive or absorptive mechanisms.

By combining effective silencing with separation in one vessel, it
is possible to produce a very compact unit and simultaneously solve
two problems.

8.9.4 Filter Silencers

Another example where silencing can be combined with another requirement is the filter silencer. Here the silencer shell can be designed to include a dry and/or wet filter, as illustrated in Fig. 8.25.
The acoustical performance of these combinations is quite good, usually
in the range of 10 to 20 dB in the critical low-frequency range, below 1000 Hz. Probably the most common applications are on the inlet

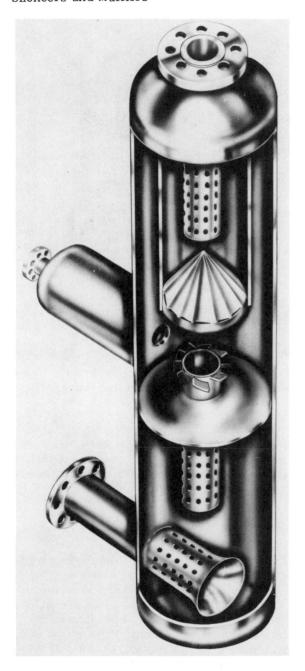

Figure 8.24 Basic design features for a spark arrest or water separator silencer. (Courtesy of Burgess Industries, Dallas, Tex.)

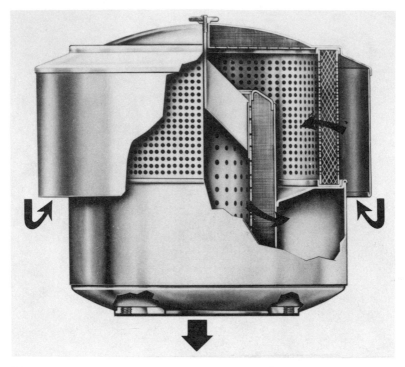

Figure 8.25 Filter silencer. (Courtesy of Burgess Industries, Dallas, Tex.)

to reciprocating compressors, internal combustion engines, vacuum pumps, rotary positive displacement blowers, etc. The acoustical energy of these machines is often concentrated in the 125- to 1000-Hz range.

8.9.5 Motor and Burner Silencers

The noise from electric motors is dominantly due to the cooling air fans. This is especially true for the explosionproof, totally enclosed fan-cooled (TEFC) types whose rotational speeds are typically about 3500 rpm. To control the fan noise, an absorptive-type silencer can usually be mounted on the *end bell* of the motor, as illustrated in Fig. 8.26. Here the cooling air enters the silencer through the end screen and passes through the silencer and exits over the motor-cooling *fins*. These air passages are typically constructed of absorbing materials which reduce the fan noise as it passes through the silencer.

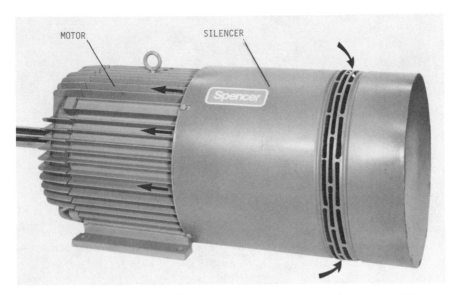

Figure 8.26 Motor with a fan silencer installed. (Courtesy of Spencer Turbine Company, Windsor, Conn.)

The attenuation of these devices is typically in the range of 6 to 10 dB at frequencies of 1000 to 8000 Hz, where the fan noise is concentrated. More specifically, resultant sound pressure levels at 5 ft with the silencers installed are typically in the range of 85 to 88 dBA for motors up to 100 hp. Other common applications of motors in this horsepower range include drive motors for pumps, compressors, blowers, fans, expanders, centrifuges, etc.

Silencers of a similar design can be applied to the inlets of gas or oil burner intakes such as are common in the petrochemical industry and for drying oven applications in the areas of food processing.

8.9.6 Dispersive Silencers

Dispersive-type silencers are used dominantly in the control of noise associated with small high-pressure relief or exhaust valves such as found on pneumatic controls, solenoid valves, air cylinders, clutches, brakes, air-driven hand tools, shop air wands, etc. The noise source is usually high-velocity air exhausted as a burst to the atmosphere. Thus the character of the noise is impulsive in duration and with a spectrum rich in broadband high-frequency noise energy.

With a dispersive silencer installed over the exhaust vent, the high-velocity air is released into an expansion chamber and subse-

(a)

(b)

Figure 8.27 (a) Examples of dispersive silencers and (b) a typical installation on pneumatic control. (Courtesy of C. W. Morris Co., Livonia, Mich.)

quently dispersed at a much lower velocity through porous sintered metal or slots. Shown in Fig. 8.27a are some commercially available dispersive silencers. Shown in Fig. 8.27b is a typical installation of slotted silencers in a pneumatic control.

It should be noted that these silencers are usually available with standard pipe thread connectors for easy direct installation and can be mounted in any orientation.

One problem that these silencers present is a tendency to *clog*, with a corresponding increase in *back* pressure. In most installations, the back pressure is intolerable, especially for pneumatic controls, clutches, and air-driven tools. As such, the diffuser elements require frequent cleaning with a solvent or replacement.

As an alternative approach, one can adapt a 1/2 or 3/4 in. diameter flexible hose to the threaded exhaust orifice and vent the air into an overhead 2 in. diameter pipe *header* running the length of, say, the press room or molding room. In this way, dozens of exhaust ports can be vented into a common expansion chamber. The ends of the **header** pipe are left open, reducing back pressure buildup and eliminating further maintenance.

8.10 ACTIVE NOISE CONTROL

The concept of reducing noise by actively adding acoustical energy is not new. Paul Lueg received a U.S. patent [7] in 1936 wherein he demonstrated that noise in a duct could be reduced by introducing sound waves out of phase with the source.

It was not, however, until the early 1980's that advancements in control theory and microelectronic circuits were brought together to demonstrate the feasibility of practical active noise control.

It is well beyond the scope of this text to discuss in detail either the theory or the electronics of active noise control systems. However, the physical plausibility of the concept is easy to understand, and from there the application follows readily. Fig. 8.28 illustrates the basic concept of active noise control. The input microphone converts the varying sound pressure of the source to an analog electrical signal. The signal is then sent to the controller where a new signal is generated and adjusted for input microphone-to-loudspeaker spacing. The new signal activates the loudspeaker, producing destructive interference with the undesired sound. One might ask, "Is there time for the controller to activate the loudspeaker?" The answer is obvious when one notes that the sound waves are travelling at the speed of sound in the duct, and the electrical signals are travelling at the speed of light (almost instantaneously) between the input microphone and the loudspeaker.

Now the precise adjustment or time delay that is calculated by the controller depends to the first order on the speed of sound in the duct

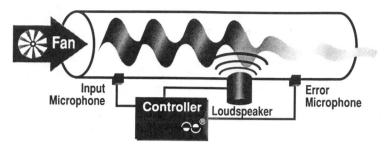

Figure 8.28 Conceptual illustration of active noise control. (Courtesy of Digisonix, Inc., Middleton, Wis.)

medium, i.e., air or other gases. Since there are generally gas flow and often dynamic changes in gas temperature that affect the speed of sound, an error microphone is included to monitor the resultant field after cancellation. This microphone essentially provides feedback to the controller to modify the time delay and optimize the cancellation process.

In summary, the active noise control system provides a powerful engineering noise reduction measure. This is especially true in the lower frequency ranges (20 to 400 Hz) where passive devices are least effective. Let us know consider some applications and actual case histories.

Shown in Fig. 8.29 is an active system installed in an exhaust fan duct. Note the loudspeakers, input, and error microphones. In this application, a discrete tone present at 120 Hz (blade-to-scroll passing frequency) was reduced by 32 dB.

Shown in Fig. 8.30 is the noise reduction achieved on dominantly broadband or random noise associated with a forward-curved centrifugal fan. Note that the attenuation was in the range of 10 to 15 dB in the critical frequency range of 60 to 200 Hz.

Shown in Fig. 8.31 is a commercially available active/passive duct silencer. The acoustical performance for the unit cannot be given, as is typically the case for passive devices. However, the commercial availability of these compact active systems clearly shows the level of reliability and the current state of the art.

In summary, practical industrial applications of active sound control are really in their infancy when compared to passive methods. However, effective applications in ducts have been demonstrated, and applications to vehicular noise are showing steady progress. The impetus here is for truck mufflers, in that back pressure loss penalties are always present with passive devices. With active control, penalties vanish and the overall engine efficiency is significantly improved, with corresponding savings of fuel. Another unique application is with headsets. Here, *selective* noise reduction of sirens on emergency

Figure 8.29 Active noise control system installed in exhaust fan duct. (Courtesy of Digisonix, Inc., Middleton, Wis.)

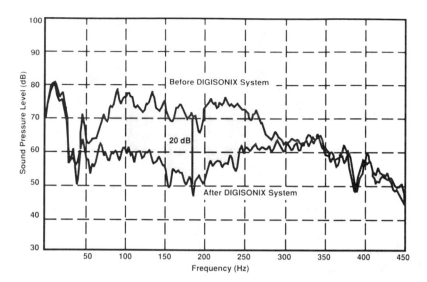

Figure 8.30 Acoustical performance of active system on broadband noise associated with centrifugal fan. (Courtesy of Digisonix, Inc., Middleton, Wis.)

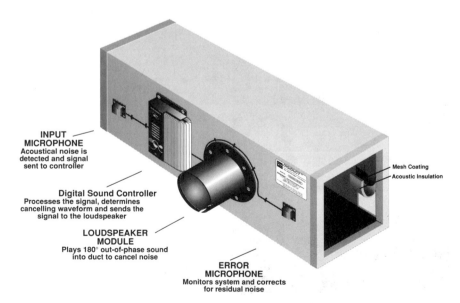

Figure 8.31 Commercially available active/passive duct silencer. (Courtesy of Digisonix, Inc., Middleton, Wis.)

vehicles has been effectively demonstrated. In short, where the geometry of the propagation is small or two-dimensional, such as in ducts, active control can be considered.

In final summary, where air or gas is either moved, conditioned, or vented, some form of noise control is usually required. Areas such as these are common not only in industry but in our daily and recreational living. As such, one recognizes that silencers, mufflers, and active noise control provide probably the most formidable weapon available to the acoustical engineer in the fight against noise.

REFERENCES

1. G. J. Sanders. Silencers: their design and application. *Sound Vib.* (Feb. 1968).
2. H. J. Sabine. The absorption of noise in ventilating ducts. *J. Acoust. Soc. Am. 12* (1940).
3. R. J. Wells. Acoustical plenum chambers. *Noise Control* 4(4) (July 1958).
4. D. D. Davis, Jr., G. M. Stokes, D. Moore, and G. L. Stevens, Jr. Theoretical and Experimental Investigation of Mufflers with Comments on Engine-Exhaust Muffler Design. *NACA Rept. 1192.* 1954.
5. L. H. Bell. *Fundamentals of Industrial Noise Control.* Harmony Publications, Trumbull, Connecticut, 1973.
6. L. E. Kinzler and A. R. Frey. *Fundamentals of Acoustics.* Wiley, New York, 1962.
7. P. Lueg. U.S. patent no. 2,043,416 (1936).

BIBLIOGRAPHY

Allie, M. C., Bremigan, C. D., Eriksson, L. J., and Greiner, R. A. Hardware and software considerations for active noise control. *Proc. ICASSP 88,* New York, 1988.

ASHRAE Guide and Data Book. ASHRAE, New York, 1988.

Beranek, L. L. (Ed.). *Noise and Vibration Control.* McGraw-Hill, New York, 1988.

Bernhard, R. Shape optimization of reactive mufflers, *Noise Control Eng. 27*(1) (1986), 10−17.

Eriksson, L. Silencers. In *Noise and Vibration Control,* D. Baxa (Ed.). Wiley, New York, 1982.

Eriksson, L. J. Active sound attenuation using adaptive digital signal processing techniques. Ph.D. thesis, University of Wisconsin−Madison, 1985.

Eriksson, L. J. Active sound attenuation system with on-line adaptive feedback cancellation. U.S. patent no. 4,677,677, June 30, 1987.

Eriksson, L. J., Allie, M. C., and Greiner, R. A. A new approach to active attenuation in ducts. *Proc. 12th ICA*, paper C5-2, 1986.

Eriksson, L. J., and Allie, M. C. System considerations for adaptive modelling applied to active noise control. *IEEE Int. Symp on Circuits and Systems*, 1988.

Faulkner, L. *Handbook of Industrial Noise Control*. Industrial Press, New York, 1975.

Harris, C. M. (Ed.). *Handbook of Acoustical Measurements and Noise Control*. McGraw-Hill, New York, 1991.

Prasad, M., and Crocker, M. Studies of Acoustical Performance of a multi-cylinder engine exhaust muffler system. *J. Sound Vib.* *90*(4) (1983), 491–508.

EXERCISES

8.1 Noise levels at 20 ft from an air-conditioning fan were 91 dB at 125 Hz, 98 dB at 250 Hz, 95 dB at 500 Hz, 88 dB at 1000 Hz, 81 dB at 2000 Hz, and 79 dB at 4000 Hz. Select the minimum length of a parallel baffle silencer from Table 8.1 which will reduce the noise level to 85 dBA.

Answer: 3 ft

8.2 Noise levels from a roof exhaust fan at the property line of an office building were 70 dB at 63 Hz, 70 dB at 125 Hz, 72 dB at 250 Hz, 78 dB at 500 Hz, 75 dB at 1000 Hz, 65 dB at 2000 Hz, 60 dB at 4000 Hz, and 55 dB at 8000 Hz. The State of New Jersey octave band noise level criteria are 72 dB at 63 Hz, 67 dB at 125 Hz, 59 dB at 250 Hz, 52 dB at 500 Hz, 46 dB at 1000 Hz, 40 dB at 2000 Hz, 34 dB at 4000 Hz, and 32 dB at 8000 Hz. Select a silencer from Table 8.1 which would meet the criteria.

Answer: 5- or 7-ft silencer

8.3 The volume flow at the inlet of a silencer is 10,000 cfm. The cross-sectional area of the silencer inlet is 8 ft^2. What is the face velocity of the silencer?

Answer: 1250 fpm

8.4 If the face velocity to a silencer is −2000 fpm, what is the dynamic insertion loss of a model 5Es silencer at
a. 250 Hz?
b. 1000 Hz?
c. 4000 Hz?
Use Table 8.3.

Answer: (a) 24 dB;
(b) 52 dB;
(c) 27 dB

8.5 What is the self-noise power level L_W for the model Es silencer with face velocity of +1000 fpm at
a. 250 Hz?
b. 1000 Hz?
c. 4000 Hz?
Use Table 8.4.

Answer: (a) 33 dB;
(b) 34 dB;
(c) 27 dB

8.6 What would be the static pressure drop for a model 10Es 24 × 36 silencer with a volume flow of 7000 cfm? Use Table 8.5.

Answer: 0.60 in. of water

8.7 Referring to Exercise 8.6, select the smallest 10Es silencer that would bring the pressure drop below 1.0 in. of water.

Answer: **10Es 24 × 30**

8.8 Calculate the attenuation at 250, 1000, and 4000 Hz of a lined duct 8 ft in length. The duct lining is acoustical-quality foam 1 in. thick with absorption coefficients of 0.48, 0.90, and 0.98 at 250, 1000, and 4000 Hz, respectively. The duct cross-sectional dimensions are 2 × 2 ft.

Answer: **6.6 dB at 250 Hz;**
15.8 at 1000 Hz;
17.8 at 4000 Hz

8.9 Estimate the sound attenuation of a 90° lined bend at
a. 125 Hz
b. 500 Hz
c. 2000 Hz
Use Fig. 8.11.

Answer: (a) 4 dB;
(b) 6 dB;
(c) 9 dB

8.10 A lined plenum chamber has the following dimensions:

$$S_w = 375 \text{ ft}^2$$

$$S_e = 8 \text{ ft}^2$$

$$d = 12 \text{ ft}$$

$$w = 10.4 \text{ ft}$$

Calculate the attenuation at 250, 1000, and 4000 Hz if the absorption coefficients of the lining are 0.53, 0.93, and 0.78 at 250, 1000, and 4000 Hz, respectively.

Answer: 15.7 dB;
         20.3 dB;
         18.6 dB;

8.11 Show that if the plenum exit area in Exercise 8.10 were reduced to 4 ft^2 the attenuation would be increased by approximately 3 dB.

8.12 The attenuation of a lined plenum is 6 dB at 63 Hz. How many identical plenums in series would be required to have an attenuation of 18 dB at 63 Hz?

Answer: 4

8.13 A compressor is located near a window in an exterior wall of a factory. The window is to be replaced by an acoustical louver to bring in outside air. What is the transmission loss of the louver model LP (low pressure loss) at
a.   250 Hz?
b.   500 Hz?
Use acoustical performance data from Fig. 8.13.

Answer: (a) 8 dB;
         (b) 9 dB

8.14 Noise from a forced draft fan emanating from a large stack creates excessive noise levels at a nearby church. If a model 24-in. SSA5 insert silencer were installed in the stack, what would be the noise reduction at the church at
a.   250 Hz?
b.   500 Hz?

Answer: (a) 16 dB;
         (b) 29 dB

8.15 Referring to Exercise 8.14, for a flow velocity of 50 ft/s what would be the pressure loss if the insert were mounted
a.   At the top of the stack?
b.   Five stack diameters lower?

Answer: (a) 0.8 in. of water;
         (b) 0.48 in. of water

8.16 The inlet noise from a compressor is to be reduced using a single expansion chamber. A minimum of 16-dB attenuation is required in the range of 125 Hz. To meet this design goal, what must be
a.   The minimum cross section ratio m?
b.   The chamber length?
Use c = 1100 ft/s.

Answer: (a) m > 16;

(b) L = 2.2 ft

8.17 To reduce the noise of a rotary blower, a single expansion chamber will be added to the exhaust. The dominant frequency is 300 Hz, and the temperature of the exhaust air is 200°F. Calculate
a. The optimum length
b. The diameter of a single chamber silencer which will reduce the noise level by 20 dB

The exaust port diameter is 6 in. Use c = 1100 ft/s

Answer: (a) L = 1.05 ft;

(b) 2.5 ft

8.18 A UR-8 silencer was selected for the exhaust of a diesel engine. The exhaust flow velocity will be approximately 100 ft/s. What will be the expected
a. Attenuation at 1000 Hz?
b. Pressure drop in inches of water?
Use table in Fig. 8.19

Answer: (a) 27.5 dB;

(b) 4.5 in. of water

8.19 A cavity resonator is to be designed to reduce a pure tone from a transformer at 120 Hz. It is convenient to have L = 10 cm and A = 10 cm^2. Determine the volume of the resonant cavity for peak ab- sorption. Use c = 344 m/s.

Answer: V = 1661 cm^3

9

REVERBERATION CONTROL

Throughout this text, we have emphasized reduction of noise at its source, which should always be the primary goal of the acoustical engineer. However, in addition to source control, significant noise reduction can be obtained by controlling the reflected noise which accumulates in enclosed interior environments. This phenomenon, commonly called reverberation buildup, is best observed in indoor swimming pools where the sound from the diving board or a shout continues on for seconds, or with a crowd present, there is often an overall *din* which makes conversation difficult.

Basically, reverberation buildup occurs when there is a lack of absorbing surfaces, and the noise levels increase until the rates of sound generation and absorption are equal. The walls, floor, and ceiling in industrial plants are typically acoustically *hard*, and as such, significant reverberation buildup is common. In this chapter, we shall discuss the basic physics of reverberation and those measures available for its control.

9.1 REVERBERATION

When sound is generated outdoors in the absence of reflecting surfaces (free field), the sound field is relatively simple and approaches the idealized concept of diverging sound waves. The sound pressure level diminishes according to the type of source, and, as we saw in Chapter 4, for a point source, each time the distance from the source is doubled, the noise level drops 6 dB. In a large room with walls, floors, ceilings, columns, tables, cabinets, etc., the sound field is far more complex. The field can, however, be broken into two parts, direct radiated sound and reflected sound, as illustrated in Fig. 9.1.

In short, the direct sound from the source reaches the listener directly along the line of sight, and the reflected sound reaches

330

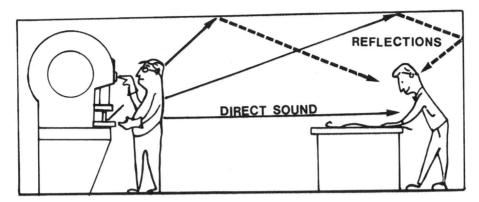

Figure 9.1 Direct radiated and reflected sound.

the listener only after reflection from the walls, floors, ceiling, etc.
Since the reflected sound generally arrives from all directions, the
resultant composite sound field is commonly called *diffuse sound*. For
example, if the sound level of the reflected sound is nearly equal to
the level of the direct sound, it may not be possible to tell the location of the source. These diffuse sound fields were studied extensively by W. C. Sabine, R. F. Norris, V. O. Knudson, and others in
conjunction with theater and concert hall design.

Now it can be shown theoretically and verified experimentally
that for a wide range of room sizes and shapes, the mean free path d
of the reflected sound waves is given by [1]

$$d = \frac{4V}{S} \quad \text{[m or ft]} \tag{9.1}$$

where

V = volume of the room (m^3 or ft^3)

S = total surface area of room (m^2 or ft^2)

The mean free path d can be interpreted to be the average distance
that sound travels between successive reflections. This is an important concept, since the magnitude and character of the diffuse and
reverberant field depend on the number of reflections and the absorption at each reflection.

For example, if each surface of the room is highly absorbent, the
intensity of the sound will be diminished at each reflection. Thus it
is easy to see that the sound level will *decay*, and the rate of decay
will depend largely on the number of reflections and the absorption
at each surface. The decay rate of reflected sound is commonly meas-

ured in terms of reverberation time. Reverberation time RT is defined as the time required for the average sound pressure level to
decay 60 dB, or in terms of sound pressure to 1/1000th of its original
value. To a close approximation, the reverberation time T_{60} at room
temperature is related to the room dimensions and total absorption by
the simple expression [2]

$$T_{60} = \frac{0.049V}{A} \quad [s] \tag{9.2}$$

where

V = room volume (ft^3 or m^3)

A = total room absorption (ft^2 or m^2)

or, in metric units,

$$T = \frac{0.161V}{A} \quad [s] \tag{9.3}$$

The total absorption A is calculated by forming the product of the
room surface area A and the corresponding absorption coefficient α
(random incidence):

$$A = S\alpha \quad [ft^2] \tag{9.4}$$

In practical situations, many of the surface areas in a room are
unalike; that is, they usually have different absorption coefficients.
Therefore, the more general and useful expression for total absorption is as follows:

$$A = S_1\alpha_1 + S_2\alpha_2 + S_3\alpha_3 + \cdots + S_k\alpha_k \quad [\text{sabins or metric sabins}] \tag{9.5}$$

where

S_k = each absorbing surface area (ft^2 or m^2)

α_k = absorption coefficient associated with each area (unitless)

Now the unit for the total absorption is obviously an area; however,
in terms of unit area, the units are referred to as sabins or metric
sabins, depending on your choice of British or metric units, respectively. As an example, in British units, 1 sabin is often likened
to the absorption of an open window having an area of 1 ft^2. Consider
an example.

Example

The acoustical treatment on the wall of a room has dimensions
4 × 5 ft. The absorption coefficient α is 0.3. What is the total
absorption A of the treatment?

Solution
From Eq. (9.5), the total absorption A is

$$S\alpha = 20 \times 0.3$$

$$= 6 \text{ sabins}$$

The 20 ft^2 of absorbing material behaves like an open window having an area of 6 ft^2.

To account for the inherent absorption of sound due to the air in a room, Eqs. (9.2) and (9.3) need to be modified slightly as follows [3–5]:

$$T = \frac{0.049V}{A + 4mV} \quad \text{(English units)} \quad [s] \tag{9.6}$$

$$T = \frac{0.161V}{A + 4 \, mV} \quad \text{(metric units)} \quad [s] \tag{9.6a}$$

where m is the air absorption coefficient which depends on relative humidity and frequency.

Except for extremely large rooms, air absorption is negligible below 1000 Hz. In addition, in most industrial environments, the relative humidity rarely varies more than from 40% to 90%, and as such the values of m listed in Table 9.1 can be used for most practical situations.

Another term which is often used to describe the absorption characteristics of a room is the average absorption coefficient $\overline{\alpha}$, which is defined as

$$\overline{\alpha} = \frac{S_1\alpha_1 + S_2\alpha_2 + S_3\alpha_3 + \cdots + S_k\alpha_k}{S} \quad \text{[unitless]} \tag{9.7}$$

where

Table 9.1 Air Absorption Coefficients for the Higher Octave Bands and a Range of Relative Humidity (RH)[a]

Octave Band Center Frequency (Hz)	Relative Humidity (%)			
	20	40	60	80
2000	0.002	0.001	0.0006	0.0005
4000	0.006	0.003	0.002	0.002
8000	0.014	0.007	0.006	0.005

[a] To convert to metric units, multiply by 3.28.

α_k = absorption coefficient (unitless)

S_k = each surface area (ft^2 or m^2)

$S = S_1 + S_2 + S_3 + \cdots + S_k$ (ft^2 or m^2)

Now since the absorption coefficient depends on frequency, it must be acknowledged that the average absorption coefficient $\bar{\alpha}$ is also frequency dependent. In fact, the reverberation time and hence the absorption coefficient are usually dealt with in terms of the conventional octave bands.

Let us now consider an example which will clarify and illustrate the use of these terms and concepts.

Example

A manufacturing area has room dimensions of $100 \times 100 \times 18$ ft. The average absorption coefficients $\bar{\alpha}$ for each octave band in the room are as follows:

Octave Band Center Frequency (Hz)

	125	250	500	1000	2000	4000	8000
$\bar{\alpha}$	0.1	0.15	0.2	0.25	0.3	0.4	0.5

What is the spectral reverberation time for the area? Assume the relative humidity is 50%.

Solution

The volume of the room is $180,000 \text{ ft}^3$, and the total internal surface area is $27,200 \text{ ft}^2$. Therefore, from Eq. (9.6) and ignoring air absorption, the reverberation time T at 125 Hz is

$$T(125) = \frac{0.049 \times 180,000}{0.1 \times 27,200}$$

$$= 3.2 \quad [\text{s}]$$

And for 250, 500, and 1000 Hz,

$$T(250) = \frac{0.049 \times 180,000}{0.15 \times 27,200}$$

$$= 2.16 \quad [\text{s}]$$

$$T(500) = \frac{0.049 \times 180,000}{0.2 \times 27,200}$$

$$= 1.6 \quad [\text{s}]$$

$$T(1000) = \frac{0.049 \times 180,000}{0.25 \times 27,200}$$

$$= 1.3 \quad [\text{s}]$$

For 2000 Hz, the air absorption term must be included in Eq.
(9.6), and from Table 9.1, m = 0.001. Thus

$$T(2000) = \frac{0.049 \times 180,000}{0.3 \times 27,200 + 4 \times 0.001 \times 180,000}$$

$$= 0.99 \quad [s]$$

For 4000 Hz, m = 0.003, and $\bar{\alpha}$ = 0.4:

$$T(4000) = \frac{0.049 \times 180,000}{0.4 \times 27,200 + 4 \times 0.003 \times 180,000}$$

$$= 0.68 \quad [s]$$

For 8000 Hz, m = 0.007, and $\bar{\alpha}$ = 0.5:

$$T(8000) = \frac{0.049 \times 180,000}{0.5 \times 27,200 + 4 \times 0.007 \times 180,000}$$

$$= 0.47 \quad [s]$$

Experimentally, reverberation time may be measured by actually
noting the time required for an interrupted or impulsive sound to
decay 60 dB from its original or peak value. One technique is to
tape-record the sound report from a pistol that fires blanks or a
one-inch firecracker. Subsequently, in the laboratory, the tape-
recorded signal is played back through an octave band filter set
into a graphic level recorder (GLR). With a logarithmic potentiom-
eter installed on the GLR and the paper speeds and writing speeds
suitably selected, a logarithmic decay curve, as illustrated in Fig.
9.2, is obtained. For the best results, the paper speed and writing
speed should be selected to yield a slope of 45° or more. In actual
practice, the decay curve will not be linear but will have some peaks
and dwells. However, a linear *eyeball* best-fit approximation can
usually be drawn through the curve. From the approximation, the
slope of the curve can be measured and the reverberation time at
each octave band center frequency calculated:

$$T_{60} = \frac{60}{\tan \theta} \quad [s] \tag{9.8}$$

where

tan θ = slope of the decay curve (referring to Fig. 9.2)

$$= \Delta Y / \Delta X \ (dB/s)$$

Example
   The measured decay (eyeball approximation) rate for sound in a
   large hall was 40 dB in 2 s. What was the reverberation time of
   the hall?
   Solution
   Utilizing Eq. (9.8), we have

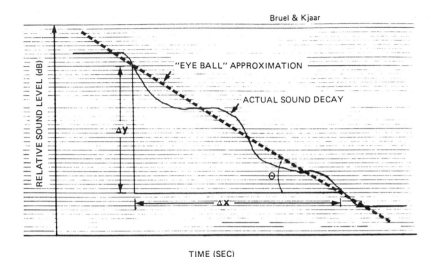

TIME (SEC)

Figure 9.2 Measured sound decay curve and calculation parameters for determining reverberation time RT.

$$T = \frac{60}{40/2} = 3.0 \text{ s}$$

Naturally, other statistical curve-fitting methods should be considered for more precise or repeatable results.

Another method uses digital electronic instrumentation, such as the sound level meter shown in Fig. 5.18, designed to measure the reverberation time. Here, the impulse is created and, nearly instantaneously, the RT appears as a numerical readout or can be retrieved from memory. To be sure, this approach is timesaving over the tape recorder approach, but the graphic level visualization has merit as a diagnostic.

The methods just outlined use basic laboratory instruments and yield good results for most applications. As a matter of interest, the concept of reverberation time and its measurement is extremely important in the design of theaters and concert halls. It is the selection of the absorption and its location that is often controversial among architectural acousticians, conductors, critics, etc. Obviously, the matter is highly subjective, and we shall discuss this aspect in a little more detail in Chapter 15.

9.2 REVERBERANT SOUND FIELD

The concept of reverberation is rather qualitative and not particularly useful in itself as applied to noise control. However, the principles

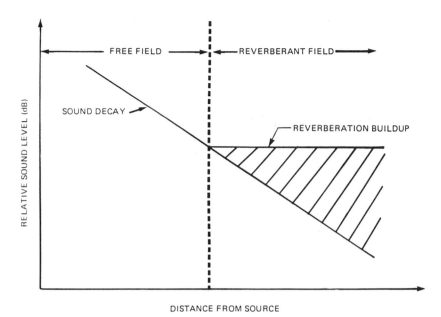

Figure 9.3 Sound fields in enclosed spaces.

are important in fully understanding the reverberant sound field in a
large room.

As mentioned earlier, as one moves away from a sound source
located in a room, the direct radiated sound will usually dominate.
However, at some transition point, the reverberant sound field will
become significant and perhaps dominate. This concept is illustrated
in Fig. 9.3. It is easy to see that the location of the transition point
and resultant level of reverberant buildup depend on the volume and
amount of absorption present in the room.

It can be shown from various references (see the end of the
chapter) that the mean-square sound pressure at a distance r from
the source is given by

$$p^2 = W \rho c \left(\frac{Q_\theta}{4\pi r^2} + \frac{4}{R} \right) \qquad [\text{Pa}^2] \qquad\qquad (9.9)$$

where

   W = source of acoustical power (W)

   ρc = characteristic impedance of medium (mks rayls)

   Q_θ = directivity factor of source (unitless)

   r = distance from source (m)

   R = room constant (m^2)

Now the last parameter R, the room constant, is very important and is defined as

$$R = \frac{S\bar{\alpha}}{1 - \bar{\alpha}} \quad [m^2] \tag{9.10}$$

where

$\bar{\alpha}$ = average absorption coefficient

S = total surface area (m^2)

In terms of the more useful logarithmic levels, Eq. (9.9) can be rewritten as

$$L_p = L_W + 10 \log_{10}\left(\frac{Q_\theta}{4\pi r^2} + \frac{4}{R}\right) \quad \text{(mks units)} \quad [dB] \tag{9.11a}$$

or

$$L_p = L_W + 10 \log_{10}\left(\frac{Q_\theta}{4\pi r^2} + \frac{4}{R}\right) + 10 \quad \text{(British units)} \quad [dB]$$

$$\tag{9.11b}$$

where

L_p = sound pressure (dB)

L_W = acoustical power level of source (dB)

Upon examining Eqs. (9.11a) and (9.11b), it is easy to see that the reverberant sound field is accounted for by the presence of the term 4/R and that the room constant R is a measure of the total absorption of the room. Note that as $\bar{\alpha}$ approaches 1 (maximum absorption), R grows very large, and thus for large values of R, the sound field approaches the free-field or outdoor results for a point source.

A useful graphic representation of Eqs. (9.11a) and (9.11b) is presented in Fig. 9.4. Here the relative difference in sound pressure level one measures as a function of distance from the noise source for various room constants is shown. Consider an example to illustrate the use of the chart in Fig. 9.4.

Example

A typical machine tool manufacturing area is 100 × 100 × 20 ft and has a room constant R of approximately 2000. From Fig. 9.4, the noise level due to the reverberant field at 5 ft from the machine is at most 2 dB above the free-field level. At 10 ft, the sound level of the reverberant field is about 4 dB above the free-field level, and at 20 ft the difference is nearly 10 dB.

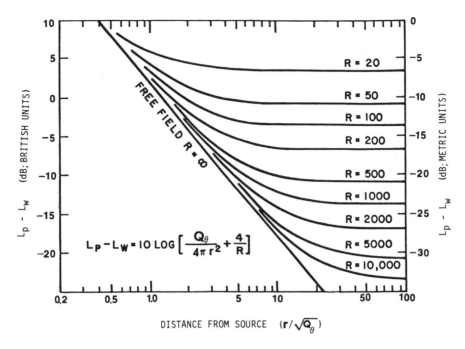

Figure 9.4 Chart for determining the sound pressure level in a large enclosure at distance r from the center of a source of directivity factor Q_θ.

Consider now the effect of increasing the room constant to 5000, i.e., adding absorption. Note that at 5 ft there is no difference between the free field and the reverberant field; at 10 ft the noise level is only 2 dB above the free field, and at 20 ft the level is only about 4 dB above the free field. If, however, the room constant were increased to 10,000 (5 times the original absorption), noise level reductions of 6 to 8 dB could be obtained for distances beyond 10 ft.

Some conclusions can now be drawn:

1. Significant noise reduction can be obtained in manufacturing, assembly, or production areas by adding absorbing materials. However, little reduction is usually achieved within 10 ft of the source, which often includes a machine operator.
2. In many factory areas, the noise sources or machines are distributed over the whole floor area. As such, moving 10 ft from a machine merely puts one into the near field of another with little or no noise reduction achieved.

In summary, reverberation control in closely packed machine environments is usually only a secondary noise reduction measure. However, in areas remote to a noisy source, say 20 ft or more, the addition of absorptive treatment may reduce the reverberant noise buildup by 10 dB or more. Often these remote areas will contain a cluster of personnel such as inspectors, clerks, etc., who will benefit significantly.

9.3 ABSORPTIVE TREATMENT

Consider now a few ways of introducing absorptive treatment into an industrial environment. In most manufacturing or heavy industrial environments, the only areas where appreciable absorptive treatment can be added is in the ceiling area above the work space. Frequently,

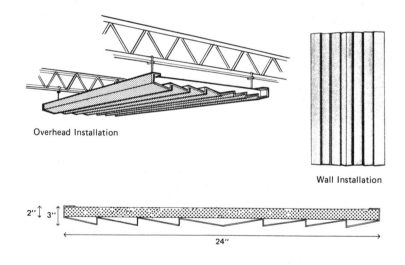

Overhead Installation

Wall Installation

2″ 3″ 24″

ACOUSTICAL PERFORMANCE

OCTAVE BAND CENTER FREQUENCY (Hz)	125	250	500	1000	2000	4000	NRC
ABSORBTION COEFFICIENT	0.68	0.84	0.86	0.87	0.95	0.89	0.85

Figure 9.5 Functional absorbing panel and typical installation methods and acoustical performance. (Courtesy of The ECKEL Industries, Inc., Cambridge, Mass.)

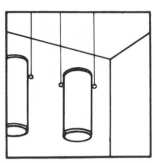

Vertical suspension

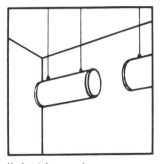

Horizontal suspension

UNITS ON 4 FT CENTERS

Mid band Frequency (Hz)	100	125	250	500	1000	2000	4000	NRC
Sabins per NOISEMASTER Unit	3.88	3.65	5.84	8.48	9.51	9.98	9.90	8.45

ASTM C423-66.

Figure 9.6 Commercially available absorptive space unit and measured acoustical performance. (Courtesy of the Proudfoot Company, Inc., Monroe, Conn.)

these overhead areas are rejected immediately since they contain myriads of plumbing lines, air-conditioning ducts, lighting, conveyors, traveling cranes, etc., which require easy accessibility. The rejection stems from the misconception that the only way of adding absorption is to install a complete acoustical tile ceiling, which would likely be intolerable. A more practical approach is to suspend absorbing panels or space units in the ceiling areas. These panels or units are commercially available and take a variety of shapes and sizes. Shown in Fig. 9.5 is an illustration of functional absorbing panels

along with installation methods and acoustical performance. The basic
construction consists simply of fibrous glass or mineral wool batts
1 or 2 ft thick sandwiched between or laid on perforated sheet metal.
Note that with these panels installed plenty of open area is available
for access. One aspect of spacing, which is often overlooked, is that
the panels do not have to be suspended symmetrically, uniformly, or
even horizontally. In fact, more effective absorption is obtained when
the panels are staggered and alternately suspended vertically and horizontally. Further, it is also desirable to treat the walls, and often
"patches" of these panels can also be installed on the walls with little
or no production interference.

Another popular approach is the use of absorbing space units.
Shown in Fig. 9.6 is a common commercially available unit along with
illustrations, typical installations, and acoustic performance. It
should be noted that often the published absorption coefficients for
units of this type exceed unity. This apparent anomaly often results
when irregularly shaped objects are tested using reverberation room
methods to measure absorption (ASTM C423). As such, the manufacturers often provide the total absorption of each space unit or
panel for a given spacing. As mentioned previously, in most practical
applications, the room constant R which appears in Eqs. (9.11a) and
(9.11b) can be approximated by the total absorption of the room:

$$R \cong S\bar{\alpha} \qquad [ft^2] \qquad\qquad\qquad (9.12)$$

Hence, for design purposes, the performance data supplied by the
manufacturers can be used directly.

Example

The total absorption at 1000 Hz in a plastic molding room is 200
sabins. How much noise reduction could be expected at 10 ft
from the molding machine if 500 units of the space units illustrated
in Fig. 9.6 were suspended uniformly over the ceiling on 4-ft
centers?

Solution

By utilizing Eq. (9.12) and from the performance table of absorption, 500 units would add

9.51 × 500 = 4755 sabins

to the room, bringing the total absorption to nearly 5000 sabins.
Referring to the chart in Fig. 9.4 and recognizing that the total
absorption is a good approximation to R, we see, comparing R =
200 and R = 5000, that at 10 ft there would be a noise reduction
of nearly 10 dB.

Another popular way of introducing absorption is to construct the
walls of hollow core masonry blocks which have slots penetrating into
the cores, as illustrated in Fig. 9.7. When these blocks are stacked,

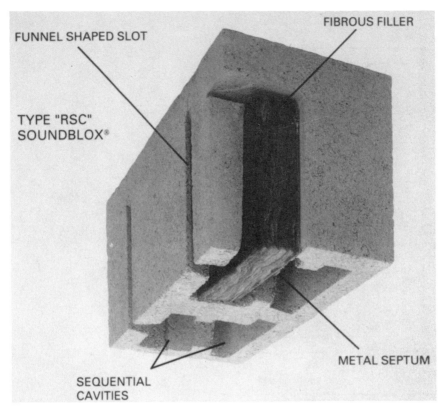

Size	Type	Surface	Exposed Slots/ Cavities	FREQUENCY – Hertz						
				125	250	500	1000	2000	4000	NRC
4"	A-1	PAINTED	2/2	.12	.85	.36	.36	.42	.45	.50
6"	A-1	PAINTED	2/2	.62	.84	.36	.43	.27	.50	.50
8"	A-1	PAINTED	2/2	.97	.44	.38	.39	.50	.60	.45
8"	Q	PAINTED	2/2	1.07	.57	.61	.37	.56	.55	.55
4"	R	PAINTED	2/2	.20	.88	.63	.65	.52	.43	.65
6"	R	PAINTED	2/2	.39	.99	.65	.58	.43	.45	.65
8"	R	PAINTED	2/2	.33	.94	.62	.60	.57	.49	.70
12"	R	PAINTED	2/2	.48	.83	.86	.54	.47	.44	.70
8"	RR	PAINTED	3/3	.61	.91	.65	.65	.42	.49	.65
6"	RSC	PAINTED	2/3	.48	1.14	.91	.76	.67	.51	.85
8"	RSC	PAINTED	2/4	.50	1.00	1.06	.66	.56	.72	.80
12"	RSC	PAINTED	2/4	.57	.76	1.09	.94	.54	.59	.85
8"	RSR	UNPAINTED SPLIT RIB	2/2	.61	.81	.57	.55	.66	.64	.65

Figure 9.7 Example of slotted hollow core masonry blocks along with acoustical performance. (Courtesy of the Proudfoot Company, Inc., Monroe, Conn.)

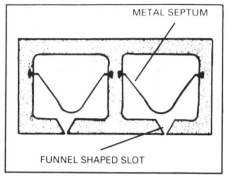

TYPE Q (8" only)

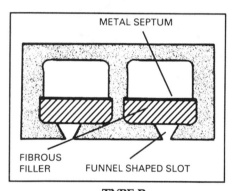

TYPE R

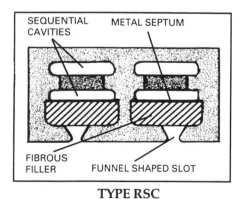

TYPE RSC

Figure 9.7   (Continued)

the cavities are closed and the slot cavity acts as a reactive absorber or resonator (Helmholtz), as discussed in Chap. 8. By varying the size of the slot or core volume and adding a batt of fibrous material to the core, a high degree of absorption over the entire audio spectral range can be obtained. Because of the reactive nature of the blocks, the peak of the absorption is in the lower frequency range, as also shown in Fig. 9.7. This is a special feature, since high absorption at low frequencies is generally difficult to obtain with resistive fibrous or porous materials. Presented in Fig. 9.8 is a photograph of an actual application of the highly absorbing hollow core blocks at an engine test faciltiy.

Alson growing in popularity is the use of hanging acoustical bats or baffles. These units are typically constructed from a 2' × 4' × 1.5" thick panel of fibrous glass core materials encased in a thin (1 or 2 mil) polyethylene bag. The polyethylene bag is permanently heat sealed around the absorbing materials, protecting it from dust, oil, and other airborne contaminants. Rolled rimmed grommets are set through the bag, providing a convenient hanging fixture. The acoustical performance of these units is extraordinary when they are

Figure 9.8 Slotted masonry blocks installed in a diesel engine test facility. (Courtesy of the Proudfoot Company, Inc., Monroe, Conn.)

suspended in rows or on "egg crate" type ceiling arrangements as
shown in Fig. 9.9. Also shown is the typical performance of thes baffle units.

Finally, spray-on type absorbing materials (see Fig. 9.10) are
also commercially available. These materials provide an attractive
textured ceiling finish with good acoustical absorption and light reflec-
tivity. The thickness of application generally ranges between ½ inch
and 2 inches. With a thickness of ½ inch, the measured noise reduction
coefficient NRC was 0.65. The spray-on material adheres readily to
typical construction materials such as concrete, steel, wood, etc., and
conforms to complex surface configurations.

Acoustical Performance (Absorption in Sabins)

Version	Frequency (Hertz)						Total
	125	250	500	1000	2000	4000	
#24G	2.52	4.58	8.75	12.24	10.70	7.33	9.07
#27G	2.77	5.04	9.68	13.22	11.95	8.26	9.97
#27E	3.58	4.40	9.81	11.79	6.99	3.31	8.25

Figure 9.9 Acoustical baffles installed on the ceiling in a wire production facility. (Courtesy of Industrial Noise Control, Inc., Addison, Ill.)

Figure 9.10 Application of spray-on sound-absorbing material. (Courtesy of International Cellulose Corporation, Houston, Texas.)

In summary, control of reverberation can be an effective noise reduction measure. However, in many industrial situations, where operators are in close proximity to the noise sources, peripheral wall or ceiling treatment will have little, if any, acoustical impact.

In final summary, control of reverberation can be an effective noise reduction measure. However, in many industrial situations, where operators are in close proximity to the noise sources, peripheral wall or ceiling treatment will have little, if any, acoustical impact.

REFERENCES

1. V. O. Knudsen. *Architectural Acoustics*. Wiley, New York, 1932.
2. W. C. Sabine. *Collected Papers on Acoustics*. Harvard University Press, Cambridge, Massachusetts, 1927.
3. R. W. Young. Sabine reverberation equation and sound power calculations. *J. Acoust. Soc. Am. 31* (1959), 912–921.
4. C. F. Eyring. Reverberation time in dead rooms. *J. Acoust. Soc. Am. 1,* (1930), 217–241.
5. L. L. Beranek. *Acoustics*. McGraw-Hill, New York, 1954.

BIBLIOGRAPHY

Beranek, L. L. *Noise and Vibration Control.* McGraw-Hill, New York, 1971.
Faulkner, L. *Handbook of Industrial Noise Control.* Industrial Press, New York, 1975.
Harris, C. M. *Handbook of Acoustical Measurements and Noise Control.* McGraw-Hill, New York, 1991.
Morse, P. M., and K. U. Ingard. *Theoretical Acoustics.* McGraw-Hill, New York, 1968.

EXERCISES

9.1 A gymnasium has the following dimensions: $200 \times 200 \times 30$ ft. What is the mean free path of sound in the gym?

Answer: 46.1 ft

9.2 The total absorption at 500 Hz in a room whose volume is 100,000 ft^3 is 2400 sabins. What is the reverberation time for sound at 500 Hz?

Answer: 2.04 s

9.3 A concert hall has a volume of 4×10^6 m^3 and a total absorption at 4000 Hz of 1×10^5 metric sabins. What is the reverberation time at 4000 Hz if the relative humidity is typically 60%?

Answer: 1.5 s

9.4 To measure the reverberation time in a theater, a blank pistol shot was fired. Using an octave band analyzer in conjunction with a graphic level recorder, the sound decay of the pistol report was found to be 30 dB in 0.75 s in the 1000-Hz frequency band. What was the reverberation time?

Answer: 1.5 s

9.5 In a typical *acoustically hard* food packaging room, the room constant is 500 in the critical frequency range above 1000 Hz. How far from a jar or can line, labeler, filler, or capper would one find the reverberant field? (Hint: Use Fig. 9.4.)

Answer: 3 ft approximately

9.6 From the results of Exercise 9.5, what must the room constant be for the reverberant field to be at least 10 ft from the acoustic center of the noise source?

Answer: 5000 approximately

9.7 In a typical power press room the operators are within 3 ft of
the presses. If the room constant were increased from 1000 to 5000
by hanging acoustic space units, how much noise reduction would
the operators receive?

Answer: Nil

9.8 Referring to Exercise 9.7, if an inspection station were located
20 ft from the presses, what noise level reduction would be achieved?

Answer: 6 dB approximately

9.9 One hundred functional panels as illustrated in Fig. 9.5 are to
be added to the walls of a machine shop. If the panels are 2 × 4 ft,
how much absorption would be added to the room in the 2000-Hz
band?

Answer: 760 sabins

9.10 Space units as illustrated in Fig. 9.6 are to be installed in a
swimming pool to reduce reverberation. If 100 units on 4 ft centers
are hung from the ceiling, how much absorption in the 1000-Hz band
would be added in the pool area?

Answer: 951 sabins

9.11 Hollow core masonry blocks are to be installed on the interior
of a transformer vault to reduce the reverberation. From the acoustical performance shown in Fig. 9.7, select a block to provide a high
degree of absorption at the dominant 120-Hz tone.

Answer: 8-in. type A-1 or Q
would be selected.

10
VIBRATION CONTROL

One of the most common sources of noise in industry is the noise generated from vibrating machinery. For example, a 5-hP motor rotating at 1800 rpm is rather quiet, but if it is bolted directly to a metal table, the noise level can rise easily to 90 dBA at a distance of 5 ft. In this instance, the metal tabletop acts as a *sounding board* and radiates the vibratory energy transmitted from some form of rotational unbalance of the motor. This sounding board effect is frequently found in industrial environments or buildings where floors or machine platforms are constructed of *diamond deck*, metal grills, hardwood, and even concrete. The excitation can originate not only from rotating equipment but from vibrators, tumblers, shakers, pressers, etc. It should be emphasized that noise is not the only undesirable effect. Floor vibration, for example, can be a serious problem to adjacent equipment, where close tolerances are maintained in machine tools or especially where optical devices are used.

The simplest and most straightforward solution to this problem is to provide some sort of mechanical isolation between the machine and the structure supporting the machine. Much has been written on vibration and shock control, and attention is usually focused on a high degree of isolation for areas extremely sensitive to vibration. However, it must be emphasized that for significant noise reduction the degree of isolation can often be much less than one would require for a high degree of vibration control. In addition, the introduction of isolation is usually easy, inexpensive, and often the complete solution to the noise problem. Before discussing the design guidelines of vibration control, it is important to consider the basic fundamentals of vibration and the transmission of vibrational energy.

10.1 FREE VIBRATION

Consider first a rigid body of mass m connected to a large inertial body on a spring, as illustrated in Fig. 10.1. Let the spring have a stiffness constant k such that when a force F acts on the spring in either compression or tension, we have

$$F = -kx \qquad [N] \qquad\qquad (10.1)$$

where

    x = displacement of the mass from equilibrium; the minus sign
        infers that the force opposes the displacement (m)
    k = spring constant (N/m)

Assume also that no other forces act on the mass. In reality, this is not a good assumption, but we shall include shortly those forces which are relevant to practical applications. Hence, under this assumption, the equation of motion of the system can be expressed from Newton's second law as

$$m \frac{d^2x}{dt^2} = m\ddot{x} = -kx \qquad [N] \qquad\qquad (10.2)$$

where

$$m \frac{d^2x}{dt^2} = \text{inertial force due to acceleration of the mass m; i.e.,}$$
$$F = ma*$$

*Hereafter we shall use the dot notation for time derivatives in accordance with the conventions of most authors.

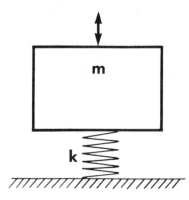

Figure 10.1 Inertial mass supported on a spring with a stiffness constant k.

$-kx$ = elastic restoring force of the spring

Rewriting Eq. (10.2), we have

$$\ddot{x} + \omega_n^2 x = 0 \qquad (10.3)$$

where

$$\omega_n = \sqrt{\frac{k}{m}} \qquad [rad/s]$$

By noting that Eq. (10.3) is a second-order homogeneous differential equation with constant coefficients, the solution will take the form

$$x(t) = A \cos \omega_n t + B \sin \omega_n t \qquad [m] \qquad (10.4)$$

where A and B are arbitrary coefficients to be determined, as we shall see from the initial conditions. First, however, note that the motion of the mass is periodic with period T:

$$\omega_n T = 2\pi$$

or

$$T = \frac{2\pi}{\omega_n} \qquad [s] \qquad (10.5)$$

Further, upon taking the reciprocal of the period T, the frequency f_n of the motion is obtained:

$$f_n = \frac{1}{2\pi} \omega_n$$

$$= \frac{1}{2\pi} \sqrt{\frac{k}{m}} \qquad [Hz] \qquad (10.6a)$$

In British units,

$$f_n = \frac{1}{2\pi} \sqrt{\frac{k'g'}{W}} \qquad [Hz] \qquad (10.6b)$$

where

   k' = spring constant (lb/in.)

   W = weight of the mass m (lb)

   g' = acceleration of gravity (386 in./s^2)

Thus we see that there is a single frequency associated with the motion called the natural frequency, and the subscript n is used conventionally as a mnemonic. Note also that for a given mass the natural frequency of motion is small for soft springs and large for stiff springs, which is somewhat intuitive from experience.

With respect to determining the values of the arbitrary coefficients A and B, suppose the mass were displaced initially at t = 0, a distance x_0, and released with velocity v_0. Then, for the first condition, t = 0, upon substitution in Eq. (10.4), we have

$$x(0) = A \cos(\omega_n 0) + B \sin(\omega_n 0) \qquad\qquad (10.7a)$$

$$= A$$

or

$$A = x_0 \qquad \text{(initial displacement)} \qquad\qquad (10.7b)$$

Differentiating Eq. (10.4) with respect to time, we get the velocity of the motion, and we have

$$\dot{x}(t) = v(t) = -x_0 \omega_n \sin(\omega_n t) + B \omega_n \cos(\omega_n t) \qquad [m/s] \qquad (10.8a)$$

and at t = 0,

$$v(0) = B \omega_n = v_0 \qquad [m/s] \qquad\qquad (10.8b)$$

Since v(0) represents the initial velocity v_0, we have

$$B = \frac{v_0}{\omega_n} \qquad\qquad (10.9)$$

Finally, with the arbitrary constants determined, we have the equation of motion of a freely vibrating single degree of freedom system:

$$x(t) = x_0 \cos(\omega_n t) + \frac{v_0}{\omega_n} \sin(\omega_n t) \qquad [m] \qquad (10.10)$$

where

x_0 = initial displacement (m)

v_0 = initial velocity (m/s)

Inspecting Eq. (10.10), we see that the motion is entirely determined from the initial conditions x_0 and v_0 and that the frequency of oscillation depends only on spring constant k and mass m.

Example

A single degree of freedom mass spring system as illustrated in Fig. 10.1 is set into motion with initial displacement of 0.01 m and zero initial velocity. What is the amplitude and frequency of the resulting oscillation? The spring constant of the spring is 400 N/m, and the mass of the body is 6 kg.

Solution

With the initial velocity $v_0 = 0$, we get the special case of Eq. (10.10):

$$x(t) = x_0 \cos(\omega_n t)$$

Thus the amplitude is just $x_0 = 0.01$ m = 1 cm and the natural frequency f_n of the motion is given by Eq. (10.6a):

$$f_n = \frac{1}{2\pi} \sqrt{\frac{400}{6}}$$

$$= 1.3 \text{ Hz}$$

10.2 FORCED VIBRATION

Returning again to the mass spring system as shown in Fig. 10.1, let the mass m be *driven* or forced into motion by a sinusoidal force F given by

$$F = F_0 \sin \omega t \qquad [\text{N}] \tag{10.11}$$

where

F_0 = amplitude of force (N)

$\omega = 2\pi f$ = angular frequency of driving force F (rad/s)

Since this is just an additional external force, the equation of motion of the system becomes

$$m\ddot{x} = -kx + F_0 \sin \omega t \qquad [\text{N}] \tag{10.12}$$

Now Eq. (10.12) is a nonhomogeneous second-order differential equation with constant coefficients. From the properties of differential equations of this type, the general or complete solution is given as the sum of the solution for the homogeneous case, called the complementary solution and a particular solution. In this case, as a general solution, we have

$$x(t) = A \cos \omega_n t + B \sin \omega_n t + \frac{(F_0/k) \sin \omega t}{1 - (\omega/\omega_n)^2} \qquad [\text{m}] \tag{10.13}$$

where the last term is the particular solution. As before, the arbitrary constants A and B will be determined from the initial conditions. For example, if x_0 is the initial displacement at t = 0, then

$$x(0) = A$$

or

$$A = x_0 \quad [m] \tag{10.14a}$$

Correspondingly, after differentiation, we obtain the velocity, and the initial velocity v_0 is

$$v(0) = v_0 = B\omega_n + \frac{F_0/k}{1 - (\omega/\omega_n)^2} \tag{10.14b}$$

and

$$B = \frac{v_0}{\omega_n} - \frac{F_0/k\omega_n}{1 - (\omega/\omega_n)^2} \tag{10.14c}$$

Now in the inherent presence of some damping forces and after numerous cycles of vibration, it can be shown that the first and second terms in Eq. 10.13 vanish or are of second order. As such we are left with the steady-state equation of motion for the system:

$$x(t) = \frac{(F_0/k) \sin \omega t}{1 - (\omega/\omega_n)^2} \quad [m] \tag{10.15}$$

We now come to one of the most important concepts in vibration. Examining Eq. (10.15), we note that if the forcing frequency ω is small compared to the natural frequency of the system, the amplitude of motion approaches a constant F_0/k. Conversely, if ω is large compared to ω_n, the amplitude of motion is small and grows smaller as ω increases. Physically, this means that as the forcing frequency grows large, the mass simply cannot follow, and the amplitude of the resultant motion is small. But now as ω approaches ω_n or the forcing frequency nears the natural frequency, the amplitude of the motion increases without bound. This condition is called *resonance*, and the concept is vital to understanding vibration and vibration control.

10.3 TRANSMISSIBILITY

Now if we define the force transmissibility T as the ratio of the force transmitted through the spring (to the support) to the force applied to the mass, utilizing Eq. (10.15, we have [1]

$$T = \frac{kx(t)}{f(t)} = \frac{(F_0 \sin \omega t)/[1 - (\omega^2/\omega_n^2)]}{F_0 \sin \omega t} \tag{10.16}$$

$$= \frac{1}{1 - (\omega^2/\omega_n^2)} \qquad \text{[unitless]} \qquad (10.17)$$

In Fig. 10.2, the transmissibility T is plotted as a function of frequency ratio $\omega/\omega_n = f/f_n$. The resonance phenomenon is clearly illustrated, when $f/f_n = 1$. Physically, one can visualize resonance as a condition when the applied external force and the motion of the mass are concurrent or *in phase*. An analogous condition exists when pushing a child on a swing. That is, if one pushes at each *down swing*, the amplitude of the motion increases easily.

Infinite extension of the spring is naturally physically impossible. However, in practical situations, the springs or vibrating structures

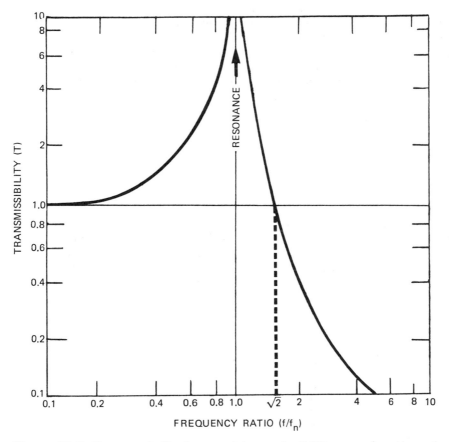

Figure 10.2 Force and displacement transmissibility as a function of the frequency ratio for a single degree of freedom mass spring system.

often are stressed to failure at resonance conditions. As such an awareness and complete understanding of the resonance phenomenon is vital to the vibration control engineer.

Inspecting Eq. (10.17) further, one notes that for $f/f_n = \sqrt{2}$, the transmissibility is unity. That is, the force transmitted and the applied force are equal. Note also that when $f > f_n \sqrt{2}$, less force is transmitted than applied, which is the basic physical property on which vibration isolation depends. We shall see applications of this concept in a later section of this chapter.

10.4 VIBRATION WITH VISCOUS DAMPING

We have thus far ignored the inherent presence of viscous damping forces in our mass spring system. When significant viscous forces are present, the resisting force is frictional in nature and is proportional to the velocity of motion. Common examples of viscous forces are the resistive forces experienced as one moves an object through a fluid such as air or water. Thus the viscous force F can be expressed as

$$F = -C_e \dot{x} \quad [N] \tag{10.18}$$

where

$\dot{x} = \dfrac{dx}{dt}$ is the velocity associated with motion (m/s)

C_e = resistance coefficient or viscous damping constant (N-s/m)

The mass spring system in Fig. 10.1 can now be further extended to include a viscous damping force by adding a *dashpot*, as illustrated in Fig. 10.3. With the dashpot (considered to have no mass), we can add the viscous damping term to the external forces, and the more general differential equation of forces becomes

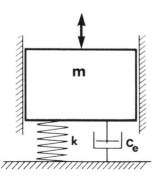

Figure 10.3 Damped spring mass system.

$$m\ddot{x} = -C_e\dot{x} - kx - f(t) \qquad [N] \tag{10.19}$$

where $f(t)$ is an external driving force.

The solution of Eq. (10.19) is considerably more complex than the previous and takes two forms depending on whether the damping coefficient C_e is greater than or less than the critical damping coefficient C_c, defined as

$$C_c = 2\sqrt{km} = 2m\omega_n \tag{10.20}$$

Another parameter ξ, the ratio of the system damping to critical damping, is also useful and is defined as

$$\xi = \frac{C_e}{C_c} \qquad [unitless] \tag{10.21}$$

Now if we take the special case of free vibration when the forcing function $f(t) = 0$, and the damping in the system is less than critical, i.e., $\xi < 1$, the solution to Eq. (10.19) and the equation of motion of the mass takes the form

$$x(t) = e^{-C_e t/2m} (A \sin \omega_d t + B \cos \omega_d t) \qquad [m] \tag{10.22}$$

where A and B are arbitrary coefficients to be determined by initial conditions and ω_d is the *damped* natural frequency.

Now it can be shown [2] that the damped natural frequency is related to the undamped natural frequency as follows:

$$\omega_d = \omega_n \sqrt{1 - \xi^2} \qquad [rad/s] \tag{10.23}$$

From a practical standpoint, when the critical damping ratio ξ is small (<0.1), the difference between the *damped* and *undamped* natural frequency is negligible.

Example

A mass of 2 kg is supported on a spring with a spring constant $k = 36$ N/m. The damping coefficient of the system is 6 N-s/m. What is the undamped natural frequency, the critical damping coefficient, the damping ratio, and the damped natural frequency of the system?

Solution

From Eq. (10.6a), the undamped natural frequency is

$$f_n = \frac{1}{2\pi} \sqrt{\frac{36}{2}}$$

$$= 0.68 \text{ Hz}$$

From Eq. (10.20), the critical damping coefficient is

$$C_c = 2 \sqrt{2 \times 36}$$

$$= 17.0 \text{ N-s/m}$$

From Eq. (10.21), the damping ratio is

$$\xi = \frac{6.0}{17.0}$$

$$= 0.35 \qquad \text{[unitless]}$$

From Eq. (10.23), the damped natural frequency is

$$f_d = 0.68 \sqrt{1 - (0.35)^2}$$

$$= 0.64 \text{ Hz}$$

Now Eq. (10.22) is the product of a periodic function and a decreasing exponential. Shown in Fig. 10.4 is an illustration of the equation of motion or the displacement as a function of time for a damped vibration system. In this illustration, the mass was displaced initially to x_0 and released with no initial velocity. The frequency of the oscillation is ω_d as given by Eq. (10.22), and the decaylike envelope of the motion is due to the decreasing exponential factor. It should be emphasized that the motion with decreasing amplitude is not periodic by definition, and hence the term frequency of oscillation is not strictly proper. However, the term is used so universally that it has become conventional, and we shall use it hereafter.

Now it is the rate of amplitude decay which is usually of the most interest to the acoustic or vibration engineer, and the rate of decay is determined by the argument of the exponential factor. That is, for $0 \le \xi < 1$,

$$\frac{-C_e t}{2m} = \frac{- C_e \omega_n t}{C_c} = -\xi \omega_n t \qquad (10.24)$$

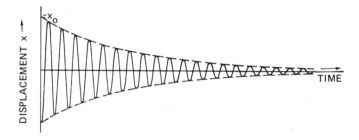

Figure 10.4 Graphic illustration of motion for damped free vibration of mass.

From the properties of exponentials, one sees that if ξ is small, which means the system damping is small compared to critical damping, the decay rate is also small. Correspondingly, the rate of decay increases as ξ approaches unity. A more analytical and quantitative method to describe the decay rate can, however, be derived in terms of the logarithmic decrement.

The logarithmic decrement of decay Δ is defined as the natural logarithm of the ratio of two successive amplitude values x_n and x_{n+1} separated by a period of oscillation. Usually, for ease of calculation and measurement, peak values are chosen. We then have

$$\Delta = \ln \frac{x_n}{x_{n+1}} \qquad [\text{unitless}] \qquad\qquad (10.25)$$

where $\ln = \log_e$ = natural logarithm.

Now, utilizing Eq. (10.22), if one takes the ratio of successive maxima, one obtains

$$\frac{x_n}{x_{n+1}} = e^{2\pi\xi/(1-\xi^2)^{1/2}} \qquad [\text{unitless}] \qquad\qquad (10.26)$$

and upon taking the natural logarithm of each side, we arrive at

$$\Delta = \frac{2\pi\xi}{\sqrt{1 - \xi^2}} \qquad [\text{unitless}] \qquad\qquad (10.27)$$

We should not be surprised that the logarithmic decrement Δ is directly and exclusively related to the damping ratio. Further, from Eq. (10.27), we see more clearly that as ξ approaches 0, Δ approaches 0, and the rate of decay is small. Correspondingly, as ξ approaches 1, Δ grows larger, and the decay rate is sharply increased. Consider an example.

Example

With a graphic level recorder, a trace of displacement versus time is recorded for a damped freely oscillating mass spring system. The ratio of two adjacent maxima was found to be 1.5. What is the damping ratio ξ of the system?

Solution

By definition,

$$\Delta = \frac{2\pi\xi}{\sqrt{1 - \xi^2}} = \ln(1.5)$$

$$= 0.405$$

Squaring and solving for ξ, we get

$$\xi = 0.064$$

Returning to Eq. (10.19), if we again consider the free vibration but where $\xi > 1$, the *overdamped* case, the form of the solution is

$$x(t) = e^{-(C_e t/2m)}\left(Ae^{\omega_n t \sqrt{\xi^2-1}} + Be^{-\omega_n t \sqrt{\xi^2-1}}\right) \quad [m]$$

(10.28)

where again A and B are arbitrary constants depending on initial conditions. However, this product has no periodic factors, and the equation of motion is just a decreasing exponential. Physically, what happens is that when the mass is displaced and released, it slowly returns to the equilibrium position without oscillation.

For the special case where $\xi = 1$, i.e., $C_e = C_c$, the solution to Eq. (10.19) takes the form [3]

$$x(t) = (A + Bt)e^{-(C_e t/2m)} \quad [m]$$

(10.29)

which also has no periodic factors, and as such the mass returns gradually to rest from displacement. At this point, the system is said to be *critically* damped.

10.5 FORCED DAMPED VIBRATION

Returning to Eq. (10.19), we must now consider the more practical case of a forced damped system. Here again, given a sinusoidal forcing function, Eq. 10.19 can be rewritten as

$$\ddot{x} + \frac{C_e \dot{x}}{m} + \omega_n x = F_0 \sin \omega t \quad [N]$$

(10.30)

This equation is still of second order but nonhomogeneous, and the previous basic method of solution applies. That is, the solution will be in two parts:

1. The first part, the complementary solution, will be the solution to the homogeneous (free damped) vibration equation and takes the form of Eq. (10.22) and (10.28), depending on the damping ratio.
2. The second part, the particular solution, takes the form

   $$A_1 \cos \omega t + B_1 \sin \omega t$$

   which is added to the complementary solution to form the general solution.

Now if there is any significant damping in the system, the system response associated with the complementary solution vanishes after a short period of time. In short, the natural free oscillating motion is transient in nature and *damps out* quickly, and we are left with the steady-state motion as described by the particular solution.

From a noise or vibration control standpoint, it is this steady-state condition that is usually of the most interest, and the equation of motion or steady-state displacement $x_s(t)$ can be expressed as

$$x_s(t) = \frac{(F_0/k)\{[1 - (\omega^2/\omega_n^2)]\sin \omega t - 2\xi(\omega/\omega_n)\cos \omega t\}}{[1 - (\omega^2/\omega_n^2)]^2 + (2\xi\omega/\omega_n)^2} \tag{10.31}$$

Here we have sinusoidal motion at the forcing frequency with constant amplitude. Further, in noise or vibration control, we are usually most concerned with the transmissibility of the vibrating forces to the platform or with the spring mass system, the spring support. Therefore, recognizing that

$$F_T = C_e \dot{x} + kx \tag{10.32}$$

and that T is the ratio of the transmitted force F_T to the applied force F_0, we finally get [3]

$$T = \frac{\sqrt{1 + [2\xi(\omega/\omega_n)]^2}}{\sqrt{[1 - (\omega/\omega_n)^2]^2 + [2\xi(\omega/\omega_n)]^2}} \quad \text{[unitless]} \tag{10.33}$$

Equation (10.33) is perhaps the most important and useful relationship in vibration control. Presented for ease of understanding in Fig. 10.5 is a graphic representation of Eq. (10.33), where the transmissibility T is given in terms of the frequency ratio $\omega/\omega_n = f/f_n$ for selected values of the damping ratio. Note that the resonance effect is present when the forcing frequency f is equal to the natural frequency f_n of the system. Note also that the amplitude of transmitted force at resonance is much lower as the system damping becomes critical, i.e., as ξ approaches 1. Finally, note that even with damping the family of curves intersects at a single point T = 1, when $f/f_n = \sqrt{2} = 1.41$. Physically, this means that when the ratio of the forcing frequency to the natural frequency exceeds $\sqrt{2}$, the force transmitted to the spring support is less than applied. Hence, from this transition point on, the spring and dashpot are providing vibration isolation. Isolation on this basis is the primary goal of the noise and vibration engineer, and only from a complete understanding of these concepts can design guidelines be presented. As we shall now see,

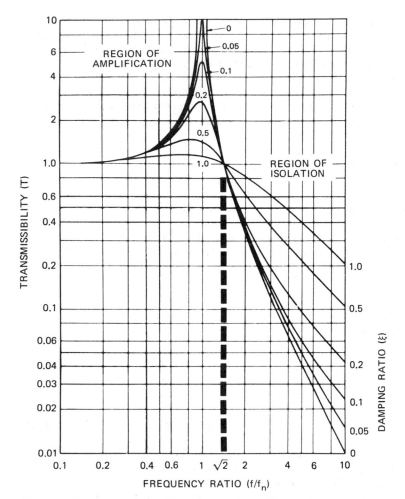

Figure 10.5 Transmissibility as a function of frequency ratio for a range of damping ratios.

many of the commercially available isolators are constructed of the same basic elements as a damped mass spring system.

10.6 VIBRATION CONTROL AT THE SOURCE

Just as emphasized in noise control, control of vibration at its source is usually the most effective measure. For example, one of the most common sources of mechanical vibration in rotating machines is due to unbalance. Unbalance may arise from a number of sources, but the solution is simply to balance the rotating parts. Often this is an easy matter, and the result can be a dynamic reduction in vibration amplitude. Further, in some instances, minor modifications such as stiffening support brackets or flanges will radically change the natural frequencies of the system, and undesirable resonance conditions can be avoided.

Often equipment such as sifters, screeners, tumblers, etc., vibrate by design at large amplitudes but very low frequencies. Generally, they are not very noisy. However, a loose sheet metal panel, door, or attachment will impact against the structure with a resounding *clatter*. The solution here is a matter of light maintenance, often with only a screwdriver. As in noise control, sufficient reduction may not always be available at the source, and hence some modifications must be made in the path between the source of vibration and the receiving element. This is best accomplished by the use of vibration isolators, which will be discussed in the following section.

10.7 VIBRATION ISOLATORS AND ISOLATION MOUNTS

There are three major types of isolators: metal springs, elastomeric mounts, and resilient pads. Each type has advantages depending on the degree of isolation required, forcing frequency, weight of equipment, temperature, etc. In short, the relative merits of each should be considered for any particular installation.

Before discussing each type of isolator, one comment must be made which is pertinent to all types. We shall assume throughout this section that we are dealing only with single degree of freedom systems. Since most systems cannot be reduced to such a simple model, the results and analysis should *not* be considered exact. Experience has proven and certainly the examples selected are such that the design methods outlined herein are reliable and conservative. It should also be emphasized that the basic design approach will be to select isolator devices such that the natural frequency of the system is well below the lowest applied forcing frequency. In this way, a high degree of isolation is assured. Referring to Fig. 10.5, one sees that isolation above 90% is possible only if the forcing frequency exceeds the system natural frequency by a factor of 4. This is true only for systems with little damping, which fortunately is the case for most commercially available isolators.

One final aspect that is relevant throughout the presentation is that for $\xi < 0.1$ the natural frequency of most isolators is related simply to their static deflection under load.

Referring, then, to Eq. (10.6a), if k is given in newtons per centimeter and m is in kilograms, we get the more useful form

$$f_n = 4.98 \sqrt{\frac{1}{\delta_{st}}} \qquad [Hz] \qquad\qquad (10.34a)$$

where δ_{st} is the static deflection under load in centimeters. Similarly, f_n is given by

$$f_n = 3.13 \sqrt{\frac{1}{\delta_{st}}} \qquad [Hz] \qquad\qquad (10.34b)$$

where δ_{st} is the static deflection under load in inches.

Example
A spring system is deflected 4 cm under a static load. What is the undamped natural frequency of the system?
Solution
From Eq. (10.34a), we have

$$f_n = 4.98 \sqrt{\frac{1}{4}}$$

$$= 2.49 \text{ Hz}$$

Therefore, for most isolators, the natural frequency of the system can be calculated from the static deflection under load.

Presented in Fig. 10.6 is another useful chart for design purposes which yields the degree of isolation or, in this case, the percent of isolation in terms of the forcing frequency f_d and natural frequency f_n of the system. Here percent of isolation is defined as

$$\text{Percent of isolation} = (1 - \text{transmissibility}) \times 100 \qquad [\text{unitless}]$$

$$(10.35)$$

This chart was obtained from Eq. (10.33) and is valid only in the isolation region where the frequency ratio exceeds $\sqrt{2}$, i.e., $f_d/f_n > \sqrt{2}$, and for systems with little damping, i.e., $\xi < 0.1$.

Table 10.1 provides a useful guide to the amount of static deflection required for various percentages of isolation at common machine running speeds. As can be seen, to obtain significant isolation at low running speeds, large static deflections are required. Likewise, for high-speed machinery, only small static deflections are required.

10.7.1 Spring Mounts

Metal springs have perhaps the widest use of all isolation types. They are particularly applicable where large heavy equipment is to be isolated. The metal springs allow for large deflections and as such are especially effective where large loads and very low forcing frequencies

Table 10.1 Vibration Isolation Deflection Chart for Common Rotational Speeds

Equipment speed (rpm)	Effective static deflection (in.) required to obtain various percentages of vibration isolation							
	99%	97%	95%	90%	85%	80%	75%	60%
3600	0.27	0.09	0.06	0.03	0.02	0.02	0.01	0.01
2400	0.62	0.21	0.13	0.07	0.05	0.04	0.03	0.02
1800	1.1	0.37	0.23	0.12	0.08	0.07	0.05	0.04
1600	1.4	0.47	0.29	0.15	0.11	0.08	0.07	0.05
1400	1.8	0.62	0.38	0.20	0.14	0.11	0.09	0.06
1200	2.5	0.84	0.51	0.27	0.19	0.15	0.12	0.09
1100	2.9	1.0	0.61	0.32	0.22	0.17	0.15	0.10
1000	3.6	1.2	0.74	0.39	0.27	0.21	0.18	0.12
900	4.4	1.5	0.91	0.48	0.33	0.26	0.22	0.15
800	5.6	1.9	1.2	0.61	0.42	0.33	0.28	0.19
700	7.3	2.5	1.5	0.79	0.55	0.43	0.36	0.25
600	9.9	3.4	2.1	1.1	0.75	0.59	0.49	0.34
550		4.0	2.4	1.3	0.89	0.70	0.58	0.41
500		4.8	3.0	1.6	1.1	0.85	0.71	0.49
400		7.6	4.6	2.4	1.7	1.3	1.1	0.77
350		9.9	6.0	3.2	2.2	1.7	1.4	1.0
300			8.2	4.3	3.0	2.4	2.0	1.4
250				6.2	4.3	3.4	2.8	2.0

are present. Shown in Fig. 10.7 are illustrations of a variety of commercially available spring mounts and hangers. Basically, the design consists of helical coil springs, which are fixtured in *end* caps. Generally, there is a central bolt which provides a tie-in or fastener to the equipment being isolated. In some cases, the springs are housed or constrained to minimize any lateral vibrating motion which may occur.

The most important feature of spring mounts is their ability to withstand relatively large deflections and as such provide good low-frequency isolation. This conclusion follows from Eqs. (10.34a) and (10.34b). Here one sees that the system natural frequency f_n is an inverse function of the deflection, or as the deflection increases, the natural frequency decreases. An example will illustrate these concepts.

Example
A large electric motor is mounted to the floor with four spring mounts at the corners. The static deflection of each mount is 1/4 in. What is the natural frequency of the system?

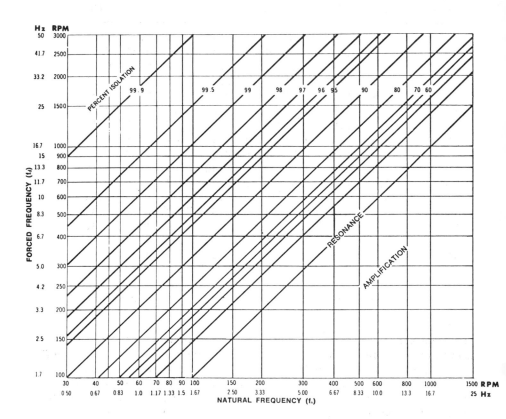

Figure 10.6 Design chart showing isolation in terms of the system natural frequency and the forced or excitation frequency.

Figure 10.7 Examples of commercially available spring mounts. (Courtesy of Mason Industries, Inc., Hauppauge, N.Y.)

Solution
From Eq (10.34b)

$$f_n = 3.13 \sqrt{\frac{1}{0.25}}$$

$$= 6.26 \text{ Hz}$$

The natural frequency of the system is then approximately 6 Hz.
    If the deflection in the example were 1.0 in., the natural
frequency would be

$$f_n = 3.13 \sqrt{\frac{1}{1.0}}$$

$$= 3.13 \text{ Hz}$$

or one-half the value for 1/4-in. deflection.
    From this example, it is easy to see why spring mounts are
especially good low-frequency isolators.

Conversely, if the forcing frequency of the motor in the example
were in the range of 6 Hz, a resonance condition would exist (referring
to Fig. 10.6) with resultant large oscillatory amplitudes and a correspondingly high transmission of vibratory forces to the floor. Naturally, this resonance condition is to be avoided. If, however, the forcing
frequency were 30 Hz, typical of large electric motors, a high degree
of isolation above 90% (from Fig. 10.6) could be expected.
    To simplify the spring mount selection method, we shall use Table
10.1 which yields static deflection as a function of applied forcing frequency for several transmissibility criteria commonly used by vibration
engineers.
    Let us now outline a systematic step-by-step procedure for select-
ing spring mounts for any degree of isolation:

Step 1. Establish the total weight and lowest expected forcing fre-
    quency of the equipment to be isolated. If a motor operating at
    1800 rpm drives a pump at 1200 rpm, the lowest frequency is
    associated with the pump (1200 rpm or 20 Hz).
Step 2. From Table 10.1, determine the static deflection required
    to provide the degree of isolation desired.
Step 3. From the listed spring constants of the mounts (supplied by
    the manufacturer), select the appropriate spring mounts.

    An example will clearly illustrate the procedure.

Example
    A motor-generator unit weighing 700 and 500 lb, respectively,
    is to be installed in a vibration-sensitive area. The motor turns
    at 3600 rpm and the generator at 1800 rpm. Select four spring
    mounts which will provide a minimum of 95% isolation.

Table 10.2   Excerpt from Spring Mount Catalog

TYPE 'SLF' MOUNT
STANDARD BASE WITHOUT BOLT HOLES

Mount No.	Mount Constant (lb/in.)	Capacity (lb)
SLF-401	50	100
SLF-402	70	140
SLF-403	90	180
SLF-404	115	230
SLF-405	150	300
SLF-406	200	400
SLF-407	305	610
SLF-408	365	730
SLF-409	500	1,000
SLF-410	650	1,300
SLF-411	940	1,880
SLF-412	1,150	2,300
SLF-413	1,500	3,000
SLF-414	2,000	4,000
SLF-415	2,680	5,360
SLF-416	3,500	7,000
SLF-417	4,750	9,500
SLF-418	6,250	12,500
SLF-419	8,200	16,400
SLF-420	11,250	22,500
SLF-421	15,000	30,000
SLF-422	20,000	40,000

Source: Courtesy of Mason Industries, Inc., Hauppauge, N.Y.

Solution
Step 1.   The total weight is 1200 lb, and the lowest forcing
frequency is associated with the generator, 1800 rpm or 30 Hz.
Step 2.   From Table 10.1, the required static deflection for
1800 rpm and 95% isolation is 0.23 in.

Step 3. Since there will be four springs supporting 1200 lb, the spring constant of each mount must be 300 lb/0.23 in. = 1300 lb/in. (approximately). From the actual catalog presented in Table 10.2, mount number SLF-412, whose mounting constant is 1150 lb/in., is selected. If additional design margin is desirable, mount number SLF-411, the next lowest capacity, would be a reasonable choice.

In addition to the selection method just outlined, several other basic design guidelines should be followed:

1. A uniform deflection should be maintained for each mount to minimize unwanted *twisting* or *rocking* motions. In some large installations, the weight may not be evenly distributed. As such the mounts should be selected accordingly to maintain uniform deflection.

2. *Driver and driven* units should be mounted on a common rigid base, or serious damage may occur to the drive-shaft bearings of either or both.

3. Where a wide range of temperatures may occur, spring mounts are especially stable. In addition, where acids, oils, solvents, etc., are present, spring mounts are more resistant to corrosion than most other mount types.

4. Because of the very low damping factors of metal springs, some kind of pad is usually also required in combination to achieve minimal damping and high-frequency isolation. In fact, many of the commercially available units have a resilient pad integrally attached to the mount base.

5. Often, the addition of an *inertia block* provides several desirable features. Shown in Fig. 10.8 is a photograph of some pumps mounted on inertia blocks which are isolated from the floor. This combination has the advantage of lowering the system center of gravity, providing a firm rigid common base for the equipment, and uniformly distributing the load. Inertia blocks are usually made from reinforced concrete and should weigh at least 1.5 times the weight of the supported equipment.

Finally, it should be mentioned that spring mounts may be used in tension but most often are designed to be used in compression.

10.7.2 Elastomeric Isolators

Over a limited range of stress, elastomeric isolators can also be used effectively for a wide variety of vibration isolation problems. This type of mount is at its best performance for applications that include small machines and relatively high excitation or forcing frequencies. The

Figure 10.8 Two motor-driven pumps mounted on massive inertia
blocks. (Courtesy of SVA Inc., Ashland, Mass.)

elastomeric medium can be easily molded into practically any size or
shape, and a range of stiffness can be controlled over rather wide
limits. Shown in Fig. 10.9 are some examples of commercially available
elastomeric mounts. The most common materials selected for elastomeric
mediums are natural rubber, neoprene, butyl, silicone, and combinations of each. A typical mount utilizing these materiasl generally employs the medium in shear but may also employ a compressive design.
By combining both shear and compression in the design, a relatively
linear deflection versus load response can be obtained, as was the
case for helical springs.

Presented in Fig. 10.10 is a typical load deflection curve for a
line of commercially available rubber mounts. The stiffness or spring
constant for this family of mounts is varied by controlling the hardness or durometer of the rubber. Intuitively, a softer rubber deflects more under load than a hard rubber. With these deflection
versus load curves, one can apply essentially the same method for
mount selection as with springs. Consider an example.

Figure 10.9 Examples of commercially available elastomeric isolators. Bold crosshatch denotes elastomeric medium. (Courtesy of Barry Controls Division, Barry Wright Corporation, Watertown, Mass.)

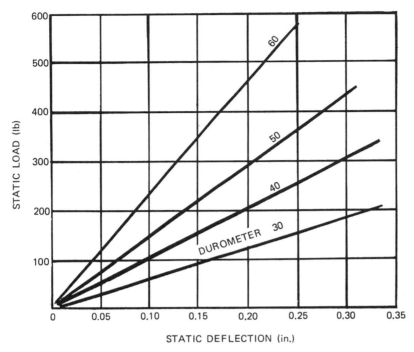

Figure 10.10   Typical load deflection curves for commercially available rubber isolation mounts.

Example

   It is desired to isolate a small internal combustion engine weighing
   900 lb whose normal rotational speed is 2400 rpm. An isolation
   transmissibility of 3% or less with four elastomeric mounts is the
   design goal. What deflection and durometer are required?
   Solution
   Recall the method outlined for spring mounts:
   Step 1. The lowest forcing frequency will be 2400 rpm = 40 Hz,
   and the weight is 800 lb.
   Step 2. From Table 10.1, the required static deflection for 97%
   isolation is 0.21 in.
   Step 3. Since the load will be distributed over four mounts, a
   mount whose deflection is 0.21 in. for 900/4 = 225 lb is required.
   Looking to Fig. 10.10, a mount with durometer 40 will do nicely
   with a little design margin.

   Elastomeric-type mounts have many advantages over springs and
others, especially where space and weight are important. This type
of mount is used almost exclusively in the aerospace industry for
engine and delicate electronic equipment installations. Another popular

application is for packaging delicate equipment in shipping containers. There seems to be no end to the configurations in which the active rubber section can be molded. Hence, this feature gives the design engineer great latitude for innovation.

The most serious limitation of elastomeric mounts is with the life and endurance characteristics. It is pretty well acknowledged that static deflection must be kept under 0.5 in., and serious changes in stiffness can be expected if operation is outside the temperature range of $-30°F$ to $150°F$. In addition, many of the rubber compounds used will deteriorate rapidly when exposed to some highly corrosive acids, oils, and solvents commonly found in industrial environments.

10.7.3 Isolation Pads

Probably the simplest and most often-used isolation mount is the *pad* type. These pads are available in natural rubber, synthetic rubber, or blocks of cork, felt, or fibrous glass and combinations thereof. Several advantages of pad-type isolations are that they

1. Are easily inserted under equipment, eliminating tearing up the floor or inserting bolts
2. Are available in sheets of various thickness
3. Can be stacked to obtain large deflections and corresponding high levels of low-frequency isolation

Let us consider briefly the basic selection criteria for the most common pad types available.

10.7.4 Rubber Pads

Shown in Fig. 10.11 is an example of commercially available *ribbed* or *waffled* synthetic rubber pads and typical load versus deflection curves. The design selection for rubber pads follows basically the method established in the preceding section on rubber mounts. That is, load versus deflection curves are always available from the man-ufacturer for a variety of rubber stiffnesses (durometers) or layer thicknesses, and the same approach to selection can be followed. Generally, these pads can be bought in sheets and cut to whatever length and width desired, and a table of recommended loading per unit area is usually available from the manufacturer, which also simplifies the selection. Also, these pads stack easily to achieve larger overall deflections than is recommended for a single sheet. For a single sheet, the upper load should not exceed 70 psi. Other-wise, most of the limitations listed for rubber mounts also apply to rubber pads.

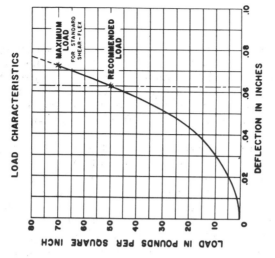

Figure 10.11 Example of commercially available rubber isolation pad and load characteristics. (Courtesy of Vibration Mounting and Controls, Inc., Butler, N.J.)

377

Example

A centerless grinding machine is to be mounted on synthetic
rubber pads for vibration isolation. The machine weighs 4000
lb, and the rotational frequency of the motor, i.e., the wheel,
is 3600 rpm. Select isolation pads from the type shown in Fig.
10.11 such that maximum load on each pad does not exceed 70
lb/ft^2. An isolation transmissibility of less than 10% is required.

Solution

The isolators in the example are available in pads 4 in. square,
i.e., (2 × 2 in.), 9 in. square (3 × 3 in.), 16 in. square (4 × 4 in.),
25 in. square (5 × 5 in.), 36 in. square (6 × 6 in.), etc. To
assure that the load limit of 70 psi is not exceeded, note that there
are four mounting points and that the machine weighs 4000 lb. Hence,
each mount will support 4000 lb/4 = 1000 lb. For a pad 4 × 4 in. =
16 in.^{2}, the load will be 1000 lb/16 in.^{2} = 62.5 psi, which is within
the limit and will allow for nearly maximum deflection.

Now from the load versus deflection chart, the deflection for
62.5 psi is approximately 0.070 in. Referring to Table 10.1, we
see that for a speed of 3600 rpm and a deflection of 0.07 in. the
transmissibility is less than 5% and hence meets the design criterion of 10% with a small margin of safety.

10.7.5 Cork Pads

Cork is perhaps the oldest isolation material and one of the most effective. The cork pads are generally available commercially in blocks
6 to 10 in. square and in thicknesses of 1/2 and 1 in. A range of
densities is available from some suppliers and provides a range of
stiffnesses for selection.

Commercially available cork pads are made by compressing cork
particles under high pressure and steam. Cork pads differ from
rubber pads in that the material is porous, containing many minute
air cells which become reduced in volume under compression. As
such, cork exhibits little lateral expansion in vertical compression.

The stiffness of cork is relatively high for commercially available cork pads. Unfortunately, cork does not possess linear deflection versus load characteristics and varies dynamically with density.
Therefore, most suppliers provide recommended loads for given densities, as shown in Table 10.3. Consider an example.

Example

A tumbling drum directly driven by a motor creates excessive
vibration to the mounting platform. The drum, motor, and
support weigh 800 lb. Select cork pad isolators for the four
mounting points.

Table 10.3 Recommended Loading for Cork Isolation Pads

Density of Cork (lb/bd-ft)[a]	Loading (psi)
0.75	Up to 5
1.25	5 to 30
1.50	30 to 50

[a] A board foot (bd-ft) is the volume occupied by a solid 1 ft square and 1 in. thick.

Solution

With four mounting points, each pad must support $800/4 = 200$ lb. Therefore, a pad 9 in. square (3×3 in.) would, if evenly distributed, support $200/9 = 22.2$ psi. From Table 10.3, a cork pad of density 1.25 lb/bd-ft would provide a high degree of isolation.

From this example, one sees that exact measures of isolation transmissibility cannot usually be calculated for cork pads. For a first-order estimate, however, commercial cork behaves much like rubber with a durometer in the range of 50 to 60.

The disadvantages of cork are the following:

1. Rather small range of available stiffness compared to rubber
2. Takes a permanent compression *set* at temperatures above 100°F
3. Not especially resistive to industrial solvents, acids, oils, etc.
4. Cannot be molded; hence is available only in slab form

On the positive side, cork possesses a rather high internal damping coefficient, typically $C_e/C_c = \xi = 0.06$, and as such can be used near resonance or for transient excursions through resonance.

10.7.6 Felt Pads

Hairfelt pads have also been used for years as vibration isolators. Used exclusively in compression, these pads are commercially available in 1/2- to 3-in. thicknesses. The most common thickness is 1 in., and where extremely high loads are anticipated, stacks of 1-in. pads are formed to provide the required deflection.

Felt pads used as isolators are generally graded in terms of the density, i.e., hard, medium, and soft. The soft grades typically can be loaded up to 50 psi, medium grades to 100 psi, and the hard, most dense pads to 200 psi. Shown in Table 10.4 is a typical selection table for commercially available felt pads. The deflection versus load curve is fairly linear up to a compression of about 25%. At this point, the

Table 10.4 Recommended Loading for Hairfelt Isolation Pads

Type	Density (lb/bd-ft)	Loading (psi)
Light	1	0–50
Medium	1.2	50–100
Heavy	1.5	100–200

stiffness increases sharply with a corresponding increase in transmissibility. Selecting felt pads for effective isolation again follows the method outlined for rubber pads. An example best illustrates the method.

Example
    A 9000-lb press is to be isolated from the floor. The press has
    six support feet, and each measures 4 × 5 in. Select a felt pad
    mount to provide a high degree of isolation.
    Solution
    Each support foot has a bearing load of 9000/6 = 1500 lb. Since
    there are 20 in.^{2} of bearing surface, the bearing load is 1500/20 =
    75 psi. Looking to Table 10.4, 75 psi falls into the medium density
    range of 50 to 100 psi. Therefore, the type medium is selected.

Many hairfelt materials are organic and therefore deteriorate rapidly in the presence of typical industrial oils or solvents. Some suppliers have industrial-quality felt, but some caution must be given this matter for each proposed felt pad installation.

10.7.7 Fibrous Glass Pads

Fibrous glass pads have vibration characteristics much like felt pads. The deflection versus stated load curves are nearly linear, up to compressions of 20% to 30%, where further compression sharply increases the stiffness. The pads are commercially available typically in 1/2- and 1-in. thicknesses and can also be stacked for low-frequency isolation. One manufacturer suggests stacks up to 20 layers (10 in. thick) to achieve 1% transmissibility at 40 Hz. Again, the stiffness of the mount is a function of density, but effective isolation below 20 Hz does not appear possible with the fibrous glass pads. The fibrous glass materials are basically inert and highly resistive to industrial oils, solvents, acids, etc.

10.7.8 Helical Isolators

One other form of isolator is the helical or steel coil isolator. These
isolators are heavy-duty assemblies of stranded stainless wire cable
attached between metal retainers. Shown in Fig. 10.12 are examples
of some commercially available helical isolators along with a photograph
inset of a typical installation. Each isolator type has its own stiffness
characteristc depending on cable diameter, number of strands per
cable, cable length, twist, and number of cables per section. They
are especially well suited to heavy-duty high-displacement applications
such as use on shipboard and for shipping containers.

Excellent isolation is available down to the range of 5 to 7 Hz
with inherent damping provided by flexural hysteresis due to rubbing
and sliding friction between the strands. As such the level of damping is usually in the range of 15% to 20% of critical, i.e., $\xi = 0.15$ to
0.20, which yields very low amplification factors at resonance. Another
positive feature is that they can be used in any orientation and are

Figure 10.12 Examples of commercially available helical isolation
mounts. Insert shows typical installation. (Courtesy of Aeroflex
Laboratories, Inc., Isolator Products Division, Plainview, N.J.)

highly resistant to industrial or other harsh environments. They
are, however recommended only for applications in compression.

10.7.9 Pneumatic Isolators

For equipment subject to low-frequency vibration or *shock* such as
power presses, drop or forge hammers, tumbling barrels, etc., pneu-
matic-type mounts are especially applicable. We have not discussed
the physical concept of shock or shock-excited systems. Analytical
methods for dealing with shock are somewhat complex, especially when
applied to system response. As such we shall say only that shock
can be considered an impulsivelike force and that system response
resembles to the first order free vibration. Hence, an isolation system that includes a high level of damping or extremely low natural
frequency or both has the primary design factors for good shock
isolation.

Returning to the pneumatic isolators, their construction consists
basically of an air-filled reinforced bellows with mounting plates attached to the top and bottom. Shown in Fig. 10.13 is a commercially
available pneumatic isolator along with construction details.

Since the stiffness of the mounts comes from an *air spring*, we
can have *zero* static deflection with a degree of isolation down to 0.5
Hz. This is particularly noteworthy since a steel spring would require deflection in excess of 3 ft to provide the same degree of
isolation.

To determine the natural frequency f_n of a mass supported on an
air column, we have [4]

$$f_n = \frac{1}{2\pi} \sqrt{\frac{\gamma A g}{V}} \quad [Hz] \tag{10.36}$$

where

γ = gas constant, 1.4 for air

A = cross-sectional area of air piston (in.^{2})

g = acceleration due to gravity, 386 in./s^2

V = volume of air column (in.^{3})

The only disadvantages of these mounts are that they must be
used in compression, and only a single, nearly absolute vertical
orientation is possible.

10.7.10 Inertia Bases or Frames

Before leaving the subject of isolation, it should be mentioned that
prefabricated inertia bases or frames which often simplify equipment
installation are readily available commercially.

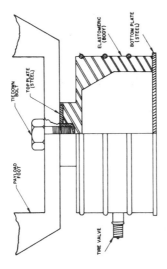

Figure 10.13 Examples of pneumatic isolation mount and construction detail. (Courtesy of Barry . Controls Division, Barry Wright Corp., Watertown, Mass.)

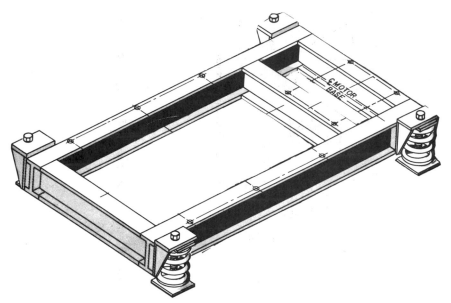

Figure 10.14 Commercially available inertia frame or base with spring isolation mounts. (Courtesy of Mason Industries, Inc., Hauppauge, N.Y.)

An inertia base generally consists of a steel frame with isolation mounts attached, as illustrated in Fig. 10.14. The frames are generally of steel beam construction designed to be used as is or to accept poured concrete, serving also as an inertial block. In either case, they (1) provide for easy installation; (2) prevent differential motion between driving and driven members; (3) reduce the system center of gravity, thus reducing undesirable *rocking;* and (4) with concrete added, act to lower the system natural frequency.

It is not possible to give many specific design guidelines for the use of inertia bases or blocks, since the requirements of space, load, etc., are unique for each installation. However, as mentioned earlier in the chapter, if the inertia base or block is a minimum 1.5 times heavier than the object to be isolated, most of the positive features will be present.

10.8 FLEXIBLE COUPLINGS

Flexible connectors used dominantly to isolate rotating or vibrating machinery from ducts or pipes play an important role in vibration and noise control. What happens here is that ducts or pipes are con-

Figure 10.15 Composite showing high-pressure rubber connectors to control structure-borne vibration in pipes. (Courtesy of Mason Industries, Inc., Hauppauge, N.Y.)

nected directly to vibrating machinery, and the vibratory energy is transmitted in a structure-borne manner to the ducts or plumbing with a high degree of efficiency. Three major problems may then arise:

1. The vibratory energy tends to produce vibratory motion at joints, couplings, or seals with resultant failure at these critical locations.
2. The vibratory energy is reradiated as noise throughout the pipe or duct system.
3. The vibrating pipes and ducts impact adjacent supports, pipes, walls, etc., with resultant clatter.

In high-pressure fluid systems, steel-reinforced rubber connectors, as shown in Fig. 10.15, are generally the best isolators. These connectors are usually flanged or threaded for ease of installation and are available in various sizes of straight or elbow sections.

Design guidelines are simple:

1. The connector isolators should be as close to the pump, blower, etc., as possible.
2. Two connectors oriented 90° to each other perform much better than a single isolator.

In low-pressure air-handling systems, flexible sleeves or hoses work well and are usually easy to install. These connector isolators also allow for expansion and contraction due to temperature changes.

10.9 DYNAMIC ABSORBERS

In some cases, the control of vibration and the resultant radiated noise cannot be achieved using the previously described methods of vibration isolation. However, when vibratory forces are a constant single frequency, dynamic absorbers can be used to control vibration. A dynamic absorber consists of a comparatively small mechanical system that has a natural frequency that is coincident with the frequency of the exciting forces of the vibrating source. The resultant mechanical system is schematically illustrated in Fig. 10.16. By selecting k and m properly such that $\sqrt{k/m}$ is equal to ω, the resultant system then has a unique dynamic property. The dynamic absorber will naturally vibrate in such a way that its spring force is at all instants equal and opposite to the exciting force $P_0 \sin(\omega t)$. Obviously, since these two forces are equal and opposite, the net force on mass M is zero and thus the original system will not vibrate.

Most practical dynamic adsorbers consist of weights that are attached to small beams. By adjusting the length of the beam and the size of the weight, the dynamic absorber's natural frequency is *tuned*

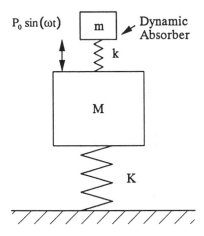

Figure 10.16 Schematic illustration of the dynamic model of a machine with a dynamic absorber attached.

to the exciting frequency. The actual process of selecting dynamic absorber sizes, mounting locations, and mechanical properties is beyond the scope of this text. However, there are many practical situations where noise and vibration control can be achieved by applying this approach. Figure 10.17 illustrates the use of two dynamic absorbers that have been attached to an elevator cable sheave. A 10-dB reduction of the radiated noise was achieved by the application of these absorbers to the sheave.

10.10 VIBRATION MEASUREMENT AND ANALYSIS

Vibration measurements are very similar to sound measurements. However, spectral analysis and related diagnostic methods closely follow the techniques employed for sound. The basic vibration measurement system includes a transducer, preamplifier, and some means of analyzing, displaying, or measuring the transducer electrical output.

10.11 TRANSDUCERS

Just as in sound measurements, the transducer is the most critical element in any vibration measurement system. By far the most popular vibration transducer is the accelerometer, followed closely by the velocity pickup. We shall discuss each in detail in this section.

Figure 10.17 An elevator sheave with two dynamic absorbers attached. (Courtesy of Timothy J. Foulkes, Cavanaugh Tocci Associates, Inc., Sudbury, Mass.)

10.11.1 Accelerometers

The accelerometer produces an electrical signal, the voltage of which is proportional to the acceleration experienced by the device.

Shown in Fig. 10.16 are several commercially available accelerometers along with a schematic showing basic construction and active elements. Essentially, we have a mass suspended on a spring between two piezoelectric crystals. These components are encased in a metal housing and attached to a base. When the assembly is mounted to a vibrating surface, the mass exerts an inertial force F on the crystals that is proportional to the acceleration. From the piezoelectric effect of the crystals, a voltage proportional to the acceleration is produced and is generally brought out through a connector to small, easily attached cables. For accelerometers, the output voltage sensitivity is usually rated in terms of millivolts per g, where g is the acceleration due to the earth's gravitational field. Typical sensitivities for commercially available piezoelectric accelerometers range from 10 to 1000 mV/g over a wide frequency range.

(a)

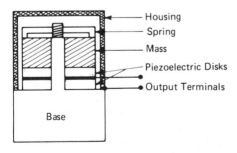

(b)

Figure 10.18 (a) Examples of commercially available accelerometers and (b) schematic showing basic construction and active elements. (Courtesy of Brüel and Kjaer Instruments, Inc., Marlborough, Mass.)

Now it must be emphasized that if the measured voltage from an accelerometer is the root-mean-square (rms) value, which is most often the case, the amplitude of the acceleration will also be the rms value. Likewise, if peak voltage is measured, peak values of acceleration will be obtained.

Example

The rms voltage from an accelerometer mounted on a large gear box is 25 mV (rms). If the signal is sinusoidal and the sensitivity of the accelerometer is 50 mV/g, what is the rms value of the acceleration and the peak value of the acceleration?

Solution

With a sensitivity of 50 mV/g and an output of 25 mV$_{rms}$, the rms value of the acceleration A$_{rms}$ is

$$A_{rms} = \frac{25 \text{ mV}}{50 \text{ mV}/g}$$

$$= 0.5g_{rms}$$

Since g is nominally taken as 9.81 m/s^2, we have

$$A_{rms} = 0.5 \times 9.81$$

$$= 4.9 \text{ m/s}^2$$

Since we have a sinusoidal signal whose voltage is proportional to acceleration, the peak value of the acceleration is related to the rms value by

$$A_{peak} = \sqrt{2} \, A_{rms}$$

We derived this expression dealing with rms values of sound levels in Chap. 1. Hence,

$$A_{peak} = 1.414 \times 4.9$$

$$= 6.9 \text{ m/s}^2$$

With respect to frequency response, these devices are extremely linear over the range of most interest to the noise and vibration control engineer. Shown in Fig. 10.17 is a typical frequency response curve or calibration chart for a piezoelectric accelerometer. Note that the response is *flat* to ±2 dB up to 10,000 Hz.

These devices are extremely stable and can be used over a wide range of temperature, nominally −100°C to 250°C. They are insensitive to magnetic fields and can be exposed without harm to most meteorological elements and harsh industrial solvents, oils, etc. The calibration applies to a single direction of motion, generally along the vertical axis. Sensitivity along the transverse axis is typically more than 30 dB *down* for most commercial equipment.

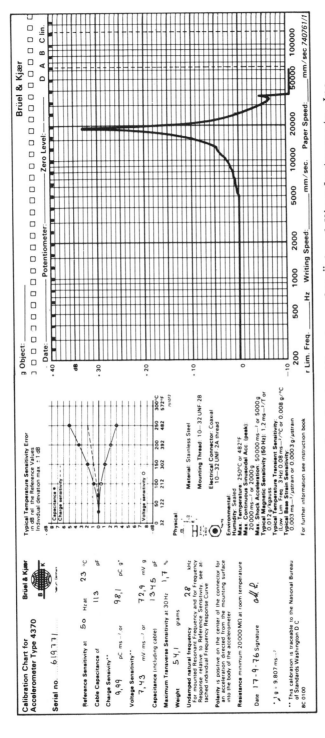

Figure 10.19 Typical accelerometer calibration chart. (Courtesy of Brüel and Kjaer Instruments, Inc., Marlborough, Mass.)

Perhaps the most critical aspect of the use of accelerometers is in the mounting. Clearly, if the accelerometer is not solidly mounted to the vibrating surface, gross errors can be expected. Shown in Fig. 10.18 are several common mounting methods. The merits and limitations of each will now be described [5].

Type 1 mounting is the best in terms of frequency response, approaching a condition corresponding to the actual calibration curve supplied with the accelerometer. If the mounting surface is not quite smooth, it is a good idea to apply a thin layer of silicon grease to the surface before threading down the accelerometer. This increases the mounting stiffness. It is essential whenever using a mounting screw not to tighten it fully, as one may introduce base bending affecting the sensitivity of the accelerometer.

Type 2 mounting is convenient when electrical isolation between the accelerometer and vibrating body is necessary. It employs the isolated stud and a thin mica washer. The frequency response is good due to the hardness of the mica. Make sure that the washer is as thin as possible (it can easily be split up into thinner layers).

Type 3 mounting employs a thin layer of wax for *sticking* the accelerometer onto the vibrating surface. Surprisingly, this method yields a very good frequency response due to the stiffness of the wax. Caution is required, however; above room temperature, the wax may melt and the method fails.

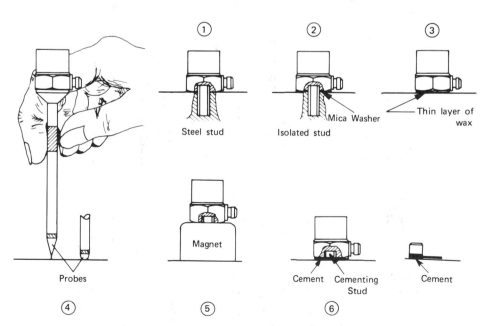

Figure 10.20 Common methods for mounting accelerometers.

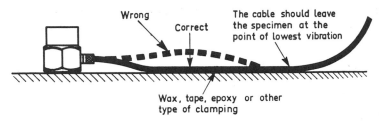

Figure 10.21 Location of cable clamping to avoid cable whip. (Courtesy of Brüel and Kjaer Instruments, Inc., Marlborough, Mass.)

Type 4 mounting employs a probe with interchangeable round and pointed tips. The method may be convenient for certain applications but should not be used for frequencies much higher than 1000 Hz.

Type 5 mounting employs a permanent magnet which also gives electrical isolation from the vibrating specimen. A closed magnetic path is used, and there is virtually no magnetic field at the accelerometer position. This mounting should not be used for acceleration amplitudes higher than about 200*g* or temperatures above 100°C.

Type 6 mounting is convenient when a cementing technique is appropriate. If the bond is good, the frequency response of the transducer is also good.

Finally, some care must also be given to avoid cable *whip*. Here some method of clamping the electrical output cable to the surface must be found to avoid a high-level whipping motion of the cable relative to the accelerometer. If not attached securely, the cable will often experience fatigue failure at the terminal. Usually either wax, tape, or glue, as illustrated in Fig. 10.19, will be adequate.

10.11.2 Velocity Pickups

Velocity pickups provide an electrical signal whose voltage is proportional to the velocity of the vibrating surface to which it is mounted. These devices consist primarily of a seismic coil surrounded by a permanent magnet. As the coil moves due to the motion of the pickup, a voltage is induced which is proportional to the velocity. Here again the signal is usually taken from the pickup itself through small cables. Since these devices are large compared to accelerometers and extremely sensitive to the magnetic fields common in industrial environments, they are used sparingly. As such we shall say only that, if utilized, most of the mounting guidelines given for accelerometers apply also to velocity pickups.

Finally, it should be mentioned that if integrating networks (usually available from the manufacturer) are employed in conjunction with accelerometers, the velocity and displacement of the vibration can be obtained as well. For example, if the vibration is periodic, the signal can be broken into pure sinusoids of the form

$$a(t) = A_0 \sin \omega t \qquad [m/s^2] \tag{10.37}$$

where

A_0 = amplitude of acceleration (m/s^2)

ω = angular frequency = $2\pi f$ (rad/s)

Then, in the case of accelerometers, the velocity $v(t)$ of the vibrating surface is

$$v(t) = \int A_0 \sin \omega t \, dt$$

$$= \frac{-A_0}{\omega} \cos (\omega t) \qquad [m/s] \tag{10.38}$$

The negative sign denotes opposite phase.

Further, the displacement $d(t)$ is then just

$$d(t) = \int v(t) \, dt$$

$$= \iint a(t) \, dt$$

$$= \frac{A_0}{\omega^2} \sin (\omega t) \qquad [m] \tag{10.39}$$

Example

The amplitude of acceleration for sinusoidal vibration of a large fan was measured to be 100 m/s^2 at 30 Hz. What are the peak values of the velocity and displacement?

Solution

From Eq. (10.38), the amplitude $|V_0|$ of the vibration is

$$|V_0| = \frac{A_0}{\omega}$$

$$= \frac{100}{2\pi \times 30}$$

$$= 0.53 \ m/s$$

From Eq. (10.39), the displacement V_0 is

$$D_0 = \frac{A_0}{\omega^2}$$

$$= \frac{100}{(2\pi \times 30)^2}$$

$$= 0.0028 \text{ m}$$

Constants of integration were ignored, since they relate only to initial velocity and displacement.

Thus from either type of transducer, accelerometer or vibration pickup, the important parameters which characterize the vibration can be determined.

10.12 PREAMPLIFIER

The second element in the measurement system is the preamplifier. When using a velocity pickup, a preamplifier is not required. However, when using a piezoelectric accelerometer, a preamplifier is always required. Basically, the preamplifier serves two useful purposes: it increases the level of the transducer signal, which is small, and it provides an impedance-matching device between the transducer and the signal-processing equipment. Most transducer manufacturers supply an assortment of preamplifiers as optional equipment. In many cases, the preamplifier is directly integrated into the accelerometer packaging.

10.13 PROCESSING AND ANALYSIS EQUIPMENT

The processing and analysis equipment for vibration is essentially the same as the equipment for sound measurement that was discussed in Chapter 5. The same equipment can be used since we are using the vibration transducer signal to measure the absolute level of the vibration. Similarly, a detailed spectrum analysis of the signal can be used for diagnostic purposes. Dedicated vibration measurement instrumentation is commercially available. Figure 10.20 presents two types of vibration meters. Tape recording of vibration signals is also often used, and detailed laboratory analysis of the signals using a spectrum analyzer can be performed. Calibration of vibration measurement systems can be achieved by using portable shakers which oscillate at a reference amplitude, typically 1 g rms. In this way, the vibration

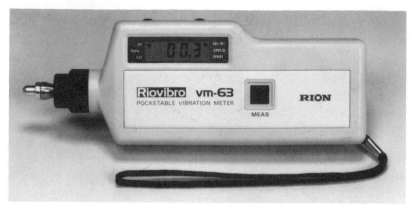

(a)

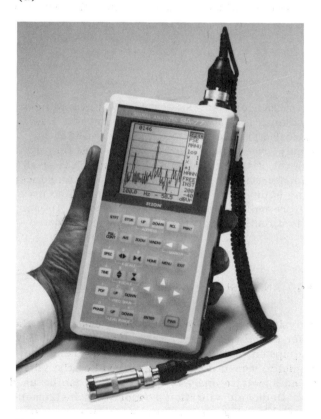

(b)

Figure 10.22 Field vibration measurement instrumentation. (a) Utilizes integrators to measure overall acceleration, velocity, or displacement. (b) Hand-held FFT spectrum analyzer with an accelerometer for signal input. (Courtesy of Scantek Inc., Silver Spring, Md.)

sensor is attached to the shaker, and the signal level is measured and adjusted to meet the level associated with the vibration calibrator.

REFERENCES

1. C. E. Crede. *Vibration and Shock Isolation*. Wiley, New York, 1951.
2. J. P. Den Hartog. *Mechanical Vibrations*. McGraw-Hill, New York, 1956.
3. C. M. Harris and C. E. Crede. *Shock and Vibration Handbook*. McGraw-Hill, New York, 1961.
4. *Application Selection Guide*. Barry Division, Barry Wright Corporation,
5. J. T. Broch. *Mechanical Vibrations and Shock Measurements*. Brüel and Kjaer,

BIBLIOGRAPHY

Faulkner, L. *Handbook of Industrial Noise Control*. Industrial Press, New York, 1975.

EXERCISES

10.1 A mass of 2 kg is supported on a spring with a stiffness constant of 100 N/m. What is the undamped natural frequency of the system?

Answer: 1.12 Hz

10.2 A mass weighing 10 lb is supported on a spring with a spring constant of 30 lb/in. What is the undamped natural frequency of the system?

Answer: 5.4 Hz

10.3 A mass of 10 kg is supported on a spring with a spring constant of 1000 N/m. If the viscous damping coefficient is 5 N-s/m, what is the
a. Critical damping ratio?
b. Damped natural frequency of the system?

Answer: (a) $\xi = 0.025$;
            (b) 1.59 Hz

10.4 The measured logarithmic decrement for a damped spring mass system is 0.5. What is the damping ratio of the system?

Answer: $\xi = 0.08$

10.5 Show that the transmissibility of an undamped system is a special case of the damped system. (Hint: Let the damping ratio approach 0.)

10.6 The damping ratio of a spring mass system is 0.90. What is the
a. Logarithmic decrement?
b. Ratio of successive maximum values?

Answer: (a)   12.97
(b) 4.3×10^5

10.7 The natural frequency of a mass spring system is 20 Hz. What is the transmissibility if an external force is applies at
a. 50 Hz?
b. 20 Hz?
c. 10 Hz?
The damping ratio is 0.10.

Answer: (a) 0.21;
(b) 5.1;
(c) 1.33

10.8 A spring isolator is deflected 1.0 in. under a static load from a large fan. What is the undamped natural frequency of the system?

Answer: 3.13 Hz

10.9 Referring to Exercise 10.8, if the excitation frequency due to rotational unbalance is 10 Hz, what is the transmissiblity of the system?

Answer: 0.11 approximately

10.10 A machine rotates at 1200 rpm. Calculate the undamped natural frequency of a spring system that would provide a transmissibility of 0.10 or less.

Answer: 6 Hz

10.11 Referring to Exercise 10.10, calculate the required deflection of the spring system to limit the transmissibility to 0.10.

Answer: 0.27 in.

10.12 A large pump weighing 2000 lb and rotating at 1100 rpm is to be isolated from a platform with four spring mounts. What deflection is required to assure a transmissibility of 0.05 or 5%. (Hint: Use Table 10.1.)

Answer: 0.61 in.

10.13 Referring to Exercise 10.12, select the proper mounts from the catalog excerpt in Table 10.2.

Answer: SLF-410

10.14 A hydraulic pump is to be mounted on a large machine tool. The unit is driven directly with an electric motor at 3450 rpm. The pump and motor combined weigh 200 lb. If four isolation pads such as shown in Fig. 10.11 are selected,
a. What would be the static deflection of the mounts?
b. What would be the percent of isolation?

Answer: (a) .06 in;
(b) 95%

10.15 A pneumatic isolator is to be used to isolate vibration-sensitive optical equipment. If the natural frequency of the mounted equipment is 1.5 Hz, what is the percent of isolation at
a. 5 Hz?
b. 10 Hz?
c. 25 Hz?

Answer: (a) 90%
(b) 97.5%;
(c) greater than 99.5%

10.16 Referring to Exercise 10.15, if the volume of the air column is 100 in.^{3}, what must the volume be to lower the natural frequency to 0.75 Hz?

Answer: 400 in^3.

10.17 An accelerometer with sensitivity of 72.9 mV/g is mounted on a support bracket of a jet engine running at takeoff power. If the output voltage is 500 mV, what is the vibratory acceleration of the bracket?

Answer: 6.86g

10.18 Referring to Exercise 10.17,
a. What is the metric acceleration?
b. If the frequency of vibration is 100 Hz, what is the displacement?

Answer: (a) 67.3 m/s^2;
(b) 0.00017 m

III
BASIC SOURCES OF NOISE: CHARACTER AND TREATMENT

It is not possible to list or discuss all the sources of noise. As we shall see, however, there are relatively few actual basic noise-producing mechanisms. For example, a power press is generally considered a noisy machine. The press noise, however, originates from numerous basic sources such as high-velocity air, gears, metal-to-metal impact, etc. Plastic molding equipment is another example; here the noise usually originates from air, hydraulic pumps, cooling fans, etc. In short, the noise from industrial machinery can usually be broken down into a relatively few noise sources or sound-producing mechanisms. Recognizing this, the acoustical engineer can apply a systematic approach to controlling the noise.

In Part III of this text, we shall identify and characterize the major sources or sound mechanisms. Further, methods for predicting the acoustical power or sound levels of these sources will be given, and, finally, methods of control or treatment including alternatives will be presented.

11
FANS AND BLOWERS

Throughout industry, the need to move large quantities of air or convey products is ever present. To meet the air- and product-handling requirements, a variety of fans and blowers is most often employed. It is rare not to find one or more fans or blowers in each department of an industrial or manufacturing complex. In many ways fans and blower noise is one of the easiest and most straightforward acoustical problems to solve. We shall now develop a systematic approach to controlling fan and blower noise. Fans will be considered first, followed by blowers. In this text, blowers will refer to the high-pressure rotary positive displacement type which can better be described as compressors.

The fans we shall focus attention on are those used to move large volumes of air for ventilation, dust or oil mist collection, drying operations, etc., which are relatively low-speed low-static-pressure units. The majority of fans can be classified as either axial or centrifugal, and we shall deal with each on an individual basis.

11.1 CENTRIFUGAL FANS

There are two basic types of centrifugal fans, backward or forward curved and radial, as illustrated in Fig. 11.1. There are numerous types designed for a wide variety of applications; however, they usually can be considered variations and/or combinations of these basic types. Now the noise from centrifugal fans is dominantly a superposition of discrete tones at the impeller or blade passing frequency and broadband aerodynamic noise. The frequency of the discrete tones is given by

$$f_n = \frac{nBN}{60} \quad [Hz] \qquad (11.1)$$

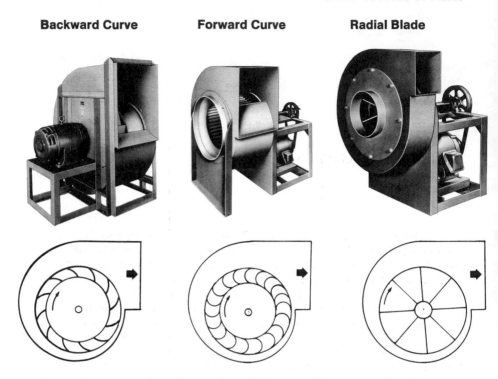

Backward Curve   **Forward Curve**   **Radial Blade**

Figure 11.1 Examples of centrifugal fans. (Courtesy of Peerless Electric Fans and Blowers, Electrical Division, H. K. Porter Company, Inc., Jersey City, N.J.)

where

> B = number of blades
>
> N = fan rotational speed (rpm)
>
> n = harmonic; i.e., n = 1 (fundamental), n = 2 (second harmonic), etc.

The origin of the discrete tones is from two sources. First, each time a blade passes a point in space, a pressure fluctuation is created due to the displacement of air and/or aerodynamic *lift* if the blade is of an airfoil configuration. Second, as the blades pass the cutoff point in the *scroll* (see Fig. 11.2), abrupt pressure changes or pulses also occur at the blade passing frequency and higher integer-ordered harmonics.

The amplitude of the discrete noise is difficult to predict; however, some design guidelines are available:

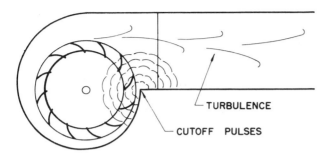

Figure 11.2 Noise generation at the scroll cutoff point.

1. The cutoff clearance is critical, i.e., the clearance between the
   blade and scroll at the cutoff point. A clearance of 5% to 10% of
   the wheel diameter is considered optimum by most manufacturers.
2. Backward-inclined blades are generally quieter than forward-
   inclined blades.

The broadband aerodynamic noise originates from vortices created at
the leading and/or trailing edge of the blades and turbulence im-
parted to the fluid, usually in the form of eddylike flow. Here again
the accurate prediction of noise levels for these fans is at best very
difficult, but an empirical approximation which provides good first-
order results for the average sound power level in the range of 500
to 4000 Hz is [1]

$$\bar{L}_W = 10 \log Q + 20 \log p_t + K \qquad [dB] \qquad\qquad (11.2)$$

where

Q = volume flow rate (cfm)

p_t = static pressure (in. of H_2O)

K = constant depending on fan type, 35 for forward- or backward-
curved blades and 43 for radial types

Consider an example.

Example

A centrifugal fan with a static pressure of 2.0 in. of water produces a volume flow of 6200 cfm. Estimate the average sound
power level for the fan.

Solution

From Eq. (11.2), selecting $K = 35$, we have

$$\bar{L}_W = 10 \log 6200 + 20 \log 2 + 35$$

$$= 37.9 + 6 + 35$$

$$= 78.9 \qquad [dB]$$

For a radial fan, taking K = 43, the average power level would be

$$\overline{L}_W = 37.9 + 6 + 43$$
$$= 86.9 \quad [dB]$$

The usefulness of Eq. (11.2) can be further extended by utilizing the correction factors given in Table 11.1. Here octave band power levels can be estimated by applying these factors to the average power level as given by Eq. (11.2). An example will illustrate the procedure.

Example

Estimate the octave band power levels for a forward curved centrifugal fan whose average power level $\overline{L}_W$ is 80 dB. The fan has 30 blades and turns at 600 rpm.

Solution

First, the average power level of 80 dB is displayed in the first line of Table 11.2 as shown. Second, the appropriate octave band corrections are listed in the second line as shown. Here the corrections selected are for the forward-curved blades. The corrections are then added algebraically to the first line, yielding line 3, the octave band power levels.

Many engineers also include a correction for the discrete noise at the fundamental blade passing frequency. Here the fundamental frequency is calculated from Eq. (11.1), and the correction given in the last column of Table 11.1, BPFI (blade passing frequency increment), is added to the appropriate octave band. In the previous example, from Eq. (11.1), the fundamental frequency is

$$\frac{30 \times 600}{60} = 300 \text{ Hz}$$

and 300 Hz falls into the 250-Hz band. Hence, a correction of +3 dB is added to the power level in the 250-Hz band, as shown in line 4 of Table 11.2.

Table 11.1 Spectral Sound Power Corrections for Estimating Fan Noise

	Octave Band Center Frequency (Hz)								
	63	125	250	500	1000	2000	4000	8000	BPFI[a]
Forward curve	15	10	5	0	−5	−8	−10	−13	3 dB
Backward curve	4	4	3	1	0	−5	−13	−20	2 dB
Radial blade	10	7	5	5	0	5	−8	−9	7 dB

[a]BPFI: blade passing frequency increment.

Table 11.2 Computation Table

	Octave Band Center Frequency (Hz)								
	63	125	250	500	1000	2000	4000	8000
1. Average power level L_W (dB)	80	80	80	80	80	80	80	80
2. Correction for forward-curved fan (dB)	+15	+10	+5	0	−5	−8	−10	−13
3. Line 1 plus line 2 = octave band power level	95	90	85	80	75	72	70	67
4. Blade passing frequency increment			+3					
5. Octave band power level including discrete blade passing noise	95	90	88	80	75	72	70	67

From this discussion, it must be emphasized that the method must be considered crude and, at best, a first-order approximation. The sound power of centrifugal fans varies dynamically with operating efficiency or performance, wheel speed, scroll design, etc. Fortunately, rather reliable spectral sound power levels are provided today by most of the manufacturers with elaborate correction tables for performance, speeds, etc. As such the estimation method just outlined should be used only. in the absence of commercially tested or measured data.

11.2 NOISE CONTROL OF CENTRIFUGAL FANS

Effective noise control of centrifugal fans is not complicated by a variety of measures or approaches. Centrifugal fans are basically low-pressure high-flow-volume devices. Therefore, the fundamental approach is the utilization of absorptive, parallel, or circular baffle-type silencers. As we saw in an earlier chapter, the features of this type of silencer are good high-frequency attenuation and minimal aerodynamic pressure loss. Often the problem of pressure loss is more difficult to resolve than the noise reduction because of the extreme sensitivity of centrifugal fans to upstream or downstream flow restriction.

Illustrated in Fig. 11.3 is a simple approach which is most often taken in industrial installations. Here a tubular silencer is installed

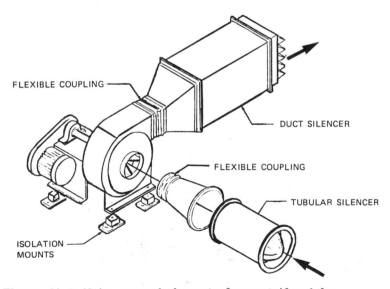

FLEXIBLE COUPLING

DUCT SILENCER

FLEXIBLE COUPLING

TUBULAR SILENCER

ISOLATION MOUNTS

Figure 11.3 Noise control elements for centrifugal fan.

on the inlet of the fan through an adapter section, and a parallel
baffle duct-type silencer is installed on the exhaust. A flexible
coupling of dense material adapts the silencers to the fan and provides vibration isolation between the fan and the ductwork. Naturally, vibration mounts are provided to isolate the fan from the floor
or support platform. We discussed the vibration isolation in detail in
Chap. 10.

   In air-conditioning or ventilating installations, the fan is often
installed in a *mechanical* room, as was illustrated in Fig. 8.6. Here
the intake silencer is a modular assembly of parallel baffles which
extends from the floor to the ceiling of the room. The fresh air,
after passing through the silencers, enters the heat exchanger and
into the inlet of the fan. The exhaust treatment is again typically
a parallel baffle duct silencer. Note again that the flexible couplings
and vibration isolators should be used to mechanically decouple the
rotating machinery from the duct system and floor.

11.3 AXIAL FANS

Axial fans take their name from the fact that the airflow is along the
axis of the fan. Shown in Fig. 11.4 is an example of an axial fan.
To avoid a circular flow pattern and to increase performance, guide
vanes are usually installed downstream of the rotor. Axial fans with
exit guide vanes are called vane axial and those without, tube axial.

Figure 11.4 Example of belt-driven vane axial fan. (Courtesy of
Peerless Electric Fans and Blowers, Electrical Division, H. K. Porter
Company, Inc., Jersey City, N.J.)

Axial fans generally operate at higher pressures than centrifugal fans and usually are considerably noisier. Common applications include heating, ventilating, and air-conditioning systems and forced or induced draft fans for steam boilers.

Because of the number of blades (8 to 30 are typical) and relatively high rotational speeds, the noise from axial fans is generally characterized by strong discrete blade passing tones. Equation (11.2) can be used to estimate the average sound power level for the four octave bands, 500, 1000, 2000, and 4000 Hz, but a constant K = 48 must be used [1]:

$$\overline{L}_W = 10 \log Q + 20 \log p_t + 48 \qquad [dB] \qquad\qquad (11.3)$$

where

Q = volume flow rate (cfm)

p_t = static pressure (in. of H_2O)

The procedure for estimating the average power level follows identically the method as given for centrifugal fans and hence will not be repeated.

Further, to obtain a spectral power level for design purposes, the corrections for the octave bands are given in Table 11.3. Again, the method of calculation of octave band power levels proceeds identically as outlined for centrifugal fans.

It must be emphasized again that these estimates of sound power should be considered, at best, first order, and the use of commercially tested or manufacturers' data is strongly preferred.

Another question that frequently arises is "What is the change in sound power level if a fan speed is changed?" First-order estimates can be obtained in conjunction with Eq. (11.2) and Eq. (11.3) from the following fan laws:

1. Volume flow (Q) varies linearly as fan speed.
2. Static pressure (P_t) varies as the square of fan speed.

Table 11.3 Spectral Sound Power Corrections for Estimating Fan Noise

	Octave Band Center Frequency (Hz)								
	63	125	250	500	1000	2000	4000	8000	BPFI[a]
Vane axial	−6	−4	0	0	0	−2	−5	−10	6
Tube axial	−7	−6	0	−1	−3	−4	−10	−12	7

[a]BPFI: blade passing frequency increments.

11.3.1 Blade-Vane Interaction Noise

In addition to the aerodynamic and discrete blade noise, another noise mechanism due to the blade-vane wake interaction may also be present in axial fans. Before discussing the interaction noise, it is essential to understand a few fundamentals of the aerodynamics and noise generation mechanism of the *rotor-only* noise. To visualize these phenomena, consider first a single-stage axial-type fan which has four blades and is rotating in a homogeneous uniform fluid medium such as air.

The aerodynamic forces or pressures associated with each blade can be resolved into two components:

1. A drag force due primarily to the thickness of the blades
2. A lift force due to the aerodynamic (airfoil) shape of the blades

Now, as the fan turns, a rotating pressure pattern is formed with pressure *lobes* at each blade. If a microphone is placed near the rotor, say a few inches upstream, it is easy to see that periodic pressure fluctuations would be felt each time a blade passed. Further, as previously emphasized, from the periodic nature of the disturbance, the rotating lobed pressure pattern can be visualized as a superposition of harmonically related sinusoidal patterns rotating at shaft speed. The rotor-only pressure pattern model is now established. With a narrow-band analyzer, the spectral Fourier components can be separated into a fundamental tone and integer-order higher harmonics.

If one were to be positioned on a blade of this fan, it would be easy to see that no dynamic pressure fluctuations would be felt during rotation. However, if the airflow to the fan were distorted, say due to the presence of an upstream guide vane or bearing support, then every time a blade passed through the wake, a pressure fluctuation would be felt by the riding observer. This fluctuation is often erroneously thought of as a *slap* as the blade impacts the wake. Actually, the wake presents velocity gradients which affect the aerodynamic performance of the blade. That is, there is a fluctuation of the lift force. To be sure, the magnitude of the fluctuation depends on the size, proximity, and aerodynamic influence of the obstacle, but it has the common property of recurring every time the blade completes a revolution. From this conceptual model, one might expect that the resultant interaction noise would be discrete tones at the vane or wake passing frequency. Such is *not* so. It can, however, be shown [2] that the resultant noise is at the usual blade passing frequency and integer harmonics thereof. In short, no new tones are produced. As such, it is difficult to separate and determine the magnitude of the rotor-only or interaction noise.

It is beyond the scope of this text to analyze further this interaction phenomenon, and the interested reader is encouraged to refer to Ref. 2.

Qualitatively, however, it is important to note that the resultant interaction pressure patterns are also *lobe* type in character, with rotation related to shaft speed in such a way that discrete tones are generated only at the blade passing frequency and integer harmonics. The number of lobes m_k of the interaction patterns can be calculated as follows [2]:

$$m_k = nB \pm kV \tag{11.4}$$

where

B = number of blades

V = number of vanes

n = harmonic number = 1, 2, 3, . . .

k = ±1, ±2, ±3, . . .

The corresponding phase rotational speed N_m associated with the m lobed interaction pattern is

$$N_m = \frac{nBN}{m_k} \qquad (n = 1, 2, 3) \qquad [unitless] \tag{11.5}$$

Example

How many lobes m would the interaction noise (pressure pattern) have in a fan having six blades and five vanes? Further, what is the rotational speed of the spinning patterns?
Solution
From Eq. (11.4), for n = 1 (fundamental), k = ±1, we have

$$m_{-1} = B - V = 6 - 5 = 1 \; (k = 1)$$

$$m_1 = B + V = 6 + 5 = 11 \qquad (k = -1)$$

Thus the interaction patterns would have 1 and 11 lobes for k = +1 and −1, respectively. Additional patterns for k = ±2, ±3, etc., exist, and the number of lobes determined similarly. Further, for the higher harmonics (n = 2, 3, . . .) the character of the spinning modes can also be obtained.
Now for the rotational speed N_m of the spinning patterns, we have from Eq. (11.5),

$$N_1 = \frac{6N}{1} \qquad for \; n = 1, \; m_{-1} = 1, \; and \; B = 6$$

Thus the 1-lobed pattern is rotating (phasewise) 6 times rotor speed. For the 11-lobed pattern m_1 we have

$$N_{11} = \frac{6N}{11} \qquad for \; n = 1, \; m_1 = 11, \; and \; B = 6$$

Thus the 11-lobed pattern is rotating or 6/11 times rotor speed.

From this example, the concept that the interaction noise is at the same frequencies as the rotor is now plausible.

The most important conclusion of this discussion can now be stated. Certain combinations of blades and vanes produce rotating patterns which rotate at speeds many times the original shaft rotational speed. Now it can be shown [1] that as the interaction pattern phase velocity increases, the radiated sound power typically increases. In fact, as the circumferential linear speed of a pattern approaches Mach 1, a rapid increase in noise level is frequently observed. At this critical point, the interaction pattern is said to be at *cutoff*, and generally maximum acoustical power is radiated. This phenomenon is well documented for axial flow compressors and propellers, and recent investigations have shown similar effects on vane axial fans.

11.3.2 Kármán Vortex Noise

Another noise mechanism which is a dominant source of noise in axial machines is Kármán vortex noise. The mechanism behind this source of noise has its origin in one of the earliest recorded references to sound. As the wind blew through the pillars of the ancient Greek temples, eerie discrete tones or whistles were produced. The Greeks felt that these whistles were the voice of the god of the wind, Aeolus, and hence the tones were called Aeolian tones. Every sailor knows these tones well as the wind blows through the rigging.

The origin of the periodic sound can be visualized by considering the flow of air over a cylinder. At a given velocity, the flow pattern behind the cylinder (downstream) exhibits an oscillatory motion as vortices are *shed* alternately on one side and then the other. The trail of *eddies* form what is called a *Kármán vortex street* which possesses strong periodic components. As a result, sound of a nearly pure tone quality is produced. From empirical data, the frequency f_a of the Aeolian tone can be calculated from

$$f_a = \frac{\alpha V}{d} \quad [Hz] \tag{11.6}$$

where

$\quad$ V = velocity of air (m/s)

$\quad$ d = diameter of cylinder (m)

$\quad$ α = Strouhal number, approximately 0.2

The connection of this phenomenon to the axial fans and propellers can now be illustrated. Consider a small segment of blade near the hub of an axial fan. Clearly, as the segment rotates through the air, a vortex tone would be produced whose frequency could be calculated from Eq. (11.6). An adjacent segment farther out radially would also

produce a pure tone but at a slightly higher frequency due to the
higher linear velocity. Continuing this reasoning over the entire
blade to the tip, and for each blade of the fan, it is easy to see that
the resultant noise would be broadband in character. In actual
practice, vortex noise from fan blades is broadband, but the bulk of
acoustical energy is concentrated in a narrow frequency range. This
is especially true for relatively low-speed fans.

11.4 NOISE CONTROL OF AXIAL FANS

Since axial fans are mounted in ducts with cylindrical geometry, one
basic approach to noise control is generally followed. A tubular absorptive silencer is installed on the inlet and exhaust. The silencer
selection methods follow those guidelines presented in the chapter
dealing with silencers and will not be repeated except to say that the
fan should be decoupled from the ductwork and the floor, ceiling, or
platform, etc. In that axial fans generally operate at higher rotational
speeds than centrifugal fans, vibration isolation is inherently easier.
     Silencer selection guidelines with respect to aerodynamic con-
straints (pressure loss) can be summarized simply as follows:

1. Calculate the volume flow or face velocity, and compare it to the
    silencer manufacturer's pressure loss specifications.
2. A good rule of thumb is to size the silencer such that the open
    flow-through cross section is 1.25 to 1.5 times the fan duct
    cross-sectional area.

11.5 BLOWERS

Manufacturers make no distinction between fans and blowers. However,
in industry the term blower is usually reserved for high-pressure devices most often used for conveying materials or products. There are
two basic types:

1. Centrifugal fans with a radial wheel construction operated at high
    speeds
2. Rotary positive displacement blowers

     We shall not discuss further the high-speed centrifugal fan types,
since the same noise control measures can be applied as presented in
detail in an earlier section of this chapter. However, the second
type is distinctly different, and we shall deal with these machines
in detail.
     Basically rotary displacement blowers consist of two counter-
rotational lobelike gear impellers, as illustrated in Fig. 11.5. As each

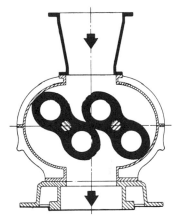

Figure 11.5 Section view of rotary positive displacement blower showing lobed impellers on dual shafts.

impeller lobe passes the blower inlet, it traps a definite volume of air or gas and carries it around the case to the blower outlet. This cycle repeats itself 4 times with every complete revolution of the driving shaft. Hence, the character of the noise is dominantly periodic with discrete tones at the compression frequency and integer-ordered Fourier harmonics thereof.

For the two-impeller configuration of Fig. 11.5 the frequencies of the discrete tones f_n can be calculated from

$$f_n = \frac{n \times 4N}{60} \qquad (n = 1, 2, 3, \ldots) \text{ [Hz]} \qquad (11.7)$$

where

N = shaft rotational speed (rpm)

The pressure pulses from these blowers are quite severe, and sound level measurements near the inlet and exhaust ports often exceed 140 dB. Hence, the character of the noise is dominantly periodic with discrete tones at the compression frequency and integer-ordered Fourier harmonics thereof.

11.6 NOISE CONTROL OF BLOWERS

Because of the discrete and dominantly low-frequency character of blower noise, reactive chamber-type silencers are especially effective. In most installations, a silencer will be required on both the inlet and

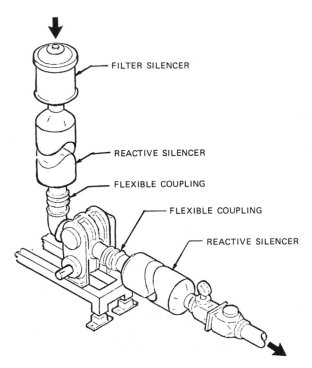

Figure 11.6 Typical rotary positive displacement blower showing recommended silencer installations.

exhaust, as illustrated in Fig. 11.6. In addition, an inlet filter may also be a requirement. Fortunately, many filter manufacturers offer a line of combination filter-silencers. As such, including a filter-silencer in series with the inlet silencer provides additional noise reduction.

Well-designed reactive silencers properly sized to the blower will have pressure losses on the order of a few inches of H_2O. These losses are usually trivial since most blowers provide static pressures on the order of 10 to 20 psi. As shown in Fig. 11.6, it is extremely important to provide vibration isolation between the blower and pipe ducts and also between the blower-motor unit and the floor or support platform. Isolation between the blower and the connecting pipe ducts is best achieved by including a high-pressure reinforced flexible coupling in the duct 3 to 5 diameters from the blower. A note of caution here: Frequently the flexible coupling provides a major acoustical leak since the transmission loss through the coupling is lower than the heavy walled pipe. However, this problem can be overcome by simply wrapping the coupling area with a composite of 1-in. polyurethane foam and $1\text{-lb}/\text{ft}^2$ dense vinyl.

Finally, even with all the noise reduction measures included as outlined, noise levels in close proximity will likely be in the range of 85 to 95 dBA. The resultant levels are due to general mechanical noise associated with gears and bearings and noise transmitted through the blower case itself. Therefore, for critical noise-sensitive installations, serious consideration should be given to enclosing the blowers. Since they generally require little monitoring or maintenance, a room-type enclosure is best, especially when several blowers (often the case) must be similarly enclosed.

In final summary, it should be emphasized that control of fan and blower noise is relatively straightforward. With the design guidelines presented here, even the most stringent design criteria can be met or exceeded with readily available commercial materials and equipment. It should also be noted that the leading manufacturers of fans and blowers are members of or subscribe to the policies of the Air Moving and Conditioning Association (AMCA) or the American Society of Heating, Refrigerating, and Air Conditioning Engineers (ASHRAE). In both organizations, noise measurement techniques for fans and blowers are strictly controlled by standard (see App. C), and, as such, reliable sound power data are usually available.

REFERENCES

1. J. B. Graham. How to estimate fan noise. *Sound Vib.* (May 1972), 24−27.
2. J. T. Tyler and T. G. Sofrin. *Axial Flow Compressor Noise Studies.* SAE, 1961.
3. *Industrial Ventilation*, 17th Edition. ACGIH, 1982.

BIBLIOGRAPHY

ASHRAE Handbook, HVAC Systems and Applications, Chap. 52, "Sound and Vibration Control." American Society of Heating, Refrigerating and Air-Conditioning Engineers, Inc., Atlanta, 1987.

Beranek, L. L. *Noise and Vibration Control,* revised edition, Chap. 12, pp. 391−393. Institute of Noise Control Engineering, Poughkeepsie, New York, 1988

Diehl, G. M. Centrifugal compressor noise reduction. In *Reduction*

of Machinery Noise, M. J. Crocker (Ed.). Purdue University
   Press, Lafayette, Indiana, 1974.
Faulkner, L., Handbook of Industrial Noise Control. Industrial
   Press, New York, 1975.
Harris, C. M. Handbook of Acoustical Measurements and Noise Con-
   trol. McGraw-Hill, New York, 1991.
Hoover, R. M., and C. O. Wood. Noise control for induced draft
   fans. Sound Vib. (April 1970).
Morse, P. M., and U. K. Ingard. Theoretical Acoustics. McGraw-
   Hill, New York, 1968.
Ostergaard, P. B. Industrial noise sources and control. In Pro-
   ceedings of the Inter-Noise 72 Conference: Tutorial Papers on
   Noise Control. Washington, D.C., October 1972).

EXERCISES

11.1 An axial fan has six blades and rotates at 3450 rpm. Calculate
the frequencies associated with the
a. Fundamental harmonic
b. Second harmonic
c. Third harmonic

> *Answer:* (a) 345 Hz;
>   (b) 690 Hz;
>   (c) 1035 Hz

11.2 Estimate the average sound power level for a radial fan which
produces 6000 cfm at a static pressure of 2.5 in. of H_2O.

> *Answer:* $L_W = 88.7$ dB

11.3 Estimate the average sound power level for an axial fan which
delivers 7000 cfm of air with a static pressure of 3.0 in. of H_2O.

> *Answer:* $L_W = 96.0$ dB

11.4 Calculate the frequency of an Aeolian tone produced by a ski
rack atop an automobile traveling at 60 mi/hr. Diameter of rack is
0.5 in.

> *Answer:* 422 Hz approximately

11.5 Calculate the frequency of the fundamental and second harmonics for a two-impeller rotary positive displacement blower at
2400 rpm.

> *Answer:* $f_1 = 160$ Hz
> $f_2 = 320$ Hz

12
GAS-JET NOISE

One of the most troublesome and perhaps the most common noise source
in industrial environments is the gas jet. This basic noise mechanism
is often referred to as aerodynamic noise, and some examples include
blowoff nozzles, steam valves, pneumatic control discharge vents,
gas or oil burners, etc. To emphasize the word troublesome, the
sound pressure level 3 ft from a typical 1/4 in. diameter *shop* air
blowoff nozzle is often in the range of 105 to 107 dBA. As has been
the practice throughout this text, we shall first consider briefly the
basic physical parameters which determine the level and character
of the noise mechanisms and then proceed to the measures for noise
control.

12.1 GAS JETS

The simplest example of a gas jet is the high-velocity airflow emanating
from a reservoir through a nozzle, as illustrated in Fig. 12.1. The
gas accelerates from near zero velocity in the reservoir to the peak
velocity in the core at the exit of the nozzle. The peak velocity of a
gas jet is a strong function of pressure difference between the reservoir pressure p_r and the external ambient pressure p_a. In short, as
the pressure ratio p_r/p_a between reservoir and ambient discharge is increased, the velocity of the gas at the discharge nozzle increases.
However, when a pressure ratio of approximately 1.9 is reached, the
flow velocity through the nozzle becomes sonic, i.e., reaches the
speed of sound, and further increase in reservoir pressure does not
significantly increase the flow velocity. When this critical pressure
ratio of 1.9 is reached, the nozzle is said to be *choked*.

The actual generation of the noise from gas jets results from the
creation of fluctuating pressures due to turbulence and shearing
stresses as the high-velocity gas interacts with the ambient gas.

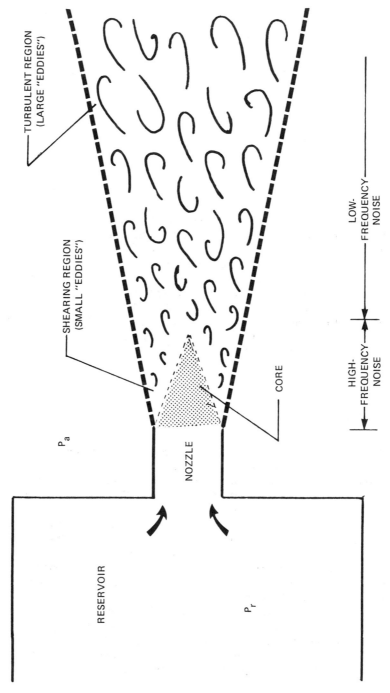

Figure 12.1 Gas jet showing mixing regions and noise sources.

Complex radiating sources, called *eddies*, are then formed, with the high-frequency noise being generated near the nozzle in the mixing shearing region and the lower-frequency noise being generated downstream in the region of large-scale turbulence. With only turbulent flow present, the pressure fluctuations are random functions of space and time, and only statistical methods can be used to describe the character of the noise. Therefore, the spectral character of gas-jet noise is generally broadband.

Shown in Fig. 12.2 is an octave band spectral analysis of an air jet from a 1/4 in. diameter copper tube (nozzle) *crimped* to blow parts from the die area of a power press. Since the reservoir or shop air pressure is usually in the range of 45 to 90 psi, the gas nozzle can be considered choked and the peak velocity in the core of the jet near Mach 1. Subjectively, the character of the noise is a broadband

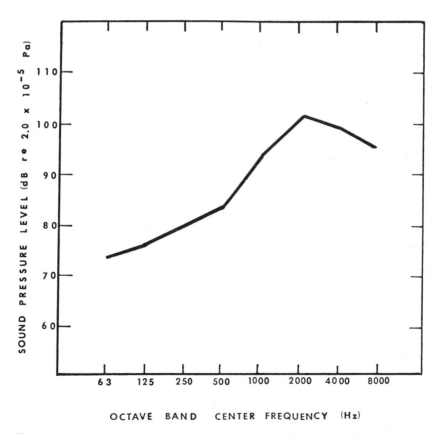

Figure 12.2 Octave band spectral analysis of a part ejection nozzle at approximately 1 m.

hiss with peak levels near 101 dB in the octave band whose center is 2000 Hz.

The magnitude and spectral character of the noise from jets cannot be accurately predicted. This is due to the complex nature of the jet itself and uncertainties associated with such factors as turbulence, nozzle configuration, temperature, etc. However, from the monumental work of Lighthill [1] and others and reams of empirical data generated dominantly by the aerospace industry, first-order estimates of the acoustical power spectral character can be obtained. The analytical background is well beyond the scope of this text; hence it will suffice to say that the overall sound power W from a subsonic or sonic jet takes the following form:

$$W = \frac{K \rho A v^8}{c^5} \quad [W] \tag{12.1}$$

where

A = area of the jet nozzle (m^2)

ρ = density of ambient air (kg/m^3)

v = jet flow velocity (m/s)

c = speed of sound (m/s)

K = constant of proportionality (unitless)

The velocity factor v in Eq. (12.1) is the fluctuating velocity and varies throughout the jet stream. As such the factor v is normally neither easy to measure nor easy to calculate. Therefore, by considering the average velocity V and assuming that the energy-carrying eddies are of the same size as the jet diameter, one can show [2] that the total radiated acoustical power W is proportional to the kinetic energy of the jet stream. Under these conditions, the total radiated power is then just a fraction of the total power discharged from the nozzle. In simpler engineering terms, the radiated power W is

$$W = \frac{e M^5 \rho V^3 A}{2} \quad [w] \tag{12.2}$$

where

V = average flow velocity through the nozzle (m/s)

M = Mach number of flow (V/c) (unitless)

ρ = density of ambient air (kg/m^3)

A = nozzle area (m^2)

e = constant of proportionality of the order of 10^{-4} [2]

Again it must be emphasized that this expression is a first-order approximation, and it is applicable to many industrial situations where the average velocity V of the jet is in the range of $0.15c < V \leq c$, where c is the speed of sound in the jet exhaust medium. Now the factor eM^5 can be considered an efficiency factor, and, from empirical data, it is plotted in Fig. 12.3 over a range of applicable average flow velocities V or, better, Mach numbers. It should be noted that the efficiency factor is given as a range which reflects the uncertainty associated with turbulence, temperature, etc. An example will add clarification.

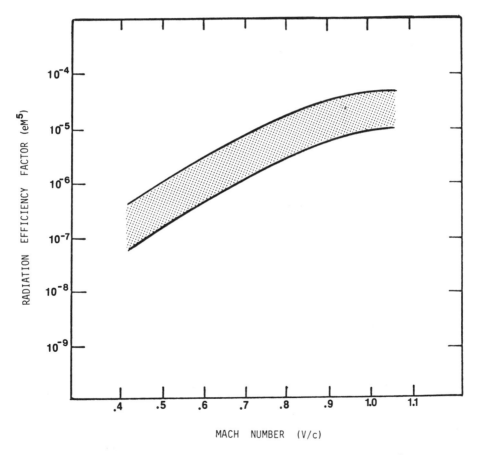

Figure 12.3 Gas-jet radiation efficiency factor for a range of Mach numbers.

Example

Calculate the overall acoustical power W of a choked 1/4 in. diameter
air jet exhausting to the atmosphere. Calculate also the sound
power level L_W.

Solution

Since the nozzle is choked, we can take the average velocity
of the nozzle to be the speed of sound, or V = 344 m/s. Further,
since we are exhausting to the atmosphere, the density of air is
ρ = 1.18 kg/m^3. Now, in metric units, the area of the nozzle
is

$$A = \frac{\pi D^2}{4}$$

$$= 3.16 \times 10^{-5} \text{ m}^2$$

From Fig. 12.3, the radiation efficiency factor (eM^5) for Mach 1
is

$$eM^5 = 3 \times 10^{-5} \qquad \text{(approximately, using the center of the range)}$$

Upon substitution in Eq. (12.2), we obtain

$$W = \frac{3 \times 10^{-5} \times 1.18 \times 344^3 \times 3.16 \times 10^{-5}}{2}$$

$$= 0.023 \text{ W}$$

In terms of the sound power level L_W, we have

$$L_W = 10 \log \left(\frac{W}{10^{-12}} \right)$$

$$= 10 \log \left(\frac{0.023}{10^{-12}} \right)$$

$$= 103.6 \text{ dB}$$

Therefore, the sound power and sound power level of the jet is
0.023 W and 103.6 dB, respectively.

From this example, we see that given the nozzle dimensions and
exit velocity, the acoustical power of the jet can be calculated to the
first order. In terms of the overall acoustical power level, the range
of accuracy is, at best, probably ±5 dB. When the gas jet is hot and
highly turbulent, such as from a gas burner, another first-order correction can be applied:

$$\text{Temperature correction} = 20 \log \left(\frac{T}{T_a} \right) \qquad [\text{dB}] \qquad (12.3)$$

where

> T = absolute temperature of the gas jet
>
> T_a = absolute temperature of ambient air

Consider an example.

Example

What would be the total radiated acoustical power level L_W for the choked 1/4 in. diameter air jet of the last example if the tempera-ture of the jet were raised from 60°F to 500°F?

Solution

Since the power level L_W for 60°F was calculated to be 103.6 dB re 10^{-12}, the correction for a temperature is given by Eq. (12.3):

$$\text{Temperature correction} = 20 \log \left(\frac{T}{T_a} \right)$$

In this example,

> T = 460° + 500° = 960°R
>
> T_a = 460° + 60° = 520°R

Therefore,

$$\text{Temperature correction} = 20 \log \frac{960}{520}$$

$$= 5.3 \text{ dB} \qquad \text{(approximately)}$$

The temperature-corrected power level is then

> L_W = 103.6 + 5.3
>
>   = 108.9   [dB]

Given the overall power level of a jet, one can proceed to calculate the overall sound pressure level at any point in space using the methods outlined in the chapter on sound propagation. In most cases of interest, the jet can usually be considered a point source with typ-ical directionality as shown in Fig. 12.4.

From Fig. 12.4 one notes that peak levels typically occur in the angular range of 15° to 45° from the axis of the jet. The relative sound pressure level as given in the figure can also be considered the directivity index DI_θ term.

Consider an example to illustrate the use of Fig. 12.4.

Example

Calculate the radiated overall sound pressure level at a radial distance of 10 m from the nozzle of the last example at angular positions of 0°, 45°, 90°, and 180°.

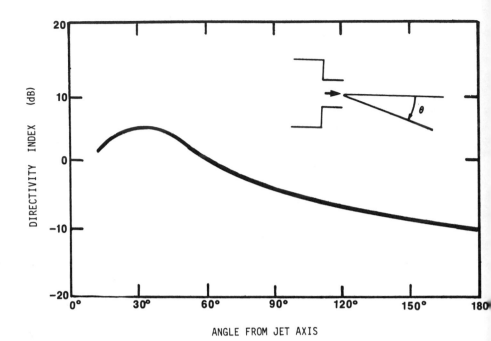

Figure 12.4 Typical directivity pattern for a small subsonic gas jet.

Solution

From a previous example, we calculated the overall sound power level of the jet to be approximately 103.6 dB. From Eq. (4.7), the sound pressure level L_p at 10 m is

$$L_p = L_W - 20 \log_{10}(r) + DI_\theta - 11$$

$$= 103.6 - 20 \log_{10}(10) - 11 + DI_\theta$$

$$= 72.6 + DI_\theta \qquad [dB]$$

Now applying the directivity correction from Fig. 12.4, at 0° we see that the DI_θ is 0 dB; hence, the sound pressure level at 0° $L_p(0°)$ is

$$L_p(0°) = 72.6 + 0$$

$$= 72.6 \qquad [dB]$$

At 45°, the $DI_{45°}$ is (from Fig. 12.4) +3 dB approximately, and we get

$$L_p(45°) = 72.6 + 3$$

$$= 75.6 \quad [dB]$$

At 90°, the $DI_{90°}$ is -5 dB approximately; hence,

$$L_p(90°) = 72.6 - 5$$

$$= 67.6 \quad [dB]$$

At 180°, the $DI_{180°}$ is -10 dB approximately; hence,

$$L_p(180°) = 72.6 - 10$$

$$= 62.6 \quad [dB]$$

The strong directional character from the jets can clearly be seen from this example and must always be accounted for in practice.

Gas jets also exhibit a strong spectral dependence, and first-order approximations of the peak frequency f_{max} can be obtained given the overall power level from the following empirical expression:

$$f_{max} = \frac{S_0 V}{D} \quad [Hz] \tag{12.4}$$

where

S_0 = Strouhal number (a constant) = 0.15 approximately for a wide range of nozzle diameters and conditions

V = average nozzle exit velocity (m/s)

D = nozzle diameter (m)

Consider an example to illustrate Eq. (12.4).

Example
What is the frequency of peak noise levels from an air blowoff nozzle (1/4 in. diameter copper tubing) whose average exit velocity is 1100 ft/s?
Solution
From Eq. (12.4),

$$f_{max} = \frac{0.15 \times 1100 \text{ ft/s} \times 12 \text{ in.}}{1/4 \text{ in.}} = 7920 \text{ Hz}$$

Hence, the peak noise level can be expected in the octave band whose center frequency is 8000 Hz.

For design analysis, the octave band distribution of acoustical power can also be calculated to the first order. The calculation method follows from experimental data which suggest that the spectrum *falls off* at an average rate of -6.0 dB/octave above the peak frequency and -7 dB below the peak frequency. Shown in Fig. 12.5 is the graphic representation of the spectral shape of jet noise in terms of dimension-

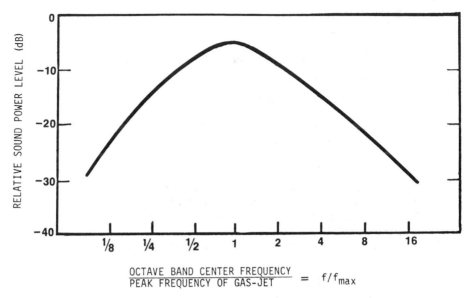

$$\frac{\text{OCTAVE BAND CENTER FREQUENCY}}{\text{PEAK FREQUENCY OF GAS-JET}} = f/f_{max}$$

Figure 12.5 Generalized power spectrum for a small gas jet.

less octave band geometric ratios. Given the peak frequency f_{max} of the jet noise, the octave band spectral power levels can be obtained from the distribution as shown in the figure. An example will again illustrate the method.

Example

The overall power level of a 1 in. diameter jet with an exit velocity of 1100 ft/s is 120 dB. What is the octave band spectral distribution of the noise?

Solution

The peak frequency is

$$f_{max} = \frac{0.15 \times 1100 \text{ ft/s} \times 12 \text{ in.}}{1 \text{ in.}}$$

$$= 1980 \text{ Hz}$$

Since 1980 Hz falls near the center of the conventional 2000 Hz band, f will be selected to be 2000 Hz. Then, from Fig. 12.5, the correction for the 2000-Hz band corresponding to $f/f_{max} = 1$ is approximately -4 dB. The level for 2000-Hz band is then

$$L_W(2000) = 120 - 4 = 116 \text{ dB}$$

Continuing on, the correction for the 4000-Hz band where $f/f_{max} = 2$ is -8 dB. The level for the 4000-Hz band is then

$L_W(4000) = 120 - 8 = 112$ dB

For the 8000-Hz band, $f/f_{max} = 4$, and the correction is -15 dB; hence,

$L_W(8000) = 120 - 15 = 105$ dB

The same method is applied for the lower frequencies; i.e., for the 1000-Hz band, $f/f_{max} = 1/2$, and the correction is -8 dB. The level for the 1000-Hz band is then

$L_p(1000) = 120 - 8 = 112$ dB

For the 500-Hz band, $f/f_{max} = 1/4$, the correction is -16 dB, and

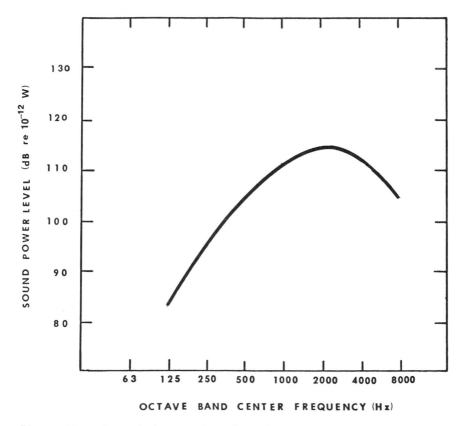

Figure 12.6 Graphic illustration of estimated octave band sound power levels of a 1 in. diameter gas jet.

$$L_p(500) = 120 - 16 = 104 \text{ dB}$$

For the 250-Hz band, $f/f_{max} = 1/8$, the correction is -23 dB, and

$$L_p(250) = 120 - 23 = 97 \text{ dB}$$

For the 125-Hz band, $f/f_{max} = 1/16$, the correction is -34 dB, and

$$L_p(125) = 120 - 34 = 86 \text{ dB}$$

These results can be seen more clearly by referring to Fig. 12.6, where the results of the calculations are displayed graphically. Here we have the estimated octave band spectral power levels of the gas jet from the previous example.

In summary, the magnitude and spectral character of gas jet noise can be estimated only crudely or at best to the first order. Because of many uncertainties, it is strongly recommended that considerable design margin be included when using these estimates. Let us now discuss the measures of gas-jet noise control.

12.2 NOISE CONTROL MEASURES FOR GAS JETS

For large high-temperature, high-velocity, and highly turbulent gas jets such as found on jet engines or gas turbines, sophisticated engineering noise control measures are required. These machines are extremely sensitive to upstream and downstream pressure losses and velocity gradients. Hence, a careful system-type design approach is necessary to meet the rigorous aerodynamic constraints. In short, the aerodynamic constraints for these test facilities or installations are often the *pacing items* and somewhat beyond the scope of this text. However, the basic approach to controlling noise is generally the use of absorptive silencers, as discussed and illustrated previously in the chapter dealing with silencers. At this point, we shall terminate the discussion of these large gas-jet facilities with the admonition that the detail design and fabrication of these silencer packages be left to commercial suppliers. Design and/or specification, however, follows the basic guidelines given for absorptive silencers.

For the simple high-velocity air jet, common in industrial environments, such as those used to eject parts, for cooling purposes, as air curtains, for pneumatic control vents, to power air tools, etc., rather straightforward noise reduction measures can be applied.

The basic guidelines to noise control, in somewhat rank order of effectiveness, are as follows:

1. Reduce the required air velocity by moving the nozzle closer to the part or scrap being ejected, thus preserving thrust.
2. Add additional nozzles, thus reducing the required velocity and again preserving thrust.
3. Install quieter diffuser and air shroud nozzles.
4. Interrupt the airflow in sequence with ejection or blowoff timing.

   With respect to measures 1 and 2, the noise reduction follows from reducing the jet velocity. As shown earlier in this chapter, the acoustical power of the radiated noise varies as the sixth to eighth power of the velocity. As such, reducing the airstream velocity should be the first consideration in any noise reduction program.
   Often the only constraint is the preservation of air-jet thrust. Now the thrust T_j of a jet is given as [3,4]

$$T_j = \frac{w'V}{g} \quad [lb] \tag{12.5}$$

where

   w' = weight flow rate (lb/s)

   g = acceleration due to gravity (ft/s^2)

   V = average jet velocity (ft/s)

Example
   What is the thrust of a choked air nozzle with a weight flow of 0.5 lb/s?
   Solution
   Since the nozzle is choked, the average velocity can be considered the speed of sound or 1100 ft/s. Upon substitution in Eq. (12.5), we have

$$T = \frac{0.5 \times 1100}{32}$$

$$= 17.2 \ lb$$

Now, from Eq. (12.5), one sees that thrust can be preserved if the weight flow rate is increased while lowering the jet velocity. Increasing the nozzle exit area and moving the nozzle closer to the ejection site will also usually allow considerable velocity reduction. Similarly, the addition of two or more nozzles will also allow air-jet velocity reduction and corresponding noise level reduction. Experimental results have shown that halving the distance between parts and nozzles or adding another nozzle will allow air velocity to be lowered 30% (preserving thrust), with an overall noise level reduction of 8 to 10 dB being typical [4] (H. R. Mull, Bell and Associates, Norwalk, Conn., unpublished data, 1978).

12.2.1 Multiple-Jet and Restrictive Flow Silencer Nozzles

Shown in Fig. 12.7 is a commercially available multiple-jet diffuser nozzle. The noise reduction occurs mainly from the reduction in jet core size and to some extent the reduction of large-scale turbulence in the mixing regions. Further, the smaller *inner* jets flow along with the outer layer of high-velocity air, reducing the shearing action to the nonmoving ambient air.

In the restrictive flow diffuser nozzles, as shown in Fig. 12.7b, the high-velocity core is also minimized by the sintered metal restrictor, typically inserted into the nozzle exit. Also, because of the restrictor, the flow velocity is usually reduced somewhat, with corresponding reduction in radiated noise. In both of these nozzle types, the reduction in noise is usually at the expense of jet thrust, and some augmentation in the form of additional nozzles may be required to restore the original untreated jet thrust.

12.2.2 Air Shroud Silencer Nozzles

Shown in Fig. 12.8 is a commercially available air shroud silencer nozzle with illustrated airflow pattern. Here the noise reduction is obtained from bypassing some of the primary air around and over the nozzle. The bypassed air lowers the velocity gradients between the

(a)

(b)

Figure 12.7 Examples of (a) multiple-jet diffuser nozzle and (b) restrictive diffuser nozzle. [Courtesy of (a) Allied Witan Company, and (b) Vlier Engineering, Burbank, Calif.]

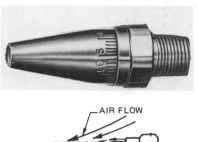

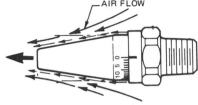

Figure 12.8 Example and illustration of flow pattern for an air
shroud silencer nozzle. (Courtesy of Vortec Corporation, Cincinnati,
Ohio.)

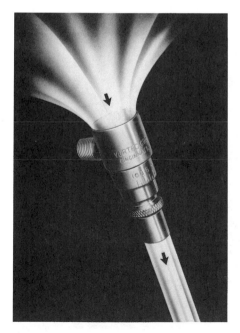

Figure 12.9 Transvector silencer nozzle. (Courtesy of Vortec Corporation, Cincinnati, Ohio.)

(a)

(b)

434

primary jet and the ambient air, thus reducing the shearing action
and lowering radiated noise. Since there is little change in weight
flow rate for these silencers, jet thrust is generally preserved. In
the model shown, the amount of bypassed air is controlled by adjusting the micrometer dial.

In summary, the noise reduction of air shroud silencer nozzles
is typically 10 to 20 dB in the critical high-frequency range of 2000
to 8000 Hz for small high-velocity air jets.

Another recently developed nozzle device is the transvector jet.
See Fig. 12.9 for an example of a commercially available unit. Here
ambient air is drawn into the nozzle throat, where it mixes with the
high-velocity gas stream. The resultant gas stream is substantially
quieter. These nozzles are designed especially for blowoff applications.

Shown in Fig. 12.10 are typical applications of these silencer nozzles. Here again, noise level reductions of 10–20 dBA in the 2000–8000
Hz frequency range can be anticipated.

12.3 IMPINGEMENT NOISE

Another source of noise which is associated with gas jets is impingement noise. Impingement noise gets its name from the sharp increase
in noise level one notes as a gas jet is brought close to and impinges
upon a solid surface or object. Shown in Fig. 12.11, for comparison,
are the octave band levels from a 1/8 in. diameter air jet in free air
and then impinging upon a hard metal surface. Note the noise level
increase, especially in the higher-frequency range of the spectrum.

Studies of this phenomenon have shown that when the gas jet
impinges upon a surface, additional unsteady forces are produced.
These pressure fluctuations take the form of aerodynamic dipoles [2],
which can be described as a pair of point sources of equal magnitude
separated by a small distance and oscillating with an angular phase
difference of π rad or 180°. In short, they are *out* of phase; that is,
when one of the point sources is positive, the other is negative. Now
it is beyond the scope of this text to proceed with further analysis
except to say that these dipole sources are the basic mathematical
models used to describe the noise and radiation of many common noise
sources including propellor noise, loudspeakers, many musical instruments, valves, air duct diffusers, grills, etc. Further, there are
usually strong directivity patterns present when the noise originates
from dipole source.

Figure 12.10 (a) Transvector jet installed on power press as blowoff
nozzle. (b) Air shroud nozzle installed on machine tool for cooling
purposes. (Courtesy of Vortec Corporation, Cincinnati, Ohio.)

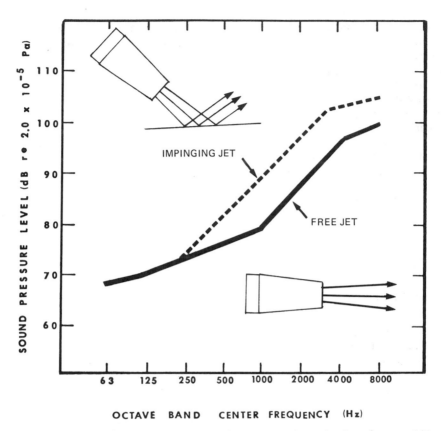

Figure 12.11 Noise levels measured at approximately 1 m from a 1/8 in. diameter jet in free air and impinging upon a hard surface.

Returning to impingement noise, the amplitude and spectral character of this noise mechanism, like jet noise, is very difficult to estimate. However, from both analytical and experimental studies, the radiated sound power for impinging subsonic jets depends to the first order on the fifth or sixth power of the flow velocity. Hence, again as for the free jet, even small reductions in flow velocity yield dynamic reductions in impingement noise.

From experience, as a jet flows over a sharp edge or discontinuity, additional noise is typically produced. In some cases, *whistlelike* edge tones will also result. The periodic components, as well as impingement noise, can be minimized by reducing the flow turbulence created as the jet flows over a cavity or obstruction. Illustrated in Fig. 12.12

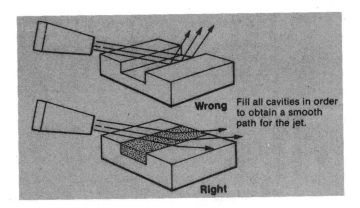

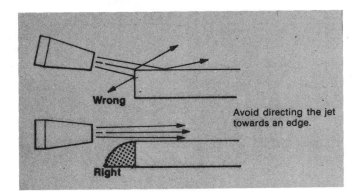

Figure 12.12 Methods for educing jet impingement noise. (Courtesy of Sunnex, Needham, Mass.)

are some basic, almost intuitive methods for reducing impingement noise.

12.4 GASEOUS FLOW NOISE IN PIPES OR DUCTS

Generally, the velocity of gas or steam in pipes or ducts is in the very low subsonic range. However, in some industries where valves or vents are used to regulate high gas pressure or flows, extremely high noise levels are common. Examples of these sources of noise are often found in the chemical, petrochemical, and electric power generation industries. Measured noise levels downstream of reducing valves in large steam lines often reach the 130–140 dB range. In most cases, the pressure drop across a control or regulator valve is sufficient to *choke* the flow at the discharge. As such the flow of the gas jet is sonic or near sonic velocity with the corresponding generation of high-intensity aerodynamic noise. The resultant high-intensity noise is then transmitted through the pipe walls to adjacent surroundings, and even worse, conducted downstream with little or no attenuation.

Prediction of the magnitude of aerodynamic noise in pipes is very difficult because the source mechanism is complex and there are other uncertainties, such as the transmission loss of the pipes or ducts. In addition, extensive service conditions such as pressure, pressure ratios, and temperature are required, demanding tedious calculations

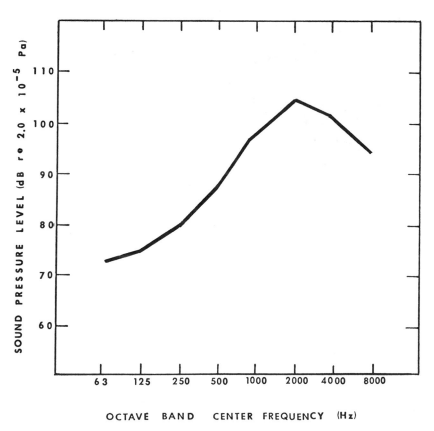

Figure 12.13 Octave band spectral analysis of noise levels from a high-pressure control valve measured 1 m downstream of the valve and 1 m from the pipe (schedule 40).

which are beyond the scope of this text. The reader is referred to excellent references at the end of this chapter, particularly [11, 13]. However, some guidelines from empirical data do exist which can be used for first-order estimates in noise control.

It should be noted that the principal area of noise generation is in the turbulent mixing region immediately downstream of the valve. Now if the pipe downstream is light walled (schedule 40 or less), overall noise levels from a choked steam valve will be in the range of 110 to 120 dBA at 1 m from the pipe. For heavy walled pipe (schedule 80 or heavier), the levels are typically 6 to 10 dB lower.

The spectral character of the noise resembles high-velocity gas jets, i.e., broadband with peak levels in the range of 2000 to 8000 Hz. Shown in Fig. 12.13 is an octave band spectral analysis of the noise from a control valve measured at 1 m downstream of the valve and 1 m from the pipe (schedule 40).

In summary, where high-pressure steam and gas are regulated or discharged through valves, noise levels in excess of 100 dB can be expected. Since the valves have relatively thick metal housings, the piping system itself downstream of the valve is usually the primary source of externally radiated noise. As such the propagation of noise generally follows the characteristic of a line source with cylindrical divergence, i.e., 3 dB per doubling the distance.

12.5 NOISE CONTROL MEASURES FOR NOISE IN PIPES

There are three basic approaches to reducing the noise radiated from these control valve areas:

1. Changing the dynamics of flow
2. Including an in-line silencer to absorb acoustic energy
3. Increasing the transmission loss of the pipe walls

12.5.1 Expansion Plates-Diffusers

Of these three approaches, changing the dynamics of the flow is noise reduction at the source and hence is preferred. Changing the dynamics refers actually to reducing the flow velocity through multiple stages of pressure reduction or diffusion of the primary jet. Shown in Fig. 12.14 is an illustration of multiple stages of pressure reduction utilizing expansion plates. Here the gas flow velocity is sequentially lowered in each expansion chamber. In addition, the plates act as a diffuser, reducing turbulent mixing. With a three-plate configuration such as shown in the figure, noise level reductions up to 20 dB have been reported [5] (H. R. Mull, Bell and Associates, Norwalk, Conn., unpublished data, 1976). However, each chamber

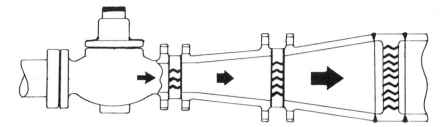

Figure 12.14 Schematic of pressure reduction throttle system showing expansion plates.

introduces *back* pressure, which is often undesirable and can be a limiting factor.

Another but similar approach is the use of diffusers to reduce flow noise. Shown in Fig. 12.15 is a schematic illustrating some commercially available diffusers. Note that the single orifice element in the valve body is replaced by a slotted *cage* which diffuses the flow into smaller interacting jets. Noise reduction in the range of 10 to 18 dB has been reported with this approach alone [6]. In combination with a downstream in-line diffuser, as illustrated, an overall noise level reduction of up to 30 dB can be expected.

In either approach, expansion plates or diffusers, careful attention must be given to proper *sizing* to assure proper valve operation. Guidelines for selections are generally simple and provided by the manufacturers.

12.5.2 In-Line Silencers

The inclusion of an in-line silencer to reduce aerodynamic noise in pipes follows closely the basic guidelines in the chapter on silencers, and the most serious design constraint is pressure loss limitations. Shown in Fig. 12.16 is an illustration of the basic construction of high-pressure, high-velocity in-line silencers. Note that at the inlet a diffuser is included and that thereafter the construction is similar to a flow-through tubular absorptive silencer. The absorbing medium in these silencers behind the perforated sheet metal is either fibrous glass or metal wool. Typically, the flow area of the diffuser is 2 to 3 times the valve discharge area, and in this way the flow velocity is sharply reduced in the silencer. The acoustical performance of these in-line silencers follows closely the performance of tubular silencers, as described in previous chapters. However, because of unusually high flow velocities, care must be taken to account for performance reduction in accordance with the manufacturer's specifications.

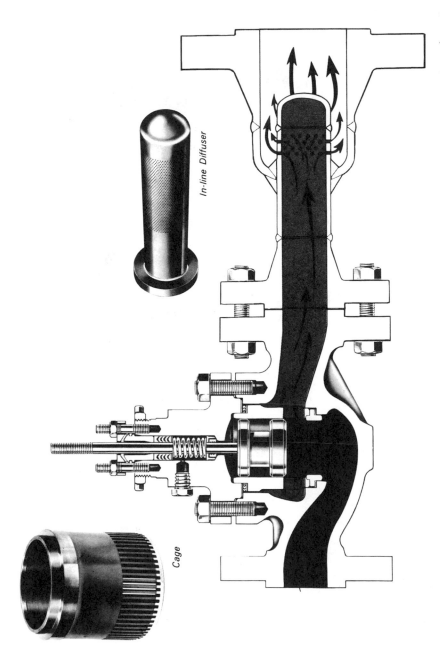

Figure 12.15 Schematic showing installed cage and in-line diffuser for reducing aerodynamic noise at control valves. (Courtesy of Fisher Controls, Marshalltown, Iowa.)

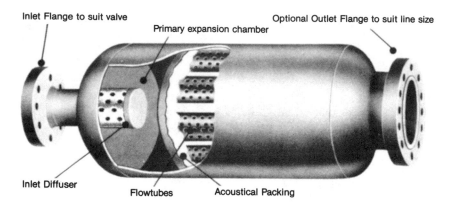

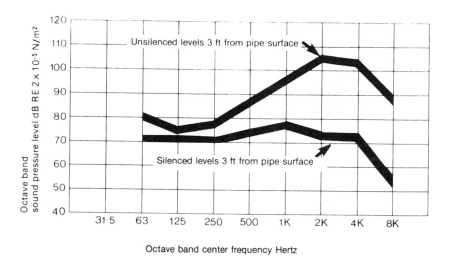

Figure 12.16 In-line silencer along with typical acoustical performance. (Courtesy of Quietflo, Division of Flaregas Corporation, Spring Valley, New York.)

Installation guidelines are simple and can be summarized as follows:

1. The in-line silencer should be installed no closer than 3 or 4 pipe diameters from the valve discharge.
2. Pipe treatment (as we shall discuss in Sec. 12.5.3) between the valve and silencer must be considered, since the silencer will only reduce noise downstream of the silencer itself.

12.5.3 Pipe or Duct Treatment

There are two basic approaches to increasing the transmission loss of pipe walls:

1. The pipe wall thickness can be increased.
2. The pipe can be wrapped or *lagged* with acoustical materials.

With respect to pipe wall thickness, no simple generality exists. for estimating the transmission loss of pipe walls. However, for carbon steel pipe, most commonly specified for high-pressure, high-velocity flow installations, schedule 40 is considered standard up to pipe sizes of 12 in. inside diameter (ID). For larger pipe sizes, standard wall thickness is constant at about 3/8 in. Now, as mentioned, because the thickness and hence the surface density of the pipe wall change with pipe size, the transmission loss of the standard pipe cannot be simply given. However, from empirical data, a table of attenuation increase, such as Table 12.1, can be very useful. From this table the increase in noise reduction due to pipe wall thickness increase (schedule) is given for a representative range of pipe sizes. An example will illustrate the use of the table.

Table 12.1 Increase in Pipe Wall Attenuation for a Representative Range of Pipe Wall Thicknesses at Various Pipe Sizes

| Pipe Size | Pipe Wall Attenuation Increase[a] (dB) at Pipe Schedules | | | | |
	20	40	80	120	160
2	—	0	6	9	12
4	—	0	7	10	13
6	—	0	8	12	15
8	—	0	9	14	18
10	—	0	10	15	18
12	—	0	10	15	20
16	0	2	10	16	20
20	0	6	12	15	20
24	0	6	12	15	20

[a] Average attenuation over the frequency range of 1000 to 8000 Hz.

Example

Noise levels downstream of a stream control valve are 94 dBA
at a machine operator's station nearby. Peak levels are in the frequency range of 2000 to 4000 Hz. The noise is radiating from a
12 in. diameter schedule 40 pipe. What schedule pipe replacement
would bring the noise level down to 85 dBA?

Solution

Schedule 40 pipe is standard for this particular installation.
Therefore, referring to Table 12.1, if 12 in. diameter schedule
40 pipe were replaced with schedule 80 pipe, 10 dB of additional
attenuation would be obtained in the high-frequency range.
Resultant levels would then be 94 -- 10 = 84 dBA, meeting the
criterion. Further, replacement with schedule 120 pipe would
reduce levels 15 dB, bringing the resultant levels to 79 dBA,
providing a small design margin.

Finally, in studies of pipe wall transmission loss, a resonancelike
condition was discovered at what is called the *ring* frequency. At the
ring frequency, the transmission loss nearly vanishes, analogous to
the coincidence frequency effect as discussed in the transmission loss
associated with barrier materials. The resonancelike performance *dip*
occurs when, for a compressional wave, a single wavelength of sound
is equal to the nominal circumference of the pipe wall. Mathematically,
the corresponding ring frequency f_r is given as

$$f_r = \frac{c_L}{D_p \pi} \quad [\text{Hz}] \tag{12.6}$$

where

D_p = nominal diameter of the pipe (m)

c_L = longitudinal speed of sound in the pipe wall (m/s)

In short, at the ring frequency, the transmission loss of pipes
drops sharply, often leaving resultant noise levels at a maximum.

Consider an example.

Example

Aerodynamic noise downstream of a throttling valve is contained
within a steel pipe with a nominal diameter of 10 in. and a standard
wall thickness of 0.365 in. (schedule 40). Calculate the ring
frequency for the pipe.

Solution

The circumference C_L of a 10 in. diameter pipe is

$\pi \times 10$ in. = 31.4 in.

The nominal speed of sound in steel C_L is 16,000 ft/s.
Hence, upon substitution in Eq. (12.6), we have

$$f_r = \frac{16,000 \times 12}{31.4}$$

$$= 6115 \text{ Hz}$$

We might therefore expect that peak noise levels radiated from the pipe would be in the vicinity of the ring frequency at 6115 Hz.

Higher-ordered resonantlike conditions at $2f_r$, $3f_r$, . . . , etc., have also been observed but are generally of little interest in noise control because of the relatively high frequency.

12.5.4 Pipe Wrapping or Lagging

For especially critical noise-sensitive areas, treatment of the noise path may also be required, and for pipes this is accomplished by wrapping or *lagging* the pipes. One very effective wrapping scheme is illustrated in Fig. 12.17. Here a 1- to 3-in. layer of acoustical-quality absorbing material such as fibrous glass, mineral wools, or polyurethane foam is wrapped next to the pipe wall. This absorbing layer is then wrapped with sheet lead, dense vinyl, or sheet metal (1 lb/ft^2 minimum). Shown in Fig. 12.18 is the typical noise reduction achieved with this wrapping scheme. For this example, the absorbing material was 1-in. thick acoustical-quality polyurethane foam, and the other layer was dense vinyl (1.5 lb/ft^2). It should be emphasized here that the outer dense layer plays the key role in achieving a high level of noise reduction.

Additional noise reduction can obviously be obtained by using a more dense outer layer or adding more composite layers. For ease of installation, various manufacturers of thermal pipe insulation have developed and made available commercially preformed lengths of ab-

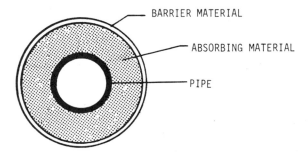

Figure 12.17 Pipe wrapping or lagging showing basic material elements.

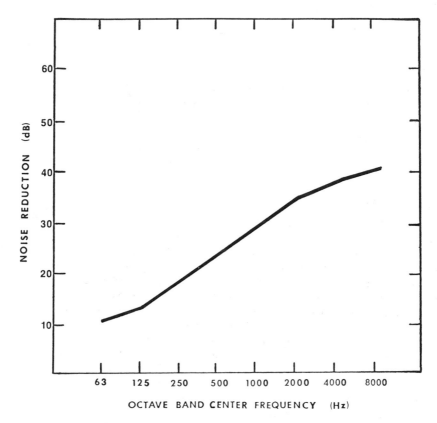

Figure 12.18 Actual noise reduction obtained by wrapping pipes as illustrated in Fig. 12.17. Here the absorbing material was 1-in. polyurethane foam, and the barrier material is dense vinyl (1.5 lb/ft^2).

sorbing materials or lagging sized to conventional or standard pipe diameters. See Fig. 12.19 for example.

Additional guidelines for wrapping include the following:

1. Avoid any mechanical coupling between the pipe wall and the outer layer of treatment.
2. Seal all edges and joints.
3. Use special materials such as fibrous glass or mineral wool for high-temperature applications.
4. Means to avoid accumulation of condensation should be included for cold piping, or *icing* will reduce the performance of the treatment.
5. Duct tape or metal packing bands are best for securing the treatment.

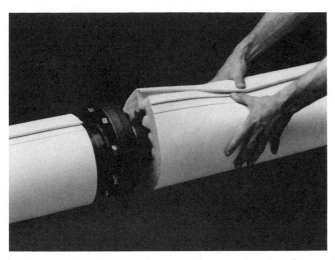

Figure 12.19 Commercially available pipe lagging. (Courtesy of Accessible Products, Co., Tempe, Arizona.)

One disadvantage of wrapping or lagging alone as the only measure for noise reduction is that long lengths of pipe will probably require treatment. Therefore, in summary, it is strongly recommended that source noise reduction be given the first priority, that is, the utilization of multiple pressure reduction stages or diffusers in conjunction with in-line silencers. If additional noise reduction is required, wrapping or lagging the pipes will usually bring levels within the most stringent criteria.

12.6 FURNACE AND BURNER NOISE

The noise from furnaces, burners, and similar combustion equipment has its origin in complex interactions associated with high-velocity flow, turbulent mixing, and combustion. The spectral character and noise levels vary dynamically with furnace or burner configurations and the fueling method. Examples of industrial applications where natural or forced draft furnaces are used include refineries, chemical plants, boilers, smelting furnaces, and heat-treating furnaces. The peak noise levels are generally low frequency in spectral character, usually below 1000 Hz, and subjectively the noise is often referred to as a *roar*. Shown in Fig. 12.20 is an actual octave band spectral analysis of a large natural draft furnace measured 2 m from the burner.

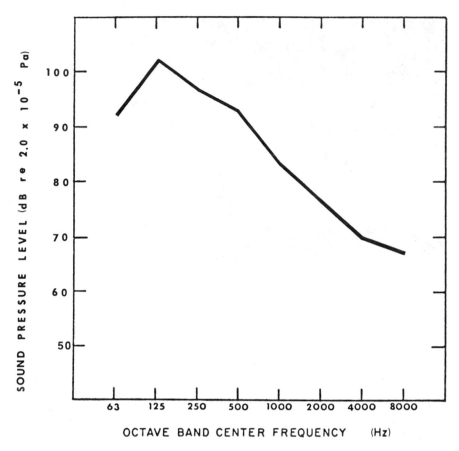

Figure 12.20 Noise levels measured at 2 m from a large natural draft furnace.

Several methods for predicting the magnitude and spectral character of burner noise have been developed, but the phenomenon is not well understood. As such we shall refer the reader to the References for detailed discussion and mention only that the flow velocity in the flame seems to be the most important parameter. Other parameters of lesser importance include mass flow rate, volumetric expansion, ratio of air to fuel mixture, etc. More important, most studies regarding burner combustion noise have yielded little encouragement for significant noise reduction at the source. In short, combustion burner noise reduction is usually followed by a sharp decrease in burner efficiency which is usually unacceptable to the operators. On a positive note, however, higher combustion noise levels are often

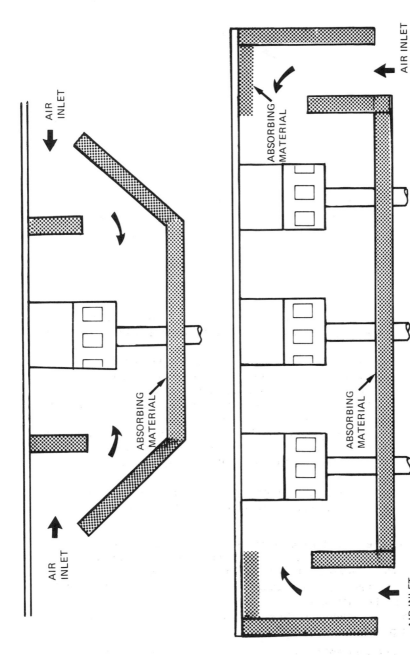

Figure 12.21 Acoustically treated plenums for controlling the noise of large natural draft furnaces or burners.

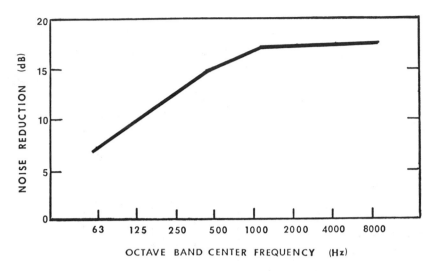

Figure 12.22 Typical noise reduction achieved for acoustically treated plenums installed on natural draft furnaces and burners.

encountered when poor flame control exists. These conditions, easily recognized by the operators, can be remedied by properly adjusting the burners.

Thus with little hope of source reduction, the acoustical engineer must look for some method of containing the noise. J. G. Seebold, with reference to Bittelich [7, 8], has shown that acoustically treated plenums have been both effective and practical. Examples of these plenum configurations are shown in Fig. 12.21. Here the interior walls are constructed of highly absorbing materials, yielding a labyrinth of parallel baffles and lined bends. The plenums are basically sound traps, and typical noise reductions achieved with these plenums are shown in Fig. 12.22.

With respect to design guidelines, walk-in-sized plenums can be constructed of typical acoustical panels with the following modifications:

1. The solid external sheet metal should be 12 gauge minimum.
2. The acoustical absorbing material must be 4 in. thick in order to have effective absorption at lower frequencies.
3. Since oil mist is often present, a 1- or 2-mil layer (maximum) of plastic film enclosing the acoustical absorbing material is recommended to prevent *wicking* or absorption of the oil, thus avoiding a possible fire hazard.

It should be emphasized that the plenums illustrated here are by no means exhaustive. The only limitations to the number of bends, baffles, switchbacks, etc., are those associated with pressure losses. Here again, each bend or switchback represents a constriction to the flow of air to the secondary registers. An adequate volume flow of air with minimal entry distortion is essential for good burner efficiency.

12.7 MIXER (SPUD) NOISE

In premix-type furnaces, another source of noise is usually encountered where the primary air is mixed with the fuel gas. Here

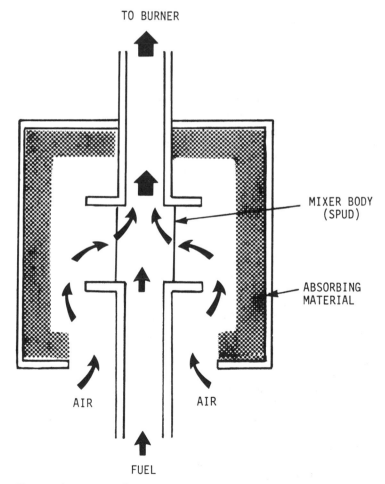

Figure 12.23 Basic design configuration for a premix *spud* silencer.

the air is inspirated into the mixing unit or *spud* by the high-velocity
fuel gas jet. The source of the noise is of course the high-velocity
gas, and since the pressure ratio is usually well above critical, a
choked condition is prevalent. As such the noise levels are extremely
intense, and the spectral character of the noise is broadband with
peak levels in the 2000- to 8000-Hz range, often above 120 dB. Because of the delicate aerodynamic balance required for proper mixing,
significant noise level reduction is rarely available at the source.
However, commercially available silencers can be obtained which
sharply lower spud noise. These silencers, shown conceptually in
Fig. 12.23, are basically absorptive types relying for the most part
on lined bends for the attenuation. Because of size and flow constraints, silencers of this configuration yield noise level reductions
of only 10 to 20 dB in the critical range of 2000 to 8000 Hz. As such,
resultant overall levels may well exceed 100 dBA within a 10-ft radius
of the spud.

Another approach which is popular, especially in new installations, is to enclose the spuds in a separate equipment room. The
spud room is a walk-in type, generally of masonry construction, extending the entire length of the furnace complex. The mixing units
require little monitoring or maintenance and hence lend themselves to
enclosure isolation. One note of caution: A generous air supply is
required to the spud room, and this can usually be achieved through
a duct penetration with absorptive parallel baffle silencer to treat the
noise.

REFERENCES

1. M. J. Lighthill. *Proc. R. Soc. London Ser. A 222* (1954).
2. P. M. Morse and K. U. Ingard. *Theoretical Acoustics.* McGraw-
    Hill, New York, 1968.
3. H. W. Lord, H. A. Evensen, and R. J. Stein. Pneumatic silencers
    for exhaust valves and parts ejectors. *Sound Vib.* (May 1977).
4. A. L. Cudworth and W. J. Hanson. Noise generation in pneumatic
    blow-off guns. *Noise Control* (1975).
5. G. Reethof. Control valve noise generation. *Noise Control Eng.*
    (Sept.-Oct. 1977).
6. H. W. Boger. Designing valves and downstream devices as low
    noise packages. *Heat. Piping Air Cond.* (Oct. 1971).
7. J. G. Seebold. *Hydraulic Processing.* Standard Oil of California,
    March 1972.
8. G. M. Bitterlich. Some findings on burner noise and its sup-
    pression. In *Symposium on Burners and Noise, 35th Mid-Year
    Meeting of the API Division of Refining,* Houston, May 1970.

9. G. Reethof. Turbulence generated noise in pipe flow. *Am. Rev. Fluid Mech. 10* (1978).
10. G. Reethof. Control valve and regulator noise generation, propagation, and reduction. *Noise Control Eng. J. 9* (1977).
11. G. Reethof. Some recent developments in valve noise prediction methods. *Noise-Con 83 Proceedings* (1983).
12. G. Reethof and A. V. Karvelis. Internal wall pressure field studies downstream from orificial-type valves. Paper No. 78-827, ISA/74 Meeting, New York, October 1974.
13. C. B. Schuder. Control valve noise—prediction and abatement. In *Noise and Vibration Control Engineering*, M. J. Crocker, Ed. Purdue University Press, West Lafayette, Ind., 1972, pp. 90-94.

BIBLIOGRAPHY

Arant, J. B. Special control valves. *Chem. Eng.* (March 6, 1972).
Faulkner, L. *Handbook of Industrial Noise Control.* Industrial Press, New York, 1975.
Seebold, J. G. Valve noise and piping system design. In *Instrument Society of America Handbook of Control Valves*, 1971.
Seebold. J. G. Reduce noise in process piping. *Hydrocarbon Processing* (October 1982).

EXERCISES

12.1 Calculate the overall sound power and level of a 1/8 in. diameter shop air jet whose average flow velocity is 900 ft/s.

$$Answer: \quad 0.001 \; W, \; and$$
$$L_W = 90 \; dB$$

12.2 A high-pressure steam relief valve is vented to the atmosphere. If the nozzle $(0.1 \; m^2)$ is choked and the temperature of the steam is 400°F, what is the acoustical power level of the vented steam? .

$$Answer: \quad L_W = 143 \; dB$$

12.3 If the steam temperature of Exercise 12.2 were raised to 600°F, what would be the increase in acoustical power level?

$$Answer: \quad 2 \; dB \; approximately$$

12.4 If the overall acoustical power level associated with the exhuast of a jet engine is 160 dB, calculate the overall sound pressure level at 100 m at an angle of 45° to the centerline of the engine.

$$Answer: \quad 111 \; dB \; approximately$$

12.5 Calculate the peak frequency of a steam jet whose velocity is
300 m/s. The exhaust nozzle diameter is 0.1 m.
Answer: 450 Hz

12.6 If the acoustical power level of the jet in Exercise 12.5 were 140
dB, what would be the octave band power levels at 125, 250, 500,
1000, 2000, and 4000 Hz?

Answer: 124 dB at 125 Hz,
132 dB at 250 Hz,
136 dB at 500 Hz,
132 dB at 1000 Hz,
125 dB at 2000 Hz,
118 dB at 4000 Hz

12.7 At a quality control station, the noise from a high-pressure 8 in.
diameter steam line (schedule 40) is 95 dBA. What schedule pipe replacement would bring the overall levels below 85 dBA.
Answer: schedule 120

12.8 Calculate the ring frequency for the steel pipe in Exercise 2.7.
Answer: 7639 Hz

12.9 A pipe containing high-velocity gas is to be lagged with 1-in.
absorbing material and 1.5-lb/ft^2 dense vinyl. If the peak noise
levels are in the range of 2000 Hz, how much noise reduction can be
anticipated? (Hint: Use Fig. 12.18).
Answer: a minimum of 30 dB

13
GEAR NOISE

In most industrial environments, the problem of gear noise is ever present. For example, machine tools, motor generators, vehicles, packaging- and product-handling equipment, etc., all include one or more gear trains. Despite the numerous and troublesome aspects, little data or design information has been published with respect to the source of gear noise or its control. Why some gear designs are quiet and others noisy has plagued mechanical engineers since their inception. Except for the identification of principal noise sources and some qualitative rank ordering of noise levels associated with gear types, the design or acoustical engineer has few guidelines to follow in control of gear noise. In short, programs of rather extensive testing involving elaborate electronic measurement and analysis systems are usually required in dealing with gear noise reduction. There are, however, some basic design guidelines, and with these guidelines, a systematic approach to gear noise control can be developed.

To establish, then, a systematic approach to gear noise control, we shall first consider some fundamentals regarding

1. The basic types of gears
2. The major noise-producing mechanisms and the spectral character of each

With this background, the available noise reduction measures can then be listed and considered in detail.

13.1 GEAR TYPES

Let us begin the discussion by considering the fundamental parts of a simple spur gear wheel.

The innermost part of the wheel is called the *hub* and accommodates the drive shaft. Extending radially outward is the flange sec-

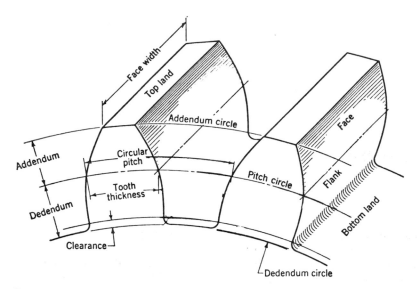

Figure 13.1 Spur-type gear with major terms used in gear design.

tion, called the *web*, and the outermost section is the *rim*, which generally includes the gear teeth. The rim and teeth can be further broken down into component parts, as illustrated in Fig. 13.1.

It is these parts which play the dominant role in gear noise generation, and listed below are the major terms used in gear design [1]:

1. The *pitch circle* is a theoretical circle on which all calculations are usually based. The pitch circles of a pair of mating gears are tangent to each other.
2. The *circular pitch* is the distance, measured on the pitch circle, from a point on one tooth to a corresponding point on an adjacent tooth.
3. The *addendum* is the radial distance between the top land and the pitch circle.
4. The *dedendem* is the radial distance from the bottom land to the pitch circle.
5. The *backlash* is the amount by which the width of a tooth space exceeds the thickness of the engaging tooth on the pitch circles.

The profile shape of gear teeth commonly used for high transmission loads and as manufactured in the United States is generally involute. In Europe, a circular-arc profile is popular.

Consider now the two basic types of gears:

1. Parallel shaft
2. Nonparallel shaft

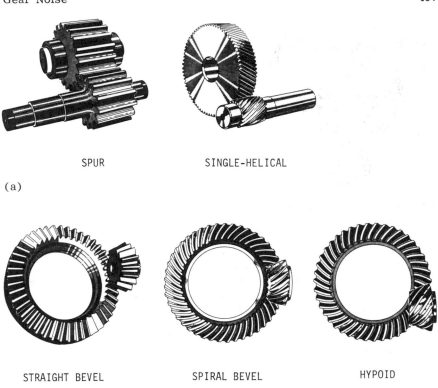

SPUR   SINGLE-HELICAL

(a)

STRAIGHT BEVEL   SPIRAL BEVEL   HYPOID

(b)

Figure 13.2 Examples of (a) parallel shaft and (b) nonparallel shaft gear sets. (From Ref. 2.)

Illustrated in Fig. 13.2 [2] are several varieties of each basic type.

It should be emphasized that only the very fundamental terms have been defined here. However, since the major extent of gear noise research deals only with first-order effects, this terminology will suffice in most instances.

The number of combinations and arrangements that gear design and application engineers commonly use is awesome. As such we shall limit the discussion to basic gear types. However, one complex arrangement, frequently encountered, that deserves mention is the planetary gear system, as illustrated in Fig. 13.3.

Rarely in gear noise problems are single gear trains involved or is there easy visibility or accessibility to the gears themselves. As such the acoustical engineer must use the characteristics of the noise as a diagnostic to identify the offending gears and, just as important, to determine the specific mechanisms or sources.

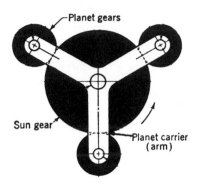

Figure 13.3 Planetary gear set showing major components.

The major mechanisms which are involved in the generation of gear noise are relatively few and rather easy to visualize and understand. Unfortunately, there are often several mechanisms present, and due to the similarity in spectral character, the resultant noise is complex and difficult to resolve electronically. In short, it is often difficult to separate and to determine the magnitude of each contributor.

As a first step in a systematic approach, let us consider the major noise mechanisms and their spectral characteristics.

13.2 MAJOR SOURCES OF GEAR NOISE

By far the dominant source of gear noise is the *meshing* or clashing of gear teeth. The complexity of gear meshing can be simplified by considering four distinct major noise component mechanisms:

1. *Engagement impulse.* On engagement, the teeth of the driven and driving or pinion gear are deflected tangentially. On disengagement, the teeth deflect back toward their original position. The corresponding noise due to the tooth deflection is periodic in character at the gear meshing frequency and integer higher harmonics.
2. *Pitch-circle impulse.* Due to load deflections during the engagement cycle, perfect rolling contact is rarely achieved. As such, fluctuating forces due to the sliding action induce radial teeth deflections. The resulting noise is also periodic at gear meshing frequency and higher harmonics.
3. *Air and/or oil jets.* During an engagement cycle, air or oil occupying space between adjacent teeth is often forced out. For high-speed gears, the air or oil jets can have escape velocities

which are nearly sonic (near the speed of sound). Local shock
waves begin to form at escape velocities of Mach 0.8 [2] with corresponding highly efficient acoustic radiation. The formation of
the jets is again periodic at the gear meshing frequency.

4. *Gear system radiation.* With periodic excitation due to the three
preceding mechanisms, the gear wheel, that is, the web and hub,
may also radiate acoustical energy. The magnitude of radiated
energy depends, to the first order, on the relationship of the excitation frequencies to the gear wheel natural frequencies. When
the excitation frequency and/or harmonics thereof correspond to
the gear wheel natural frequencies, a resonance condition is
established, and the radiated energy can be the major source of
gear noise. Furthermore, these resonance conditions must be extended to include the shafts and adjacent gear sets.

The four mechanisms listed here are not the only sources associated
with the gear mesh noise. However, these sources are first-order effects, always present to some extent, and must be considered in any
analysis.

13.3 CHARACTER OF GEAR NOISE

As mentioned before, the most powerful weapon in analyzing gear
noise is to determine the spectral character. The first step in determining the spectral character is to calculate the fundamental gear
mesh frequency. Since each gear meshes once per revolution, the
fundamental gear meshing frequency f_1 is the product of shaft rotation and the number of gear teeth:

$$f_1 = \frac{kN}{60} \quad [Hz] \tag{13.1}$$

where

   k = number of gear teeth (unitless)

   N = shaft rotation speed (rpm)

Generally, because of the periodic nature of gear meshing, strong
integer-order harmonics of the fundamental are also present. Therefore, the general expression for the fundamental gear meshing frequency and integer higher harmonics is as follows:

$$f_n = \frac{nkN}{60} \quad [Hz] \tag{13.2}$$

where n = 1, 2, 3, . . . , corresponding to the harmonics.
   An example will illustrate this concept.

Example

   Calculate the fundamental gear meshing frequency and first two
   harmonics for the set of gears 3 and 4 illustrated in Fig. 13.4.
   Solution
   From Fig. 13.4, the set of gears 1 and 2 is a step-up pair, i.e.,
   2:1, and the rotational speed N of gear 3 is

   N = 2 × 3600 rpm = 7200 rpm

Since gear 3 has 20 teeth, i.e., k = 20, upon substitution into
Eq. (13.2), we have

$$f_1 = \frac{20 \times 7200}{60} = 2400 \text{ Hz } (n = 1) \qquad \text{(fundamental harmonic)}$$

$$f_2 = \frac{2 \times 20 \times 7200}{60} = 4800 \text{ Hz} \qquad (n = 2) \qquad \text{(second harmonic)}$$

$$f_3 = \frac{3 \times 20 \times 7200}{60} = 7200 \text{ Hz } (n = 3) \qquad \text{(third harmonic)}$$

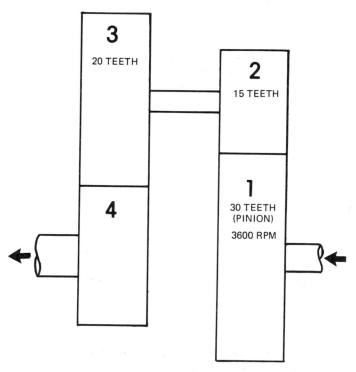

Figure 13.4 Simple spur-type gear train.

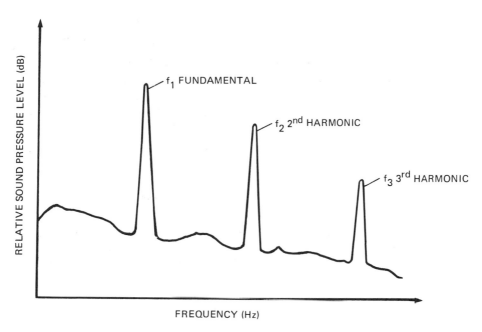

Figure 13.5 Typical narrow-band spectral analysis of gear noise.

Figure 13.5 illustrates this typical discrete tone spectral character of gear meshing noise. In the absence of other factors, the
amplitude of these discrete tones generally decreases as n increases,
or, more simply, higher harmonics are usually lower in amplitude.

To calculate the gear meshing frequencies of a planetary gear
system, Eq. (13.2) has to be modified to account for the relative motion between the sun gear, the planet gears, and the ring gear, as illustrated in Fig. 13.3. One should also recognize that the gear mesh
frequency associated with the sun and planet gears must be equal to
the gear mesh frequency associated with the planet and ring gears.
Thus:

$$f_n' = nk_s(N_s - N_c) \qquad [Hz] \qquad\qquad (13.3)$$

or

$$f_n' = nk_r(N_c - N_r) \qquad [Hz] \qquad\qquad (13.4)$$

where

k_s = number of teeth on the sung gear (unitless)

k_r = number of teeth on the ring gear (unitless)

N_s = rotational speed of the sun gear (rpm/60)

N_r = rotational speed of the ring gear (rpm/60)

N_c = rotational speed of the planet carrier system (rpm/60)

 n = 1, 2, 3, . . . corresponding to the harmonics

Note: rotation in the same direction as the sun gear is considered a positive rotational speed, and rotation counter to the direction of the sun gear is considered a negative rotational speed.

Using Eqs. (13.3 and 13.4), a very useful general relationship for various shaft rotational speeds can be derived:

$$k_s N_s + k_r N_r = N_c (k_s + k_r) \qquad (13.5)$$

An example will help to illustrate the use of these equations:

Example
Calculate the fundamental gear meshing frequency and the first two harmonics of a planetary gear set where the sun gear has 50 teeth and is being driven at a rotational speed of 1800 rpm (30 Hz) and the ring gear has 100 teeth and is fixed (not rotating).
Solution
Here k_s = 50, N_s = 30, k_r = 100, and N_r = 0, and from Eq. (13.5) we have

$$N_c = \frac{k_s N_s}{k_s + k_r}$$

$$= 10 \qquad [Hz]$$

Thus using Eq. (13.3),

f_1' = 50(30 − 10) = 1000 Hz   (fundamental harmonic)

f_1' = 2 × 50(30 − 10) = 2000 Hz (second harmonic)

f_1' = 3 × 50(30 − 10) = 3000 Hz (third harmonic)

With these methods for determining the specific frequencies of gear noise, let us now consider the specific properties and character of these discrete tones which serve as symptomatic diagnostics.

13.3.1 Gear Tooth Irregularities

Probably the easiest diagnostic to recognize is the presence of one or more irregular gear teeth. If one or more teeth are broken or worn excessively, corresponding high noise levels can be expected at the shaft rotational frequency N and/or integer-ordered harmonics thereof. Shown in Fig. 13.6 is a narrow-band spectral analysis of gear

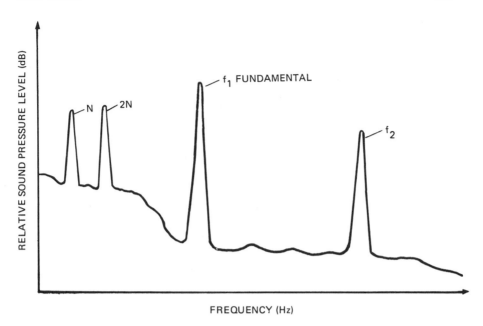

Figure 13.6 Narrow-band spectral analysis of gear noise illustrating the presence of discrete tones at shaft rotation frequencies N and 2N.

noise illustrating the presence of strong discrete tones at shaft rotation frequency N and the second harmonic 2N. Note that the gear meshing tones f_1 and f_2 are also present. In general, then, the presence of discrete tones at shaft rotation N or integer harmonics thereof (2N, 3N, . . .) is symptomatic of uneven tooth meshing loads.

13.3.2 Gear Wheel Resonance

A high harmonic with unusually large amplitudes almost always suggests a gear wheel resonance condition. Since the gear wheel has many natural frequencies (not necessarily harmonic), it is common for harmonics of gear meshing to coincide with these natural frequencies of the gear wheel with resulting high radiated noise levels. Hence, the presence of higher harmonics with unusually large amplitudes is the diagnostic *key* to recognizing a gear wheel resonance. Shown in Fig. 13.7 is a narrow-band spectral analysis of gear noise illustrating gear wheel resonance at the fourth harmonic f_4 of gear meshing frequency.

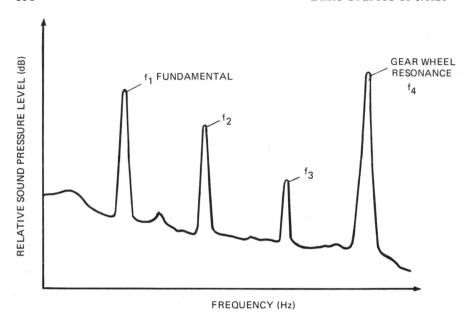

Figure 13.7 Narrow-band analysis of gear noise illustrating the presence of a gear wheel resonance.

13.3.3 Eccentricity of Base Circle: Misalignment

Frequently, in addition to the discrete tones of gear meshing, extremely narrow-band analysis of gear noise will show the presence of *side band* tones. These side band tones are generally found on either side of the gear meshing tones at plus or minus the shaft speed rotation, as illustrated in Fig. 13.8. For the upper side band, the frequency f_u is

$$f_u = f_{gm} + f_0 \qquad [Hz] \tag{13.6}$$

where

$$f_u = \text{upper side band frequency (Hz)}$$
$$f_{gm} = \text{gear meshing frequency (Hz)}$$
$$f_0 = \text{shaft rotation (r/s)}$$

and for the lower side band, the frequency f_l is

$$f_l = f_{gm} - f_0 \qquad [Hz] \tag{13.7}$$

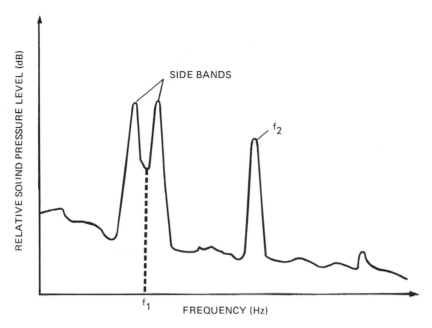

Figure 13.8 Narrow-band spectral analysis of gear noise illustrating discrete side band tones clustered around the fundamental gear meshing frequency.

The generation of side bands is due primarily to periodic fluctuating loads associated at integer-order harmonics of shaft rotation. Consider, for example, that one gear of a set is eccentric, say "egg" shaped. With this condition, it is easy to see that the fluctuating loads on the teeth will vary periodically with shaft rotation N. As such the amplitude of the gear meshing noise will be modulated periodically at a frequency corresponding to integer-ordered harmonics of shaft rotation.

To see this, consider the sound pressure associated with the fundamental discrete gear mesh tone, which can be represented as follows:

$$p(t) = C \sin \omega t \qquad [Pa] \tag{13.8a}$$

$$= C \sin(2\pi Nkt) \qquad [Pa] \tag{13.8b}$$

where

N = shaft rotation (r/s)

k = number of teeth

C = amplitude of sound pressure

If either gear is eccentric, the amplitude of the gear meshing noise C would not be constant but periodic, or without loss of generality, simple harmonic (sinusoidal), i.e.,

$$C = D \sin 2\pi Nt \qquad (13.9a)$$

and upon substitution into Eq. (13.7b), we have

$$p(t) = (D \sin 2\pi Nt) \sin 2\pi Nkt \qquad [\text{Pa}] \qquad (13.9b)$$

From the well-known identity

$$\sin A \sin B = \cos(A - B) + \cos(A + B)$$

Eq. (13.8b) can now be rewritten:

$$p(t) = D \sin (2\pi Nt) \sin(2\pi Nkt)$$

$$= \frac{D \cos 2\pi (kN + N)t + D \cos 2\pi (kN - N)t}{2} \qquad [\text{Pa}] \qquad (13.10)$$

The right-hand side of Eq. (13.9) is just a pair of discrete tones at frequencies of gear mesh plus or minus the shaft rotation:

$$kN \pm N$$

For example, with the set of gears 3 and 4 in Fig. 13.4, the fundamental gear meshing frequency kN and shaft rotational speed N is, respectively,

$$kN = 20 \times 120 = 2400 \text{ Hz}$$

$$N = 120 \text{ r/s}$$

Therefore, from Eqs. (13.6) and (13.7) the side band frequencies associated with the fundamental tone for the gear set are

$$f_l = 2400 - 120 = 2280 \text{ Hz}$$

$$f_u = 2400 + 120 = 2520 \text{ Hz}$$

In general, the side band noise level can occur at any integer order of shaft rotation and cluster around any harmonic of gear meshing frequency. Thus if one of the gears were oval shaped, it is easy to see that the fluctuating loads would occur twice per shaft revolution. The side band frequencies generated then would cluster around gear mesh frequency at plus or minus twice the shaft rotation:

$$kN \pm 2N$$

In summary, side band frequencies are generally symptomatic of gear wheel eccentricity or shaft misalignment.

13.4 GEAR NOISE REDUCTION MEASURES

There are five fundamental and distinct approaches to gear noise
reduction available to the acoustical engineer:

1. Change the basic gear type or material.
2. Modify the tooth profile.
3. Introduce damping or shock reducing material.
4. Increase the transmission loss TL of the gear box.
5. Total enclosure.

Often a combination of selected measures from each approach is
required to meet the most stringent noise criteria. With this in mind,
let us consider the different measures and factors in each approach
along with the corresponding range of noise reduction likely.

13.4.1 Basic Gear Types and Relative Noise Levels

Consider first the most common parallel gear sets, spur and helical.
In general, the loudest of all gears is the spur gear. Here gear tooth
meshing is *in line* or *rolling contact*, and all mechanisms of noise
production including impulse, resonance, and air or oil ejection are
usually present. With respect to helical gears, it has been demonstrated [2−4] (H. R. Mull, Bell and Associates, Norwalk, Conn.,
unpublished data, 1974) that helical gears are often from 5 to 15 dB
quieter in operation than spur gears for identical loading and speed.
In many cases, the noise reduction is even more for the higher harmonics of gear meshing. Intuitively, the quieter operation of helical
gears can be attributed to distributed gear tooth loading and reduced impulse flexure.

In nonparallel types of gear sets, the straight bevel gears are
generally the loudest. Here again the straight bevel gears have line
or rolling contact as opposed to the sliding contact of the spiral bevel
or hypoid variety. From experience [2,4] (H. R. Mull, Bell and Associates, Norwalk, Conn., unpublished data, 1974), spiral bevel and
hypoid gears are 5 to 15 dB quieter than straight bevel gears. The
sliding contact feature is again the element responsible for the noise
reduction.

By far the quietest of all the basic types of gear sets is the worm
gear. Noise level reductions of 15 dB or more are often achieved with
worm gears. With this configuration, the continuous sliding contact
minimizes impulse and corresponding tooth deflection. In addition,
the nature of the continuous contact adds internal damping to the gear
system.

Table 13.1 summarizes the typical range of noise reduction associated with basic types of gear configurations.

Table 13.1   Relative Range of Reduction for Parallel and Nonparallel
Gear Sets

Gear Type	Relative Noise Reduction (dB)
Parallel gears	
Spur	0
Helical	5 to 15
Nonparallel gears	
Straight bevel	0
Spiral bevel	5 to 15
Hypoid	5 to 15
Worm	15 to 20

13.4.2 Gear Contact Ratio

From the previous discussion of basic noise-generating mechanisms,
it is easy to see that gear design plays a dominant role. One of the
most significant parameters assoiciated with gear noise is the gear
contact ratio. For ordinary rotating spur gears, the number of mating tooth pairs simultaneously in contact alternates between 1 and 2.
The transverse contact ratio Q_t is defined as the time average number
of tooth pairs simultaneously in contact. Depending on the gear
geometry, this value will range between 1 and 2. However, for
high-contact-ratio spur gears (which have long, thin teeth) this
value can increase to between 2 and 3. In this case, as many as
3 teeth can be in contact simultaneously. For a given application,
as the contact ratio increases the generated noise will decrease. A
first-order approximation of this attenuation for spur gears is given
by the following equation [7]:

$$\Delta \, dB = -20 \log \frac{Q_{t2}}{Q_{t1}} \tag{13.11}$$

where

Q_{t2} = the higher contact ratio

Q_{t1} = the lower contact ratio

This implies that for a particular application, changing the spur
gear transverse contact ratio from 1.5 to 2.5 will reduce the noise
level by approximately 4 dB. One note of caution: high-contact-
ratio spur gears generally have more teeth and hence generate noise

at higher frequencies than ordinary spur gears. Therefore, the noise reduction obtained by using a higher contact ratio may be off- set by the relief associated with A-weighting.

Consider as an example a gear set with a fundamental gear mesh- ing tone at 250 Hz. If increasing the number of teeth by a factor of 2, and thus increasing the transverse contact ratio, results in a 3 dB noise reduction, the A-weighted level would actually increase by 3dBA. This is due to the A-scale weighted difference between 250 and 500 Hz of approximately 6 dB.

In contrast to spur gears, there are two contact ratios associated with a meshing pair of helical gears. The transverse contact ratio Q_t is the contact ratio previously defined for spur gears. The axial contact ratio Q_a is associated with the number of gear teeth that are intersected by a line that is parallel to the shaft at the pith circle. Depending upon the helix angle and the thickness of the gear, this value can range from below 1 to more than 10. By switching from a spur gear design to a helical gear design, reduction in noise level can be estimated by the following equation [7]:

$$\Delta dB = -20 \log \sqrt{2\pi Q_a} \tag{13.12}$$

This equation helps to explain the significant noise reduction advan- tages of helical gears in comparison with their spur gear counter- parts. For example, a helical gear set with an axial contact ratio of 2 is estimated to be 19 dB quieter than a similar spur gear set.

13.4.3 Tooth Profile

Noise level reductions of up to 5 dB have been reported [2,6] in which the gear tooth profile was modified such that most tooth contact occurs during the recess portion of the line of contact. Despite the promis- ing acoustical aspects of recess-action gears, little published data or design guidelines are currently available.

Within nominal values, backlash is a secondary factor in gear noise generation. It must be emphasized that too little backlash results in tight mesh and that excessive backlash can increase the noise level 10 dB or more [6].

13.4.4 Lubrication

With respect to lubrication, dry-running gears can be 10 to 20 dB noisier than well-lubricated gears. Studies on the effect of viscosity over a wide but realistic range of conditions indicate more viscous oil may reduce noise levels up to 2 dB [4]. In short, viscosity and pos- sibly related oil temperature appear to be secondary factors compared to the first-order factor of deficient lubrication.

Table 13.2 Typical Range of Noise Reduction Available from Gear
Tooth Modifications

Area of Modification	Range of Noise Reduction (dB)
Contact ratio	2 to 5
Tooth profile-recess action	0 to 5
Lubrication (viscosity)	0 to 2
Finish (electropolish)	0 to 2

13.4.5 Gear Finish

Studies on gear finish have shown that there is no significant difference between ground and milled gears [4]. However, refinements in
the teeth such as electropolishing indicate noise level reductions up to
2 dB. Gear tooth errors also appear to be a secondary factor, provided a minimum quality of AGMA 12 or better is maintained [2,6].

In summary, the noise reduction achievable from minor modification of the gear teeth, gear quality, or lubrication is very limited.
The only exceptions are those situations where gear operation or
selection includes deficient lubrication or excessive tolerances. Summarized in Table 13.2 are the gear tooth features and modifications
along with the corresponding range of noise reduction available.

These measures form the basis of good design guidelines for the
control of gear noise. One should not, however, expect that by including each design feature the net noise reduction is strictly cumulative.

13.5 SYNTHETIC GEAR MATERIALS

Gears constructed of synthetic materials such as nylon, plastic,
Dacron, phenolic, etc., have been utilized extensively as a gear noise
reduction measure. The obvious limitations of load, temperature, and
slide characteristics have not deterred engineers from "reaping" an
acoustical profit where noise-sensitive design goals must be met.

Where applicable, noise level reductions of 10 to 20 dB are often
achieved. With these materials, tooth impulse shock loads are easily
absorbed, and due to the high internal damping, resultant induced
vibratory response is minimal. Further, because of the high internal
viscous damping, noise levels associated with harmonics of gear mesh
frequency (above n = 3) are seldom significant.

Design limitations for the use of synthetic gears include the following:

1. Loads should not exceed 500 psi.
2. Temperatures should not exceed 100°C.

13.6 APPLICATION OF DAMPING MATERIALS

It is generally accepted that high-strength thin-web gears are extremely susceptible to high induced vibratory response or, in jargon, *ringing*. This phenomenon is simply the basic noise mechanism of resonance between gear meshing excitation and the natural frequencies of the gear itself. The obvious remedial approach is to stiffen the gear, which will generally increase the weight, or to introduce *damping*, or both. With the development of highly effective lightweight viscoelastic materials and easy-to-adhere constrained layers, the damping approach has been both popular and successful. In many applications, only a thin layer of material is circumferentially applied in narrow strips in the web and rim areas. When applied in these areas to both sides of the gear, noise reduction of up to 5 dB has been reported (H. R. Mull, Bell and Associates, Norwalk, Conn., unpublished data, 1975).

Currently available damping materials can be easily applied by spraying or troweling or adhered by using precured strips backed with pressure-sensitive adhesives. In addition, these materials have a high resistance to most oils and industrial solvents.

Schlegel et al. [2] have also reported excellent noise reduction results from introducing a compliant material, stiff rubber or plastic, between the rim and hub. This approach essentially *decouples* or

Table 13.3 Typical Range of Noise Reduction for Basic Gear Noise Approaches

Basic Noise Reduction Approach	Range of Noise Reduction (dB)
Gear type change	5 to 20
Tooth features, pitch, finish	2 to 5
Gear material, metal versus synthetic	10 to 20
Damping material	2 to 5

mechanically isolates the two units. Unfortunately, excessive dynamic and static distortion, unbalance, and wear rates were also experienced.

In final summary, the basic approaches to gear noise reduction, along with the range of anticipated noise reduction, are listed in Table 13.3. Because of the discrete frequency character of gear noise, it must be emphasized again that careful measurements utilizing narrow-band analysis with statistical signal conditioning are usually essential to detect differences and to evaluate applied measures.

13.7 GEARBOX ENCLOSURES

All too often gear tooth modification or gear type changes are either too costly, involve major design changes, or require long-term development programs. Therefore, to meet immediate design goals, consideration should be given to a total gearbox enclosure.

The design of a gearbox enclosure is approached in essentially the same manner as any other enclosure, and the usual problems of accessibility, visibility, and maintainability must be considered along with cooling air, etc. Since these problems were dealt with in detail in Chap. 7, duplicate treatment will not be made here.

A marvelous feature of this approach is that, usually, there is no lower limit to the degree of noise reduction obtainable. The practical limitations are only size and weight, and in most installations these are not too restrictive. Exceptions include, of course, road vehicles, boats, aircraft, etc.

Before leaving gears, one other noise source needs mentioning. Frequently, a resonant condition associated with the gearbox occurs when the natural frequencies of the gearbox case coincide with gear tooth meshing frequencies. Under these resonance conditions, the gearbox "rings like a bell." Gearbox natural frequencies can be changed or the response reduced significantly by adding stiffening members to the case and/or by the application of damping materials. Specific design recommendations cannot be made regarding stiffening. However, for damping material application on heavy castings, a minimum layer of 0.50 in., well distributed over 50% of the case, is a good rule of thumb.

REFERENCES

1. J. E. Shigley. *Kinematic Analysis of Mechanisms.* McGraw-Hill, New York, 1959.
2. R. G. Schlegel, R. J. King, and H. R. Mull. How to reduce gear noise. *Mach. Des.* (Feb. 27, 1964), 134–142.

3. U. Arns and F. Unterholzner. Ursachen von getriebegerauschen
    and ihre vermessung (causes of gearing noises and their meas-
    urement). *VDI-Z 102*(6) (1960), 225—230.
4. G. Niemann and M. Unterberger. Gerauschminderung bei
    zahnrädern (reduction of noise in gearing). *VDI-Z 101*(6)
    (1959), 201—212.
5. W. D. Route. Gear Design for Noise Reduction. *SAE Paper
    208E.* June 5—10, 1960.
6. R. G. Schlegel and K. C. Mard. Transmission Noise Control—
    Approaches in Helicopter Design.
7. W. D. Mark. Gear noise. In *Handbook of Acoustical Noise Mea-
    surements and Noise Control,* C. M. Harris (Ed.). McGraw-Hill,
    New York, 1991.

BIBLIOGRAPHY

Bradley, W. A. Sound gear quality. *Mech. Eng.* (Oct. 1972).
Faulkner, L., *Handbook of Industrial Noise Control.* Industrial
    Press, New York, 1975.
Gear Nomenclature American Standards. *AGMA-112.02 ASA-86.10.*
    AGMA and ASME, 1950.
Livshits, G. A. 16 dB noise level reduction in high-speed transmis-
    sion. *Standartizatsiya.*
Spotts, M. F. *Design of Machine Elements.* Prentice-Hall, Engle-
    wood Cliffs, New Jersey, 19

EXERCISES

13.1 Calculate the gear meshing frequency and two higher harmonic
frequencies for a spur gear having 36 teeth rotating at 2000 rpm.

> *Answer:* 1200, 2400, and
> 3600 Hz

13.2 Calculate the fundamental gear meshing frequency of a planetary
gear system where the sun gear has 40 teeth and is being driven at
3600 rpm. The ring gear has 80 teeth, and the corresponding rotational speed of the planets is 1200 rpm.

> *Answer:* 1600 Hz

13.3 Calculate the side band frequencies about the fundamental for a distorted oval-shaped spur gear which has 50 teeth and rotates at 1800 rpm.

Answer: f_u = 1560 Hz,
f_l = 1440 Hz

13.4 To reduce gear noise, it is possible to replace a straight bevel gear with either a spiral bevel or worm gear. Which would probably provide the most noise reduction?

Answer: worm gear

14

ADDITIONAL TOPICS AND CASE HISTORIES

In the preceding chapters of Parts II and III we have devoted whole chapters to describing the major and basic sources of industrial noise and control measures. There are, however, other sources, mechanisms, and treatment which should not be considered secondary and need to be mentioned. However, because of their uniqueness, complexity, or simplicity, it is not possible to present these topics in a continuous or related manner. Therefore, a case-history-like format will be used. The author personally feels that published case histories are rarely directly applicable to particular noise problems. However, case histories can provide the acoustical engineer with useful approaches for consideration and review.

14.1 POWER PRESSES (PUNCH PRESSES)

Of all the machine tools, the power press or punch press has over the years presented one of the most formidable challenges to the acoustical engineer. There is no single noise source present but rather many noise sources. Some of these noise mechanisms can be traced to easily identifiable singular causes, while others result from a complex combination or sequence of events. In addition, the noise paths are, in some cases, straightforward and direct, while other are insidiously confounding. However, from recent research programs, the major noise sources or mechanisms have been identified and some noise reduction measures developed. As has been our convention, we shall first identify the major sources of noise and the character of the noise and then consider those measures or design guidelines for control.

The punch press can be separated into four principal areas or functions with respect to the emanation of noise. In somewhat rank order of intensity they are as follows:

1. Die space area
2. Press equipment and controls
3. Press structure
4. Material handling

In the die space area, the noise originates from

1. The metal-to-metal impact of the tool or stripper plate
2. Part *breakthrough* or fracture
3. High-velocity air to eject parts

Press drive equipment and controls such as gears, pneumatic control vents, motors, etc., are also major contributors but usually only on a cumulative basis.

With respect to press structural members, significant noise which is difficult to control is often radiated from the frame, die sets, Pitman, flywheel coverguards, etc.

Finally, there is the metal-to-metal clatter associated with material handling which cannot usually be ignored. Here we have part-to-die-set impact at load, part-to-part impact, or part-to-bin impact after ejection and finally scrap collection. Although these areas or functions have been ranked in order generally, it must be emphasized that any of these basic sources or mechanisms can, in any given blanking or forming operation, be dominant. More often, however, it is the accumulation of all sources that renders significant noise level reduction difficult to achieve. Now, of the sources listed, specific control measures have been given in previous chapters for all but the metal-to-metal impact and fracture which occurs in the die area. As such we shall focus attention mainly in the die area. However, a few supplementary comments which are unique to presses will also be presented.

14.1.1 Die Area

Noise emanating from the die area generally originates from the following mechanisms or causes in somewhat sequential order:

1. Stripper plate (if used) impact with the workpiece
2. Tool impact with the workpiece
3. Breakthrough (fracture)
4. Restrike or knockout
5. High-velocity air to eject parts

We shall discuss each source individually.

Stripper Plate

In high-speed blanking operations, noise from the stripper plate is often the dominant source of noise. The only reported noise reduction measure is to *cushion* the plate where it strikes the workpiece

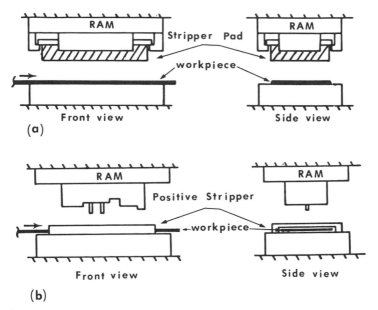

Figure 14.1 (a) Acoustically treated stripper bars and (b) positive stripper configuration. (From Ref. 1.)

with rubber or polyurethane pads or *snubbers*, as illustrated in Fig. 14.1a [1]. Since the impact of the plate is a punishing *heavy-duty cycle*, the life cycle characteristics and integrity of materials attached to or embedded in the stripper plate are typically brief and as such a high-maintenance item. In short, these materials wear or deform quickly and require frequent replacement. Treatment of this type applied to knockout pins has also been shown to be troublesome.

Application of viscoelastic damping material to the stripper plate or die set has also been widely reported. However, there is also a preponderance of data that suggests that there is little, if any, noise reduction from this superficial treatment.

As an alternative to the stripper plate, a *positive stripper* approach eliminates completely the stripper plate impact noise. The positive stripper, when applicable, is as shown conceptually in Fig. 14.1b. Here the parts are stripped with little or no impact noise.

Tool Impact and Breakthrough Fracture

To understand the tool impact and breakthrough noise, consider the following sequence of events. As the tool impacts the workpiece, a tremendous amount of energy is expended over an extremely short period of time. Even though only a small part of the expended energy

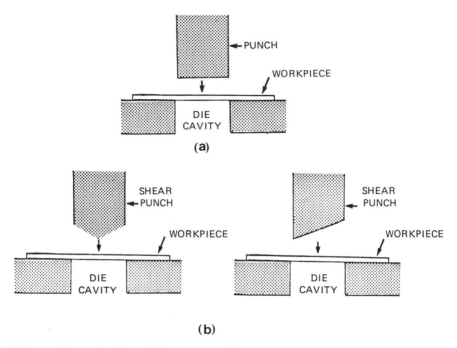

Figure 14.2 Tool and die set modifications which reduce noise. (From Ref. 1.)

is converted to sound, the result is a high-intensity, impulsivelike noise. Peak sound pressure levels from the punch or tool impact are often in the range of 130 to 150 dB.

The noise as the tool breaks through the workpiece is also impulsive in character. The magnitude of the noise level depends to the first order on workpiece hardness, thickness, etc.

Significant noise level reduction for this source is difficult to achieve, but two factors are fundamental: (1) If the duration of the impact can be extended and (2) if the speed of the tool or punch can be lowered, the noise levels are generally reduced.

To achieve these goals, the following measures have been shown to reduce tool impact noise where applicable:

1. Utilize shear, tapered, or step-cut dies or punches, as illustrated in Fig. 14.2 [1].
2. Utilize a press with sufficient or extra capacity, and minimize ram stroke speed.

With respect to punch modification, the shearing action tends to extend the impact duration and/or ease the metal cutting action. Intuitively, a slower ram stroke reduces the impact velocity and extends the impact duration.

With respect to breakthrough noise, it is acknowledged by tool designers that breakthrough noise is greatest for brittle and hard-to-machine materials. The character of the noise is subjectively described as a sharp *snap* or *crack*. We shall leave the subject here in that no noise reduction measures or research has been published dealing with control of breakthrough noise.

It is important to note that reliable studies relating to treatment of stripper plates and the use of shearing tools, etc., yielded noise level reduction in the range of only 2 to 3 dBA. However, the use of presses of larger capacity than necessary, reducing ram stroke velocity, typically reduced levels 10 dB or more [2]. Unfortunately, some production penalties are often incurred with slowing the stroke speed.

High-Velocity Air to Eject Parts

In many automatic press operations, noise associated with high-velocity air ejection nozzles is dominant. The basic approach to noise reduction is of course to reduce velocity by moving nozzles closer, using additional nozzles, or using the quieter air shroud, eductor, or diffuser nozzles. In addition, interrupting the airflow in sequence with or *timed* to part ejection is also an effective noise reduction measure. Here the high-velocity flow is present only for a short time when ejection is required. The burst of air can usually be *triggered* by a mechanically (cam) operated valve associated with the ram motion.

An alternative to air ejection is mechanical part ejection, which is virtually noiseless. There are two basic approaches to mechanical ejection. In the first approach, an armlike mechanism enters the die area, attaches itself to the part (generally with suction cups), and removes the part or scrap. This approach works well on large slowly operating presses where flat sheet metal parts are being formed.

The second approach uses an air-actuated cylinder piston to push or eject the part from the die area. A photograph showing an air cylinder installed on a press is presented in Fig. 14.3. The cylinder is usually mounted horizontally, and the ejection piston impact is automatically sequenced as the ram motion reaches top dead center or thereabout. This air-actuated piston approach does not work well if the parts are very small or flat since the area for piston impact is minimal. It must be emphasized that a vent silencer is also required for the air cylinder exhaust vent which we discussed previously.

14.1.2 Material Handling

The high velocity and subsequent impact of metal parts and scrap against tote bins are often a significant contributor to press room noise. In fact, where large presses (600 to 2000 tons) and corre-

Figure 14.3 Arrow shows air cylinder and piston for ejecting parts from press die area.

sponding large parts are being handled, the metal-to-metal clatter associated with parts handling can be the dominant source of noise. Consider, for example, the clatter a sheet of steel (soon to be a truck fender) makes when it is placed (thrown) into the press die. Next comes the rather modest *crunch* of the forming operation and then, again, the impact noise as the scrap and fender are thrown into a metal tote bin or stacked. Another example is the noise made as small parts (several ounces) are blown out of the die onto the side of a metal tote bin and subsequently fall to the bottom, impacting other similar parts.

Material handling, like die area noise, is not easy to control, but listed below are two basic guidelines:

1. Treat areas where parts impact the tote bin with rubber or polyurethane impact pads, as previously discussed.
2. Minimize the parts or scrap *drop* distances.

14.1.3 Press Room Machine Layout

Finally, significant reduction in noise exposure can be accomplished by following some common sense guidelines in press room layout.

As a first consideration, isolate (when possible) typically noisy equipment from quieter operations. In particular, it may be possible to *wall off* high-speed blanking operations, thus reducing exposure to adjacent manual press operators and inspectors. The wall may be of simple *stud* construction, acoustical panels, or flexible acoustical curtains.

Obviously, the application of these press room layout guidelines are easiest to implement and more effective when facility expansions or new machine installations are in the planning stage.

14.1.4 Summay

In summary, press room noise control must be considered as one of the most challenging areas of noise control. There are emphatically no simple gimmicks.

The measures and guidelines presented herein will, in most cases, provide measurable noise level reduction and in many cases will significantly lower operator exposure. It must, however, be emphasized that a coordinated effort among the tool designer, manufacturing engineer, setup, and the maintenance department is essential for effective results.

14.2 HYDRAULIC PUMPS

The varieties of hydraulic pumps are far too numerous to list. However, there are four basic types which are commonly found in industrial applications: centrifugal, screw, reciprocating, and gear. In all cases, discrete parcels of fluid are taken in at the inlet, compressed, and recombined at the discharge. Ideally, the flow would be steady with no fluid or pressure pulsations. In practice, however, the fluid flow and discharge pressures are not steady but contain periodic components due to the pump compression mechanism. The compression mechanisms may include vanes, pistons, gears, screws, etc. It is from these pulsations that the major sources of pump noise originates. Additional broadband noise is often present due to mechanisms such as cavitation, turbulence, etc.

These broadband sources are usually secondary in magnitude and related to poor pump design or applications.

14.2.1 Character of Pump Noise

For the periodic pulsations, the frequencies of the discrete tones f_n can be calculated from

$$f_n = nP_r N \tag{14.1}$$

where

  N = pump rotational speed (r/s)

  P_r = number of compressions or pumping events per rotation

  n = harmonic number, 1, 2, 3, . . .

The form of this equation is of course the same as for fans, gears, internal combustions engines, etc. A little more care, however, must be taken in determining the number of compressions or pumping events. Some pump manufacturers will combine the fluid delivery from several compression units in order to minimize the flow pulsation *ripple*.

For example, the flow delivery of pairs of cylinders may be combined, yielding a fundamental tone (n = 1) at half the frequency one calculates by just counting the cylinders.

With respect to the magnitude of pump noise, little can be said because of the number and diversity of basic pump designs. However, a first-order approximation given by Irwin and Graf [3] shows that the overall acoustical power L_W in the four octave bands 500, 1000, 2000, and 4000 Hz can be estimated as follows:

$$L_W = 10 \log hp + K_p \qquad [dB] \tag{14.2}$$

where K_p is the pump constant: 95 dB for centrifugal, 100 dB for screw, and 105 dB for reciprocating pumps (below rated speeds of 1600 rpm, subtract 5 dB).

Further, under the assumption of equal energy in each band, the individual octave band levels can be taken as 6 dB less than the overall power level.

Example

  A screw-type pump is driven by a 100-hp motor near peak efficiency. Calculate the overall acoustical power level L_W and the acoustical power level for the 1000-Hz octave band.
  Solution
  Using Eq. (14.2), we have

$$L_W = 10 \log hp + K_p \qquad [dB]$$

$$= 10 \log(100) + 100$$

$$= 120 \text{ dB}$$

where L_W is the overall sound power level for the four bands 500, 1000, 2000, and 4000 Hz. Here K_p was chosen as 100 for screw-type pumps. Under the assumption that the power level is the same for each band, the sound power level in the 1000-Hz band is then

$$120 - 6 = 114 \text{ dB}$$

It is pretty well acknowledged by manufacturers that reciprocating pumps with odd numbers of pistons produce less flow pulsation amplitude or ripple. For example, a nine-piston pump generates one-fifth as much ripple as an eight-piston pump of equal displacement. It is for this reason that most piston pumps are made with an odd number of cylinders. The added bonus is of course generally less noise.

14.2.2 Noise Reduction Measures

In many cases, most of the noise from pumps is radiated from the case of the pump itself. As such one of the most effective noise reduction measures is to totally enclose the pump. Often the enclosure design is rather simple since most pumps do not require air for cooling purposes, and therefore only strict attention to detail is required for acoustically sealing the plumbing and shaft penetrations.

With respect to the plumbing penetrations, rubber grommetlike seals as illustrated in Fig. 14.4 provide a good acoustical seal. In addition, if the rubber is soft, durometer 40 to 60, good vibration isolation is also obtained.

With respect to the shaft, most pumps are driven directly from the motor. As such a slot can usually be cut to allow the shaft to penetrate the enclosure. Naturally, the clearance should be minimal.

With the pump itself enclosed, noise associated with the plumbing must be given priority. Here the noise, which is fluid-borne, radiates from the pipes themselves, usually at the compression frequency and harmonics thereof. Reactive silencers designed to reduce noise at the fundamental compression frequency are quite effective. However, their range of effectiveness is rather narrow, and performance falls off rapidly for even small changes in speed. These silencers resemble the Helmholtz resonators discussed in Chap. 8, and the basic design equations are applicable to the hydraulic case.

One other interesting design approach is to split the flow equally and then recombine, as illustrated in Fig. 14.5. Now if the difference in path lengths is one-half wavelength, the fluid pulsations are 180° out of phase when they recombine, and they *cancel*. This device is called a Quincke tube and is often used in aircraft and shipboard hydraulics. The limitations are clear; the device can only be tuned to a single frequency, say the fundamental, and integer odd harmonics. For the even harmonics, the pulsations are in phase when recombined, and thus there is reinforcement.

Example
    Suppose the speed of sound of a fluid used in a hydraulic system is 5000 ft/s. What increase in length should the bypass section of a Quincke tube be to have phase cancellation at 500 Hz?
    Solution
    The wavelength of the pressure pulsations in the fluid is

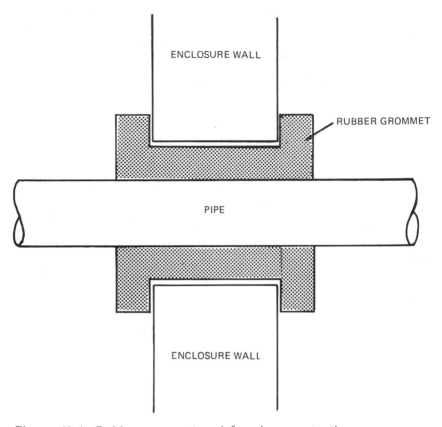

Figure 14.4 Rubber grommet seal for pipe penetrations.

$$\lambda = \frac{5000}{500}$$

$$= 10 \text{ ft}$$

The increase in path length for phase cancellation is

$$\frac{\lambda}{2} = 5 \text{ ft}$$

Cancellation will also occur at 1500 Hz, 2500 Hz, etc. Reinforcement will occur at 1000 Hz, 2000 Hz, etc.

In summary, devices to reduce fluid-borne noise are somewhat limited in overall noise reduction. As such, pipe wrapping or lagging methods as previously discussed are often the only practical approach for significant noise reduction.

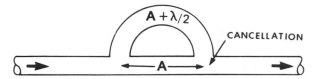

Figure 14.5 Quincke tube for reducing fluid-borne noise in pipes.

There is another source of noise related to the plumbing which always must be considered. The fluid pulsations and motion of the pump will cause the mechanically coupled pipes to vibrate. As such they often impact against bulkheads, other adjacent pipes, walls, etc., with a resultant *clatter*. This pipeline vibration can be controlled by using an abundance of clamps and supports with elastomeric or rubber *spacers* between the pipes, bulkheads, walls, etc. Shown in Fig. 14.6 are some examples of these cushion-clamp–type assemblies. In addition, a short length of flexible hose should be installed near the pump inlet and discharge to provide isolation between the pump and the pipes. In high-pressure systems, 3000 psi and up, two sections of flexible hose installed at 90° are strongly recommended for vibration-sensitive installations.

Finally, the vibration of the pump itself can transmit high-level vibratory energy to the supports, mounting table, and beyond. This structure-borne noise can be controlled by utilizing good-quality isolation mounts. Selecting these mounts follows the procedures outlined in detail in Chap. 10. One rule of thumb which applies directly to pumps is to select a mount such that the system natural frequency (fundamental) is at least 0.5 times the shaft rotational speed. At this point, transmitted energy at the shaft rotational frequency is sharply reduced, and the energy transmitted at the compression frequency and harmonics will usually be more than 30 dB down.

Guidelines for control of pump noise and vibration can be summarized as follows:

A. Isolate Pumps
   1. Use flexible couplings between pump and motor shafts.
   2. Use adequately sized hose between pump pressure port and start of pipe runs. A two- or three-foot length is generally sufficient.
   3. Avoid mechanical contact between suction and return lines and fluid reservoir.
   4. Isolation mounts between the pump and motor and the reservoir or machine frame are essential.
B. Rigidly Mount All Valves
   1. Mount all directional and flow control valves to the machine base or frame. Avoid in-line mounting. Smaller two-port

ANCHOR CHANNEL

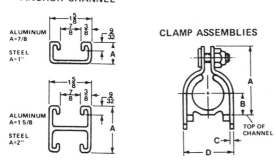

Figure 14.6 Example of cushion-clamp–type pipe isolators. (Courtesy of Hydra-Zorb Co., Auburn Hills, Mich.)

solenoid, in-line flow, and in-line check valves can be considered rigidly mounted when installed directly next to a clamp point.

C. Minimize Fluid Resistance
   1. Use adequately sized components and lines.
   2. Avoid drilled junction blocks.
   3. Use bends in the conduits rather than right-angle fittings. Straight fittings may be used to facilitate disassembly and service.

D. Secure the Fluid Conductor System
  1. Use resilient cushion clamps.
  2. Clamp spacing distance should not be more than 60 diameters. Where severe vibration is present, clamp spacing distance should not be more than 30 diameters.
  3. Provide clamps as close as possible to each side of a bend.

14.3 NOISE FROM ELECTRICAL EQUIPMENT

Noise from electrical equipment such as motors, generators, transformers, ballasts, etc., is characterized generally as a discrete low-frequency *hum*. Narrow-band spectral analysis will show, in most cases, a fundamental frequency at twice the line frequency, i.e., $2 \times 60 = 120$ Hz or $2 \times 50 = 100$ Hz, and integer-ordered harmonics thereof. The question is often asked, "Why twice the line frequency?" To see this, one must consider the origin of the noise. When there is an increase in magentic flux density in ferromagnetic materials, a mechanical strain called magnetostriction results. Most iron alloys increase in length when magentized; nickel alloys, on the other hand, decrease in length. For example, a transformer core excited by alternating current will experience an increase in flux density twice every cycle. Hence, the magnetostrictive displacement and corresponding noise occur at twice the line frequency and harmonics thereof.

These dimensional changes, though small, often less than 1 part in 10,000, can provide a high level of acoustical energy when mechanically coupled to a large radiating surface such as found in transformers.

Another discrete source, commonly referred to as pole attraction, occurs dominantly in rotating electrical equipment. Here rotors or stators are displaced due to applied magnetic stresses. The effect again at twice the line frequency is most pronounced in slotted core structures or in dual-pole single-phase motors or generators.

Thus, with the major noise sources defined and characterized, we can now consider the major electrical components and methods of noise control. It should be emphasized that other noise sources such as cooling fans, bearings, or pumps and noise due to vibration may well exceed the noise of electrical origin. We have, however, treated these subjects previously.

14.3.1 Transformers

The noise from transformers is almost exclusively due to magnetostrictive effect in the cores. As such the magnitude of the noise increases steadily with flux density in the core up to saturation. For large oil-immersed units, the most common for power distribution, the following expression can be used to obtain an estimation of the overall A weighted level L_{pA} at 30.5 cm or 12 in from the case.

$$L_{pA} = 10 \log_{10} \frac{R}{R_0} + 35 \quad [dB] \quad\quad (14.3)$$

where

R = maximum transformer rating (kVA)

R_0 = reference (1 kVA)

Consider an example.

Example

A power transformer operates at maximum power of 50,000 kVA. Estimate the overall A weighted sound level at 12 in. from the case.

Solution

From Eq. (14.3), we have

$$L_{pA} = 10 \log_{10} \frac{5 \times 10^4}{1} + 35$$

$$= 10 \log_{10} 5 + 40 + 35$$

$$= 7 + 40 + 35$$

$$= 82 \text{ dBA}$$

It should be noted that Eq. (14.3) was derived from data published by the National Electrical Manufacturers Association (NEMA) and should be considered only the average of levels measured around the transformer in accordance with the NEMA standard (TR 1-9.04).

With respect to noise reduction, few guidelines are available. Extensive studies have shown that little noise reduction is available at the source, that is, at the core. Core material variations, lamina shape modifications, core clamping, and internal barriers yielded only relatively small noise reduction. In short, the noise levels associated with currently available transformers represent, more or less, the state of the art in noise control. As such, partial barriers or total enclosures are the only practical measures available to the acoustical engineer.

For large power distribution transformers located outdoors, partial barriers of masonry construction have been used successfully. However, since the noise is dominantly low frequency, 120 to 480 Hz, rather high walls are often required to control defracted sound. In critical noise-sensitive neighborhoods, a total building-type enclosure should be considered.

Until recently, relatively large transformers were also present in rectifiers, welding units, etc. Now advances in solid-state rectifiers, which are virtually silent, are making the older types obsolete.

14.3.2 Motors and Generators

The variety of types, sizes, and models of electric motors and generators is enormous. However, since 1969, NEMA has provided lists of acoustical power levels not likely to be exceeded for several common motor types. Shown in Table 14.1 [4] are excerpts from the list which represent many common motors used for industrial applications. It must be emphasized that the overall power levels are A weighted and include the noise associated with cooling fans which in many cases is dominant. Note in particular the sharp increase in acoustical power as shaft rotational speed increases from 1800 to 3600 rpm, which is characteristic of fans. Therefore, since air-cooling fans are generally the major source of electric motor noise and electrical noise is negligible, we shall focus attention on the control of cooling fan noise.

For the cooling fans of totally enclosed fan-cooled (TEFC) motors up to 150 hp, absorptive-type silencers can be used effectively, as

Table 14.1 Sound Power Levels of Polyphase Squirrel-Cage Induction Motors[a]

Speed (rpm)	Frame Size Designation	Horse Power[b]	A Weighted Overall Sound Power Level (decibels re 10^{-12} Watt)	
			OPEN	TEFC
3600	143T, 145T	3:2	76	87
	182T, 184T	7-1/2:5	80	91
	213T, 215T	15:10	82	94
	254T, 256T	25:20	84	96
	284T, 286T	40:30	86	98
	324T, 326T	60:50	89	100
	364T, 365T	100:75	94	101
	404T, 405T	150:100	98	102
	444T, 445T	250:250	101	104

Table 14.1 (Continued)

Speed (rpm)	Frame Size Designation	Horse Power[b]	A Weighted Overall Sound Power Level (decibels re 10^{-12} Watt)	
			OPEN	TEFC
1800	143T, 145T	1/2 to 2	70	70
	182T, 184T	3 to 5	72	74
	213T, 215T	7-1/2 to 10	76	79
	254T, 256T	15 to 20	80	84
	284T, 286T	25 to 30	80	88
	324T, 326T	40 to 50	84	92
	364T, 365T	60 to 75	86	95
	404T, 405T	125:100	89	98
	444T, 445T	200:150	93	102
1200	143T, 145T	1/2 to 1	65	64
	182T, 184T	1-1/2 to 2	67	67
	213T, 215T	3 to 5	72	71
	254T, 256T	7-1/2 to 10	76	75
	284T, 186T	15 to 20	81	80
	324T, 326T	25 to 30	83	83
	364T, 365T	40 to 50	86	87
	404T, 405T	60 to 75	88	91
	444T, 445T	100 to 125	91	96
900	143T, 145T	1/2 to 3/4	67	67
	182T, 184T	1 to 1 1/2	69	69
	213T, 215T	2 to 3	70	72
	254T, 256T	5 to 7 1/2	73	76
	284T, 186T	10 to 15	76	80
	324T, 326T	20 to 25	79	83
	364T, 365T	30 to 40	81	86
	404T, 405T	50 to 60	84	89
	444T, 445T	75 to 100	87	93

[a] The no-load sound power levels of design A, B and C polyphase squirrel-cage induction motors generally do not exceed the values given in the table when measured in accordance with IEEE Std 85.
[b] Where colons occur, the first number is the upper hp rating in that frame size for OPEN motors, the second is for TEFC. In all other cases, range is representative or approximate.
Source: NEMA Standards Publication No. Mg1-1978, Copyright 1978 by the National Electrical Manufacturers Association.

discussed in Chap. 8. Typically an overall noise reduction of 6 to 10 dBA can be expected from these silencers. For larger motors, however, total enclosure with carefully designed low-pressure-loss inlets and exhausts is the only practical approach to significant noise reduction. With total enclosure, there is virtually no lower limit to acoustical isolation. Shaft misalignment or rotor unbalance can also create excessive vibration and corresponding additional noise, as discussed previously. Since these are maintenance items, little more need be mentioned. Finally, careful attention to vibration isolation should always be a design consideration for a motor installation regardless of size or shaft rotational speed.

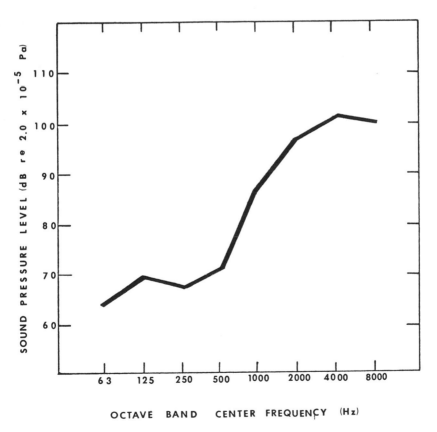

Figure 14.7 Actual octave band sound levels measured 1 in from a vibratory feeder bowl.

14.4 VIBRATORY BOWLS

A very common noise source in manufacturing areas is the vibratory *feeder* bowl. Such feeders are used extensively in assembly operations, with automatic machine tools, and as *screeners* or sorters, etc.

Basically, the feeder bowl consists of a cylindrical bowl with a spiral track attached inside of the bowl. The bowl is coupled through leaf springs to an electromagnetic base exciter unit. The angular vibratory motion of the bowl then causes the parts, generally small, to move, or, in a sense, to *march*, in an almost military fashion up the spiral tract to the exit ramp. The major source of noise is the part-to-bowl and part-to-part impact. The noise is broadband in spectral character with peak levels typically in the range of 2000 to 10,000 Hz. Shown in Fig. 14.7 is an actual octave band spectral analysis of the noise from a small vibrator bowl with the microphone located at approximately 1 m.

There are several approaches to noise reduction for these devices, and the most straightforward is total enclosure. Shown in Fig. 14.8 is an example of a commercially available total enclosure for vibrator bowls. Note the transparent upper section for good visibility and the access door at the top. Partial enclosures, such as transparent plastic lids, are relatively ineffective, with noise reduction seldom exceeding 3

Figure 14.8 Commercially available vibrator bowl enclosure. (Courtesy of Automation Devices, Inc., Fairview, Pa.)

to 5 dB. Application of polyurethane or rubberlike damping material to the inside bottom of the bowl and the track has, however, been shown to be effective. The damping material acts as a cushion between the parts and the bowl, and significant noise reduction of up to 8 to 10 dBA has been reported (H. R. Mull, Bell and Associates, Norwalk, Conn., unpublished data, 1978). One novel approach, which is often applicable, is to add water or solvent to a depth of approximately 1 in. in the bowl. Measurements evaluating this approach showed an overall noise reduction of 12 dBA (H. R. Mull, Bell and Associates, Norwalk, Conn., unpublished, 1978). Further, only minor asjustments to the excitation level or vibratory amplitude were required. To maintain part movement, it should be emphasized that damping material applied to the exterior of the bowls has been shown to be ineffective, with noise reduction often less than 2 dB.

14.5 ROUTERS

Intensely high noise levels in excess of 124 dBA are common in routing operations of both wood and metal. The noise is discrete in character at the router tool impact frequency with intense integer-ordered higher harmonics usually present. The presence of higher-ordered harmonics is due primarily to resonant conditions associated with the dimensions of the workpiece itself. The actual impact of the tool to workpiece is at shaft rotation frequency or 2, 3, or 4 times shaft rotation frequency, depending on the configuration of the tool bit. Hence, the strong higher sixth, twelfth, twentieth, etc., harmonics are effectively radiated resonances associated with the workpiece itself.

Unfortunately, even with the origin and character of the noise rather well understood, highly effective noise level reduction measures are not currently available. Because of the close visual and manual requirements of the operator as shown in Fig. 14.9, enclosure is not generally practical. Further attempts to reduce the radiated noise by using damping materials or clamping arrangements have not been encouraging.

In some sheet metal routing operations a numerical control (NC) positioning approach has been used successfully. With this approach remote monitoring and operator isolation in a control room has proved practical and effective.

A novel approach which is applicable for some metal routing is to perform the actual machining underwater. Here the actual cutting is done beneath the water surface at a depth of 1 to 2 in. In Fig. 14.10 is a water table, showing the air-driven router and routing guide. In this setup the workpiece is clamped under approximately 1.5 in. of water. Overall noise level reductions of more than 15 dB were

Figure 14.9 Typical operator station at a routing machine.

Figure 14.10 Water table to reduce noise from metal routing operations. Arrow shows an air-powered router mounted on a guide.

achieved with no loss in production quality (H. R. Mull, Bell and Associates, Norwalk, Conn., unpublished data, 1980). Further, as an added bonus, the life of router bits increased fivefold due to cooler operating temperatures.

14.6 PLANERS OR SURFACERS

The intense noise from wood planers or surfacers originates from the impact of the rotating knives and the workpiece. As such, strong discrete tones are present at the knife blade impact frequency and integer higher-ordered harmonics thereof. Further, at idle (not cutting) as the knives pass stationary surfaces such as the chip breaker, intense noise of aerodynamic origin is also produced at the knife passing frequency and higher harmonics. Noise levels in excess of 110 dBA are common.

Progress in the control of planer noise has, however, been in-couraging. Planers lend themselves rather readily to total enclosure. Here conventional acoustical panels or even stud wall construction has been used successfully. One note of caution: Noise radiated from the workpiece itself during the planing operation is considerable, if not dominant. Hence, a lengthy inlet tunnel and exit tunnel are required

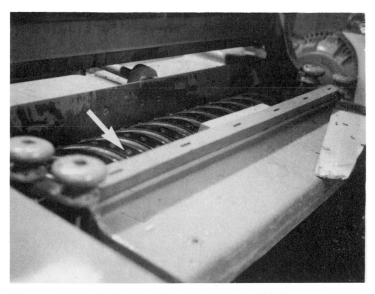

Figure 14.11 Helical cutting head installed in a wood planer or sur-facer. Arrow shows the cutting head.

for optimum noise reduction. These tunnels can be of plywood construction and should include absorbing materials on the interior surface.

The most encouraging breakthrough in planer noise control is the use of continuous helical cutter heads, as shown installed in Fig. 14.11. With these heads the cutting edges are in continuous contact with the workpiece, eliminating periodic impact. Overall noise reduction of more than 15 dBA has been reported (H. R. Mull, Bell and Associates, Norwalk, Conn., unpublished data, 1979) with these helical cutter heads. In addition, the discrete aerodynamic noise at idle essentially vanishes, further reducing operator noise exposure.

14.7 TOOL SQUEAL OR SCREECH

A major noise source in machine shops is the shrill *screech* that often occurs during metal cutting on lathes, drills, etc. The spectral character of this noise source is almost exclusively discrete and often nearly a pure sinusoidal tone. It is easiest to understand this noise mechanism itself by considering the special case of its generation applied to lathes. Shown in Fig. 14.12 is an illustration of a lathe tool cutting a *chip* as the part moves with linear velocity V. Now, as the tool "plows" along, the tool is deflected, as shown by the dashed lines, in the direction of the moving part. This deflection is due to steady forces required to force the tool through the metal part. In addition, there are dry frictional forces between the tool and part, which also tend to deflect the tool. These forces, it turns out, are not steady.

To see this, note that when the tool is deflected exclusively due to frictional forces, it *springs back* toward equilibrium when the frictional forces are overcome. At rest, frictional forces again deflect the tool, and the motion repeats itself. Now, when the tool moves in the same direction as the part, its velocity relative to the part is less than when it springs back or moves in a direction opposite to the part. Hence, we have a varying or oscillating relative velocity ΔV between the tool and the part, as also illustrated in Fig. 14.12. Now the behavior of dry frictional forces with respect to velocity is such that a negative slope or inverse relationship exists, as illustrated in Fig. 14.13. Note that as the velocity increases, the frictional force decreases. One experiences this phenomenon when moving or *pushing* furniture across a floor. Intuitively, as the velocity, of, say, a heavy chair increases, the pushing generally becomes easier. Furthermore, this is contrary to the behavior of viscous forces, as illustrated in Fig. 14.13, which we dealt with in Chap. 10.

Finally, given this relative velocity difference between the tool and part, we have a plausible argument for the oscillating frictional

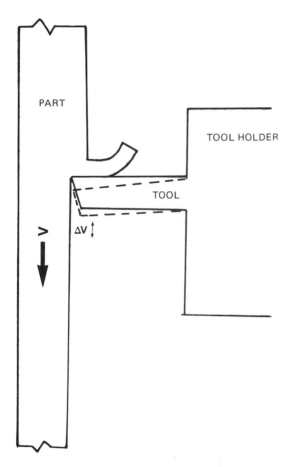

Figure 14.12 Illustration of lathe tool cutting a chip.

forces acting on the cutting tool. Further, the oscillation is regenerative or *self-exciting*, and thus we would expect the frequency of oscillation to bè strongly related to the natural frequencies of the tool and
holder system. Experimental data show this to be true (H. R. Mull,
Bell and Associates, Norwalk, Conn., unpublished data, 1981) and
explains the discrete spectral character of the noise.

This form of regenerative self-excited vibration is often referred
to as *stick slip*, and other common examples of stick slip with similar
noise-producing mechanisms are the following:

1. Moving a violin bow across a string
2. Fingernails drawn across a blackboard
3. Scraping paint

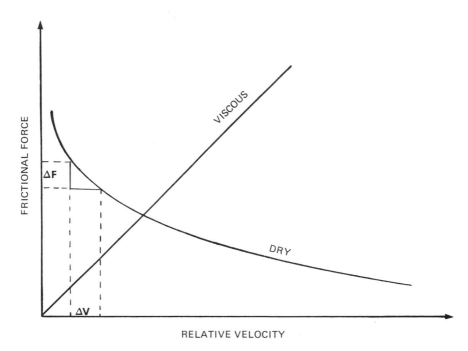

Figure 14.13 Graphic illustration of dry and viscous frictional forces as a function of relative velocity.

4. Auto brake squeal
5. Subway car wheel squeal on a curve

Returning to tool screech, from extensive measurements on vertical turret lathes over a wide range at different speeds and *feeds*, it has been shown that the screech frequency and harmonics remain constant. In short, the spectral character of tool screech is virtually independent of speed and feed.

Feed is the term commonly used to describe the rate of tool advance; it is also a measure of chip thickness. Note that this conclusion is reasonable when one considers bowing a violin. For example, it does not matter how fast one moves a bow across a violin string, say an E string; the resultant frequency of the tone is always the same, an E and harmonics. The amplitude of the sound level, however, generally increases as one applies more normal force or bows harder, which is also often the case with lathe tool noise.

The previous discussion must be considered a plausible argument for providing physical insight into the stick-slip mechanism. Another more analytical way of looking at the stick-slip phenomenon is to return to and consider again the damped vibration of a mass spring system. From Chap. 10, the differential equation of motion is

$$m\ddot{x} + c\dot{x} + kx = f(t)$$

and the form of the solution was given as

$$x(t) = e^{-ct/2m}(A \cos \omega_d t + B \sin \omega_d t)$$

which represents a decaying oscillatory motion. If, however, we took the sign of the friction term to be negative, i.e., $c < 0$, the argument of the exponential would be positive, and we would have an oscillation with ever-increasing amplitude at the natural frequency of the system. As mentioned earlier, this is the characteristic of a regenerative self-excited system. We know, however, that when the oscillation becomes sufficiently large the damping eventually becomes positive, and this effect limits further amplitude increase.

To account for the finite amplitude, the damping term probably takes the form [2]

$$-c + ax^2$$

which is an amplitude-limiting term, and the differential equation becomes

$$m\ddot{x} + (-c + ax^2)\dot{x} + kx = 0 \qquad (14.4)$$

This is a form of Van der Pol's equation, and the solution is initially an exponentially increasing oscillation to a point in time where amplitude remains constant.

In summary, the stick-slip mechanism can also be considered, for purposes of analysis, an example of regenerative or self-excited vibration due to negative damping.

With respect to noise control, little research has been expended in this area of tool squeal. This is probably due to the fact that the surface finish of parts is usually not adversely affected by the small amplitude of the vibration. However, on turret lathes where the tool squeal is a common problem, significant noise level reduction in the range of 8 to 10 dB has been achieved by wrapping the tool or tool holders with a dense vinyl or mastic (H. R. Mull, Bell and Associates, Norwalk, Conn., unpublished data, 1981). From this, one can reasonably conclude that for this situation the tool holder was the dominant radiating surface. As such the contribution of noise radiated from the part being machined can be considered secondary. This may not be the case for large disks or cone-shaped parts.

There is also experimental data that show (H. R. Mull, Bell and Associates, Norwalk, Conn., unpublished data, 1981) that lubrication plays a small role in the stick-slip phenomenon. This is probably due to the manner in which lubrication is applied. Typically the lubrication is of a water-soluble type, *flushed* over the tool. As such, little lubrication is actually at the critical tool-to-part contact and probably actually acts more as a coolant than a lubricant.

Finally, it should be emphasized that from studies (H. R. Mull, Bell and Associates, Norwalk, Conn., unpublished data, 1981) tool squeal occurs most often while machining hard-to-machine nickel or tantalum alloy steels with machinability rating indices less than 20%. Suggested areas of research into this little understood phenomenon include tool shape, tool rake, and tool holder damping, etc. Tool squeal should not be confused with tool chatter, which is usually due to poor tool design, dull tools, improper rake, or chip thickness variations.

Tool squeal associated with drilling operations is also common for hard-to-machine alloys. Unfortunately, there are no published engineering data or reduction measures in this area. The noise mechanism is, however, probably stick slip in origin and again occurs almost exclusively on hard-to-machine nickel alloys.

14.8 PRINTING PRESSROOM NOISE CONTROL

Noise levels in large, high-volume printing pressrooms typically exceed 90 dBA. The sources of noise are many, but general mechanical

Figure 14.14 Adjacent workroom enclosure for large printing press. (Courtesy of Industrial Noise Control, Inc., Addison, Ill.)

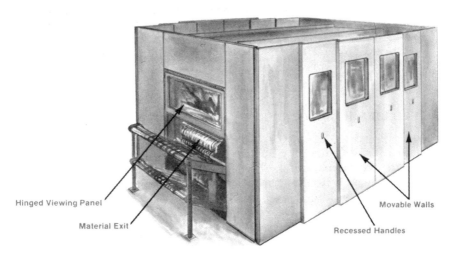

Hinged Viewing Panel

Material Exit

Movable Walls

Recessed Handles

Figure 14.15 Press folder enclosure.   (Courtesy of Industrial Noise Control, Inc., Addison, Ill.)

noise associated with gears, bearings, motor bearings, etc., is often the dominant source. As such, enclosure is often the only practical noise reduction measure. For large presses, adjacent workroom-like total enclosures have reduced operator exposure to acceptable risk of hearing loss criteria. See Fig. 14.14 for actual work room enclosure installation. In addition, ancillary equipment such as folders and scrap "hoggers" have been successfully enclosed. Figures 14.15 and 14.16

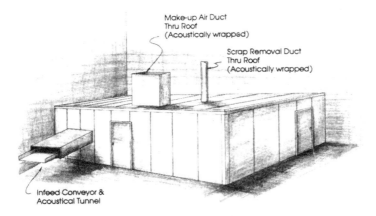

Make-up Air Duct
Thru Roof
(Acoustically wrapped)

Scrap Removal Duct
Thru Roof
(Acoustically wrapped)

Infeed Conveyor &
Acoustical Tunnel

Figure 14.16 Scrap hogger enclosure.   (Courtesy of Industrial Noise Control, Inc., Addison, Ill.)

show conceptual examples of these enclosures. Resultant sound levels at the operator and inspection stations were below 90 dBA in actual installations. In addition to the enclosures, reverberation control utilizing hanging ceiling baffles has provided measurable noise reduction at the periphery of the pressroom.

14.9 FOUNDRY NOISE CONTROL

Noise control in the foundry industry has shown considerable progress. This is especially true for ancillary equipment. Shown in Figs. 14.17 and 14.18 are conceptual examples of enclosures for the shakeout and

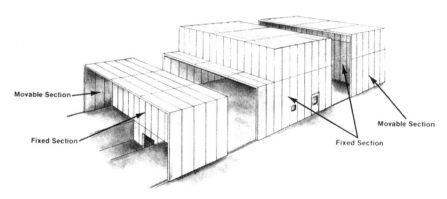

Figure 14.17 Shakeout Equipment enclosure. (Courtesy of Industrial Noise Control, Inc., Addison, Ill.)

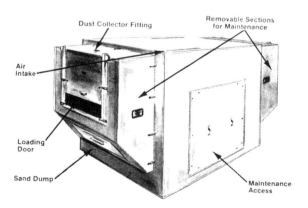

Figure 14.18 Parts cleaner enclosure. (Courtesy of Industrial Noise Control, Inc., Addison, Ill.)

parts-cleaning equipment. Here again, resultant overall levels adjacent to these enclosures was less than 90 dBA.

REFERENCES

1. L. H. Bell. Guidelines to Power Press Noise Reduction. *Society
    of Manufacturing Engineers TE80-338.*
2. C. M. Harris and C. E. Crede. *Shock and Vibration Handbook.*
    McGraw-Hill, New York, 1961.
3. J. D. Irwin and E. R. Graf. *Industrial Noise and Vibration
    Control.* Prentice-Hall, Englewood Cliffs, New Jersey, 1979.
4. J. M. Guinter. Controlling noise from electrical equipment.
    Noise Control Eng. (Nov.-Dec. 1979).

BIBLIOGRAPHY

Alfredson, R. H. Noise source identification and control of noise
    in punch presses. In *Reduction of Machinery Noise Seminar,
    Dec. 10—12, 1975.* Ray Herrick Laboratories, Purdue University

Allen, C., and R. Ison. A practical approach to punch press quieting.
    Noise Control Eng. 3(1) (July 1974), 18—20.

Bruce, R. D. Noise control for punch presses. In *21st Annual
    Technical Conference, American Metal Stamping Association,*
    New York,

Daggerhart, J. A., and A. Berger. An evaluation of mufflers to
    reduce punch press air exhaust noise. *Noise Control Eng.*
    (May-June 1975).

Faulkner, L. *Handbook of Industrial Noise Control.* Industrial
    Press, New York, 1975.

Harris, C. M., *Handbook of Acoustical Measurements and Noise
    Control.* McGraw-Hill, New York, 1991.

Locke, R. C. Automatic strip feed press noise and its reductions.
    In *Reduction of Machinery Noise Seminar, Dec. 10-12, 1975.*
    Ray M. Herrick Laboratories, Purdue University

Shinaishin, O. A. Impact induced industrial noise. *Noise Control
    Eng.* 2(1) (1974), 30—36.

Shiniashin, O. A. *Sources and Control of Noise in Punch Presses.*
    Office of Research and Development, Environmental Protection
    Agency.

Stewart, N. D., J. R. Bailey, and J. A. Daggerhart. Study of
    parameters influencing punch press noise. *Noise Control Eng.*
    5(2) (Oct. 1975), 80—86.

EXERCISES

14.1 A nine-cylinder hydraulic pump is operated at 3450 rpm. Calculate the frequency of the fundamental, second, and third harmonics of the anticipated noise.

Answer: 518, 1035, and
1553 Hz

14.2 Estimate the sound power level of a centrifugal pump driven by a 50-hp motor.

Answer: L_W = 112 dB

14.3 Estimate the sound power level for a reciprocating pump driven by a 100-hp motor at 1500 rpm.

Answer: L_W = 120 dB

14.4 Estimate the sound level 12 in. from a large transformer rated at 100,000 kVA.

Answer: 85 dBA

14.5 Estimate the overall A weighted sound power level of a TEFC motor operating at 1800 rpm with a rating of 60 hp. (Hint: Use Table 14.1.)

Answer: L_{WA} = 95 dBA

IV
ENVIRONMENTAL ACOUSTICS

Due to the public's growing awareness of noise as a nuisance and a source of health problems, brief introductory discussions on sound control in buildings (Chap. 15), community noise (Chap. 16), environmental and community noise regulations (Chap. 17), and personal hearing protection (Chap. 18) have been included in this text. Each topic is broad enough in scope to be the subject of whole volumes, but for ready reference, the fundamental aspects of each topic are presented. In no way is the presentation intended to be complete or exhaustive; however, with this introduction and the rather comprehensive list of references, a student or reader can proceed easily to further study and understanding.

15

SOUND CONTROL IN BUILDINGS

Probably no aspect of acoustical engineering touches the lives of more people than sound control in buildings. The practice of sound control in buildings began in the early 1900s with the concern for good listening conditions in churches, theaters, concert halls, auditoriums, etc. Further, additional impetus was added with emphasis on privacy and quiet conditions for apartments, offices, restaurants, commercial buildings, etc. Concurrently, the rapid growth of air conditioning added additional problems, and the public's awareness of noise followed naturally.

The subject is broad in scope but can be broken down into three distinct and separate areas:

1. Acoustical correction
2. Sound isolation
3. Mechanical equipment

Acoustical correction deals with the shaping of spaces and control of reverberation for the best possible listening conditions. The primary objective is to provide an acoustical environment that assures hearing sound uniformly and favorably reproduced throughout the space. Such spaces include auditoriums, concert halls, classrooms, meeting rooms, offices, etc.

Sound isolation deals with acoustical privacy and focuses attention on the design of building construction to reduce noise transmission through walls, floors, ceilings, etc.

With mechanical equipment, one deals primarily with air-conditioning systems which produce airborne as well as structure-borne noise. While the noise of mechanical systems cannot be eliminated at the source, it can be controlled throughout the building to even the most stringent criteria by the proper selection of noise reduction devices and construction details.

Of these three areas, more attention will be given to sound isolation, as it involves a wider variety of topics and methods. Further, in the other areas, discussions in previous chapters are directly applicable and need only to be mentioned and referenced.

15.1 ACOUSTICAL CORRECTION

As mentioned in Chap. 9, when sound originates in a room with highly reflective surfaces, a reverberant buildup occurs. Furthermore, superposition of the reflected waves creates a diffuse sound field. Now if the sound source is a human voice, this buildup interferes dramatically with speech intelligibility and may render communication nearly impossible. Further, if the source is music, excessive reverberation can cause an undesirable unbalance. On the other hand, too much absorption leaves a hall acoustically *dead*, resembling the outdoors. In the case of music halls in particular, the sound will have little *warmth* or *dimension*. For design purposes, shown in Fig. 15.1 is a chart for selecting reverberation times for rooms or halls up to 1 million ft^3 [1]. Given the room volume and application (music, speech, or a combination thereof), the chart provides the optimum reverberation time at 500 Hz to assure good listening acoustics. For example, if the hall were to be used for musical concerts exclusively, the upper values of the range would be used as design criteria. If the hall were to be used for speech exclusively, the lower values of the range would be selected. If the hall were to be used for a combina-

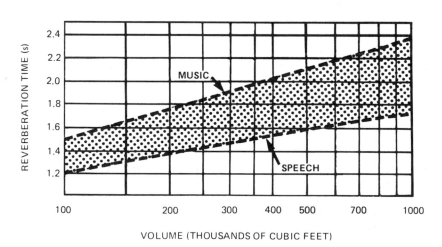

VOLUME (THOUSANDS OF CUBIC FEET)

Figure 15.1 Recommended reverberation times at 500 Hz (full hall). (From Ref. 1.)

tion of music and speech, say a church, then values near midrange
would be selected. Consider a specific example.

Example

   A room to be used as a lecture hall has a volume of 400,000 ft^3.
   To assure a high level of speech intelligibility, what reverberation time should be selected as a design criterion? If the hall
   is to be used for music, what is the optimum reverberation time
   at 500 Hz?

   Solution

   From the chart in Fig. 15.1 [1], for a volume of 400,000 ft^3, the
   recommended reverberation time is 1.5 s at 500 Hz where speech
   is the dominant application. Now if the hall were to be used for
   choral or chamber music, the design criterion would be approximately 2.0 s.

   Now the chart in Fig. 15.1 [1] yields only the optimum reverberation time at 500 Hz. To further improve the listening quality of the
room, a spectral optimization can be added to the design system.
Shown in Fig. 15.2 is a chart which provides spectral modification
factors for reverberation time [1]. Used in conjunction with Fig. 15.1,
one can now determine the optimum reverberation time for any given
room volume and application. To see this, note that in the previous
example the optimum reverberation time at 500 Hz for a room of 400,000
ft^3 used for music was 2.0 s. To enhance the musical quality of the
room, the reverberation times at 250 and 125 Hz are increased by a
factor of approximately 1.15 and 1.5, respectively, as given in Fig.
15.2 [1]. Thus the reverberation design criterion T at 250 Hz would be

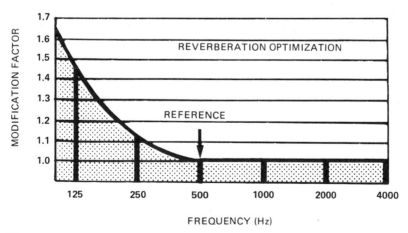

Figure 15.2 Chart for selecting spectral modification factors for
optimum reverberation times. (From Ref. 1.)

$$T = 1.15 \times 2.0 = 2.30 \text{ s}$$

and at 125 Hz it would be

$$T = 1.5 \times 2.0 = 3.0 \text{ s}$$

In summary, professional specialists in the design of concert halls, theaters, auditoriums, etc., generally agree that a little more reverberation at lower frequencies greatly improves the listening quality. Note that the modification factor is unity for the higher frequencies.

Now it should be emphasized that a difference of 0.2 s in reverberation time is probably not perceptible to a listening audience. However, it is possible to design to the nearest 0.1 s, and the reader is encouraged to do so as a good design discipline.

Now the reverberation time of a hall or room can be calculated from the sabine equation:

$$T = \frac{60V}{1.086cA} \quad [\text{s}] \tag{15.1}$$

where

V = volume of room or hall

c = speed of sound

A = total absorption

For engineering design purposes, Eq. (15.1) is often written in the more useful form as presented in Chapter 9.

$$T = 0.161 \frac{V}{S\overline{\alpha}_{sab} + 4mV} \quad \text{(mks units)} \quad [\text{s}] \tag{15.2}$$

$$T = 0.049 \frac{V}{S\overline{\alpha}_{sab} + 4mV} \quad \text{(English units)} \quad [\text{s}] \tag{15.3}$$

where

$$\overline{\alpha}_{sab} = \frac{S_1(\alpha_{sab})_1 + S_2(\alpha_{sab})_2 + \cdots + S_i(\alpha_{sab})_i}{S} \tag{15.4}$$

and

$$S = S_1 + S_2 + \cdots + S_i \quad \text{(total absorbing surface area)} \quad [\text{m}^2 \text{ or ft}^2] \tag{15.5}$$

where

$S_1, S_2, \ldots, S_i$ = areas of the individual sound absorptive surfaces (m^2 or ft^2)

$(\alpha_{sab})_1$, $(\alpha_{sab})_2$, ... , $(\alpha_{sab})_i$ = respective individual sabine absorption coefficients of the ith surface (unitless)

m = air absorption constant

Now $\bar{\alpha}_{sab}$ can be considered as a weighted average room absorption coefficient which contains the readily available published sabine absorption coefficient. For ready reference, shown in Appendix F is a listing of common construction materials and decorative accessories along with measured absorption coefficients. It should be emphasized that the manner in which materials are mounted affects dramatically the absorption performance. This is especially true for acoustical ceiling tile, and the number of standard mountings is growing daily. A complete listing is beyond the scope of this discussion; however, listed below are several standard mounting configurations, selected because they provide representative examples of common applications.

Type A: Laid directly against rigid solid surface; no airspace.
Type B: Spot cemented to gypsum board; 1/8" airspace.
Type D-20:   Fastened to wood furring strips; 3/4" × 1½".
Type E-400: Suspended over plenum; 16" airspace.

For details and a complete listing, the reader is referred to ASTM Standards C-423 and E-795.

It must be emphasized that the absorption of objects such as seats, people, and space units must be included in calculating the average absorption coefficient. In some cases, such as a concert hall, they provide the major absorbing surfaces. These objects generally are difficult to define in terms of area, and as such it is common practice to give their absorption directly in sabine units per person or per unit area, etc. The absorptive contribution of these objects is simply an additional term to be summed in the denominator of Eqs. (15.2) and (15.3).

With respect to the air absorption constant, Table 9.1 can be used to include this effect for most rooms or halls where temperatures are typically in the range of 20°C to 25°C (68°F to 77°F) and the relative humidity (RH) lies between 50% and 70%. Consider an example.

Example

A concert hall with a seating capacity of 2000 has a volume of 500,000 ft^3 and a total wall area of 50,000 ft^2. The average sabine absorption coefficient for the hall is 0.20 in the frequency range of 2000 Hz, including occupied seats but without an orchestra. Calculate the reverberation time of the hall with and without an orchestra of 100 musicians. Assume a temperature of 20°C (68°F) and a relative humidity of 50%.

Solution

By utilizing Eq. (15.3), the reverberation time T for the occupied hall is

$$T = \frac{0.049 \times 500,000}{50,000 \times 0.20 + 4 \times 0.001 \times 500,000}$$

$$= \frac{24,500}{12,000}$$

$$= 2.04 \text{ s}$$

Note that for air absorption m = 0.001 is approximate. From the table in App. F, the absorption of a musician is 13 sabins, and for 100 musicians the additional absorption is

13 × 100 = 1300 sabins

The total absorption including the orchestra is

12,000 + 1300 = 13,300 sabins

The reverberation time T with the orchestra is

$$T = \frac{24,500}{13,300}$$

$$= 1.84 \text{ s}$$

Naturally, the methods illustrated in this example can be extended to the entire spectrum. However, usually the octave band center frequencies from 125 Hz to and including 4000 Hz are taken for design purposes.

In residential rooms or office space, it is generally considered good design practice to keep reverberation time below 0.50 s. This can usually be accomplished with a suspended ceiling tile treatment. Hall-ways, in particular, generally need absorptive treatment on the ceilings as well as the floors to reduce the concentrated flow of sound along the hallway itself and to other branch rooms.

In small rooms, such as bedrooms, living rooms, etc., upholstered furniture and a carpet will usually provide sufficient absorption such that ceiling treatment in the form of acoustical tile will not be re-quired.

15.2 SOUND ISOLATION

Sound isolation in buildings deals principally with the control of sound transmission through walls, floors and ceilings, and openings. Sound is transmitted through building construction as a result of either airborne or structure-borne sound.

Sound travels in a structure-borne manner through walls, floors, and ceiling constructions or in an airborne manner by way of open air *flanking* paths. Common examples of these flanking paths are illustrated in Fig. 15.3 [2,3].

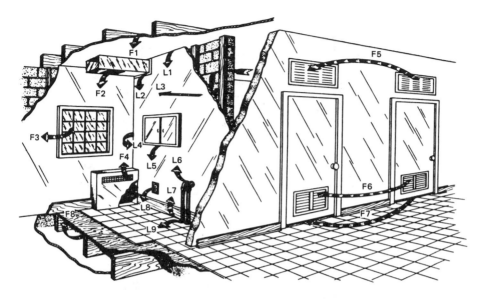

FLANKING PATHS		ACOUSTICAL LEAKS	
F1	OPEN PLENUMS	L1	POOR SEAL: CEILING
F2	DUCT RUNS	L2	DUCT PENETRATIONS
F3	WINDOW TO WINDOW	L3	POOR JOINTS
F4	HEATING UNITS	L4	POOR SEAL: WALLS
F5	OPEN VENTS	L5	BACK-TO-BACK CABINETS
F6	LOUVERED DOORS	L6	WALL PENETRATIONS
F7	UNDER-CUT DOORS	L7	POOR SEAL: FLOOR
F8	UNDER FLOORS	L8	BACK-TO-BACK OUTLETS
		L9	FLOOR PENETRATIONS

Figure 15.3 Common airborne noise transmission paths in buildings. (From Refs. 2 and 3.)

The basic approaches to controlling airborne sound transmission include the following:

1. Selecting constructions that reduce the degree of transmission
2. Eliminating paths of direct air transmission
3. Using sound-absorbing material within the free airspace of frame construction

Often a combination of these control measures is necessary to achieve suitable sound privacy between rooms.

Structure-borne sound transmission through walls, floors, ceilings, etc., results when the surfaces of the construction are set into vibration through acoustical coupling or direct mechanical contact. With the surfaces thus set into motion, the vibratory energy is carried to other parts of the building. Structure-borne sound trans-

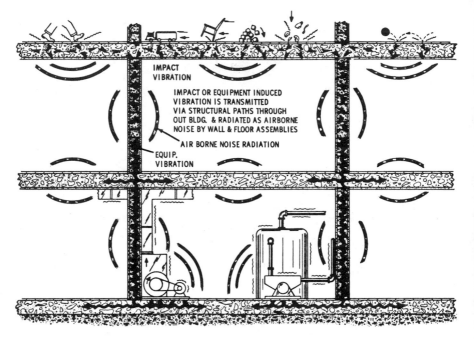

IMPACT
VIBRATION

IMPACT OR EQUIPMENT INDUCED
VIBRATION IS TRANSMITTED
VIA STRUCTURAL PATHS THROUGH
OUT BLDG. & RADIATED AS AIRBORNE
NOISE BY WALL & FLOOR ASSEMBLIES

AIR BORNE NOISE RADIATION
EQUIP.
VIBRATION

Figure 15.4 Common structure-borne noise and vibration paths in buildings. (From Refs. 2 and 3.)

mission comes from sources such as footfalls, dropped objects, doors slamming, or induced vibration from mechanical systems and equipment. Illustrated in Fig. 15.4 [2,3] are some common examples of structure-borne noise and its paths. Basic methods of control include the following:

1. Increasing the transmission loss of the wall, ceiling, or floor
2. Vibration isolation of equipment
3. Utilizing discontinuous construction assemblies

Regardless of the mode of transmission, structure-borne or airborne, the primary objective is to reduce the transmitted sound to an acceptable level. Consider now some basic design and construction guidelines.

15.2.1 Measuring Sound Transmission

By far the most accepted method for rating the performance of sound-isolating building constructions is in terms of sound transmission class or, as commonly abbreviated, STC. To determine the STC of a

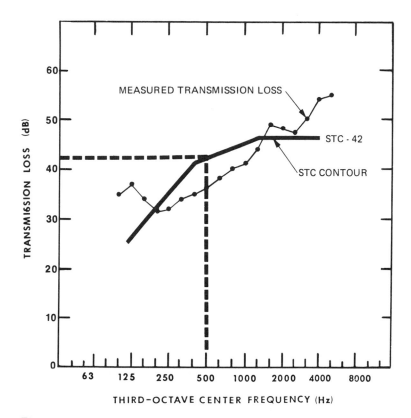

Figure 15.5 An STC rating is obtained by adjusting the standard contours to the test curve (heavy line) and reading the contour intercept at 500 Hz. (Courtesy of the National Concrete Masonry Association.)

test specimen, say a wall construction, the transmission loss TL is measured in a series of 16 test bands, usually in accordance with ASTM E90, Standard Recommended Practice for Laboratory Measurement of Airborne Sound Transmission Loss of Building Partitions. These TL values are then compared to a reference contour, as illustrated in Fig. 15.5. The reference curve, usually a transparent overlay, is moved or adjusted to a position where the sum of deficiencies and maximum deficiencies meet a given criterion of the standard. The STC value is then taken as the intercept of the reference contour and the 500-Hz ordinate. For complete details, the reader is referred to the ASTM Standard Classification for Determination of Sound Transmission Class, Designation E413.

In summary, the STC value is a single-number index that rates the specimen's (usually a wall) isolation performance. To gain some

Table 15.1 Typical Hearing Quality for a Wall of Rated Sound Transmission Class (STC)

Sound Transmission Class (STC)	Hearing Quality
25	Normal speech can be understood quite easily and distinctly
30	Loud speech can be understood fairly well; normal speech can be heard but not understood
35	Loud speech can be heard but is not intelligible
42	Loud speech is audible as a murmur
45	Loud speech is not audible
50	Very loud sounds such as musical instruments or a stereo can be faintly heard

insight into the relative value of the STC, shown in Table 15.1 are some qualitative measures of acoustical isolation in terms of STC. Note that for any "real" privacy, such as for offices, apartment walls, external walls, etc., an STC of 45 or greater is required.

15.2.2 Walls

The sound isolation efficienty of a wall, ceiling, or floor depends primarily on the following parameters:

1. The surface density (weight per square foot) of the construction
2. The separation of each barrier element
3. The addition of resilient mounting channels or material
4. The addition of sound-absorbing material between the barrier element
5. The quality of workmanship with strict attention to eliminating airborne flanking paths

Shown in Fig. 15.6 is a composite showing some very effective wall constructions. Note that absorbing materials in combination with construction discontinuities are required to assure a high level of acoustical isolation. Note also the improved acoustical performance of metal studs over wood. One note of caution, however: Metal-resilient mounts do not significantly improve the TL in combination with metal studs.

A **STC•39** Single wood studs 16″ o.c.; single layer ½″ gypsum board each side; one thickness R-11* Fiberglas insulation
STC•35 Single wood studs 16″ o.c.; single layer ½″ gypsum board each side; no insulation

*Nominal 3½″ Thick

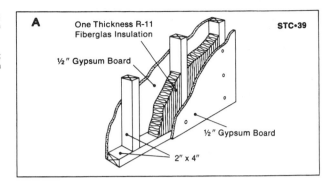

A One Thickness R-11 Fiberglas Insulation
½″ Gypsum Board
½″ Gypsum Board
2″ x 4″
STC•39

B **STC•45** Single wood studs 16″ o.c.; double layer ½″ type x gypsum board each side; one thickness R-11 Fiberglas insulation
STC•39 Single wood studs 16″ o.c.; double layer ½″ gypsum board each side; no insulation

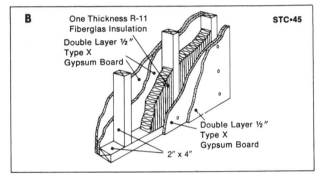

B One Thickness R-11 Fiberglas Insulation
Double Layer ½″ Type X Gypsum Board
Double Layer ½″ Type X Gypsum Board
2″ x 4″
STC•45

C **STC•46** Single wood studs with resilient channel; single layer ½″ gypsum board each side; one thickness R-11 Fiberglas insulation

STC•40 Single wood studs with resilient channel; single layer ⅝″ type x gypsum board each side; no insulation
STC•39 Single wood studs with resilient channel; single layer ½″ gypsum board each side; no insulation

STC•56 Single wood studs with resilient channel; double layer ½″ gypsum board each side; one thickness R-11 Fiberglas insulation

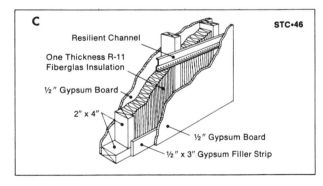

C Resilient Channel
One Thickness R-11 Fiberglas Insulation
½″ Gypsum Board
2″ x 4″
½″ Gypsum Board
½″ x 3″ Gypsum Filler Strip
STC•46

Figure 15.6 Composite showing examples of frame wall construction. STC values should be considered approximate and for design purposes only. (Courtesy of Owens-Corning Fiberglas Corporation.)

D STC•59 Double wood studs
16″ o.c.; single layer ½″ gypsum
board each side; two thicknesses
R-11 Fiberglas insulation
STC•56 Double wood studs 16″
o.c.; single layer ½″ gypsum board
each side; one thickness R-11
Fiberglas insulation

STC•47 Double wood studs 16″
o.c.; single layer ½″ gypsum board
each side; no insulation

STC•64 Double wood studs 16″
o.c.; double layer ½″ gypsum board
each side; one thickness R-11
Fiberglas insulation

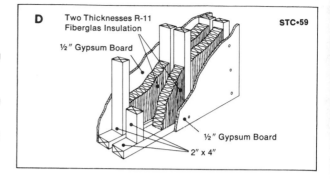

E STC•51 Staggered wood studs 16″
o.c.; single layer ½″ gypsum board
each side, two thicknesses R-11
Fiberglas insulation
STC•49 Staggered wood studs
16″ o.c.; single layer ½″ gypsum
board each side; one thickness R-11
Fiberglas insulation

STC•55 Staggered wood studs
24″ o.c.; double layer ½″ gypsum
board each side; one thickness R-11
Fiberglas insulation

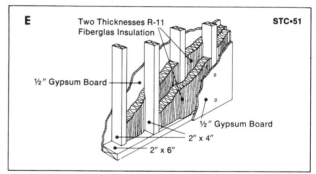

F STC • 45 2½″ metal studs 24″ o.c.;
single layer ½″ gypsum board
each side; one thickness R-8*
Fiberglas insulation
STC • 37 2½″ metal studs 24″
o.c.; single layer ½″ gypsum
board each side; no insulation

STC • 52 2½″ metal studs 24″ o.c.;
double layer ½″ type x gypsum
board each side; one thickness
R-8 Fiberglas insulation

*Nominal 2½″ Thick

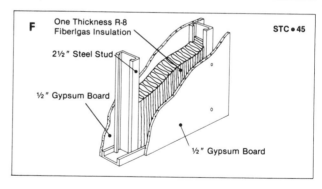

Figure 15.6 (Continued)

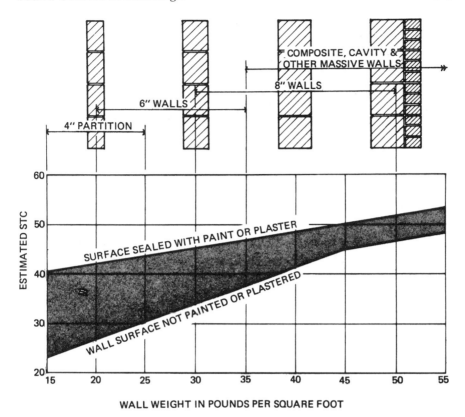

Figure 15.7 Typical STC values for sealed and porous concrete masonry walls according to the weight of the wall. (Courtesy of the National Concrete Masonry Association.)

Concrete and masonry walls, because of their relatively high surface density, are excellent construction materials for isolation. The acoustical performance of concrete and masonry walls follows closely the transmission loss as predicted by the mass law. The key to optimum performance lies in eliminating acoustical leaks, and thus masonry walls should be sealed with paint or plaster. Shown in Fig. 15.7 is the range of isolation one can expect for these massive walls in terms of surface density.

One other approach which needs mentioning is the addition of a thin sheet of lead in the wall construction. Field test reports indicate that an increase in STC of 5 to 7 dB can be expected if 1- or 2-lb/ft^2 lead sheet is included. Here the lead sheet is usually *furred* out 1 to 2 in. from the stud or masonry block, and a finished layer of gypsum board or plywood is then applied to cover the lead.

15.2.3 Floors and Ceilings

Airborne sound transmission is always a design consideration in floor and ceiling constructions, and the basic guidelines to sound isolation follow closely those for walls. However, the problem of structure-borne impact sound transmission usually is the more difficult to control. Conventional wood joist floor construction provides very little acoustical resistance to impact sound such as from footfalls. Therefore, additional design attention must be included where sensitivity to impact sound is anticipated.

The basic guidelines for control of structure-borne sound in floors and ceiling systems can be summarized as follows:

1. Cushion the actual impact with soft or resilient materials such as carpet, tile, pads, etc.
2. Include discontinuous construction by *floating* the finished floor
3. Provide discontinuous construction between the floor and the finished ceiling by using resilient channel mounting on the ceiling below or by using separate ceiling joists.
4. Include sound-absorbing materials in the cavity between the floor and ceiling.

Shown in Fig. 15.8 are several construction details for floors and ceilings showing popular construction methods utilized for controlling impact sound.

For purposes of design specifications and evaluation, several laboratory test methods have been developed for rating the acoustical isolation performance of floor and ceiling systems. The most common method utilizes a standardized *tapping* machine which delivers, with small hammers, successive impacts to the test specimen. Normalized sound pressure level measurements of the transmitted impacts are then made, and a single-number rating called the impact insulation class (IIC) is calculated. The IIC method is similar in many respects to the test method for determining the STC, and the reader is referred for further details to ANSI/ASTM E492, Standard Method of Laboratory Measurement of Impact Sound Transmission Through Floor-Ceiling Assemblies Using the Tapping Machine. For high-quality floor-ceiling systems, the IIC, like the STC, must exceed 40.

15.2.4 Doors

Doors and windows are often the "weakest link" in acoustical isolation for buildings. In particular, the use of undercut doors to accommodate rugs or carpets or to allow return air movement can destroy an effective acoustical isolation design. With respect to door construction, solid core doors are generally better sound barriers than hollow core

A. Wood Floors (all on 2″ x 10″ joists)
STC•53, IIC•73 Carpet and pad; particle board surface; plywood subfloor; single layer ½″ type x gypsum ceiling on resilient channel; one thickness R-11 Fiberglas insulation
STC•43, IIC•60 Carpet and pad; particle board surface; plywood subfloor; single layer ½″ type x gypsum ceiling attached directly to joists; one thickness R-11 Fiberglas insulation
STC•42, IIC•60 Carpet and pad; particle board surface; plywood subfloor; single layer ½″ type x gypsum ceiling attached directly to joists; no insulation

B. Lightweight (Cellular) Concrete Floors (1½″ thick over ⅝″ plywood subfloor on 2″ x 10″ joists)
STC•58, IIC•74 Carpet and pad; single layer ½″ type x gypsum ceiling attached to joists by resilient channel; one thickness R-11 Fiberglas insulation
STC•47, IIC•59 Carpet and pad; single layer ½″ type x gypsum ceiling attached directly to joists; no insulation

C. Steel Joist (7¼″ x 18 gage, @ 24″ o.c.) STC•56, IIC•71 Carpet and pad, ¾″ T & G plywood sub floor. ⅝″ type x gypsum board attached to ceiling joists by resilient channel; one thickness R-11 Fiberglas insulation
STC•43, IIC•57 Carpet and pad; ¾″ T & G plywood sub floor, ⅝″ type x gypsum board attached directly to the joists; no insulation

*The carpet and pad were selected to represent the type and quality commonly specified for new construction and was 20-oz. textured loop nylon carpet backed by a 40-oz. all-hair pad.

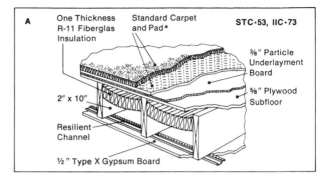

A | One Thickness R-11 Fiberglas Insulation | Standard Carpet and Pad* | **STC•53, IIC•73**
⅜″ Particle Underlayment Board
⅝″ Plywood Subfloor
2″ x 10″
Resilient Channel
½″ Type X Gypsum Board

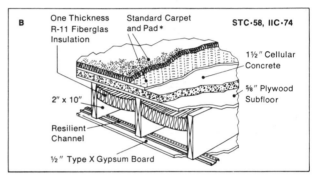

B | One Thickness R-11 Fiberglas Insulation | Standard Carpet and Pad* | **STC•58, IIC•74**
1½″ Cellular Concrete
⅝″ Plywood Subfloor
2″ x 10″
Resilient Channel
½″ Type X Gypsum Board

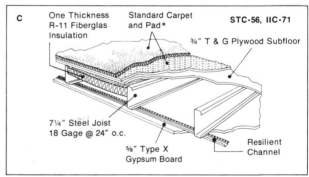

C | One Thickness R-11 Fiberglas Insulation | Standard Carpet and Pad* | **STC•56, IIC•71**
¾″ T & G Plywood Subfloor
7¼″ Steel Joist 18 Gage @ 24″ o.c.
⅝″ Type X Gypsum Board
Resilient Channel

Figure 15.8 Construction details for floor and ceiling systems that control impact sound. (Courtesy of Owens-Corning Fiberglas Corporation.)

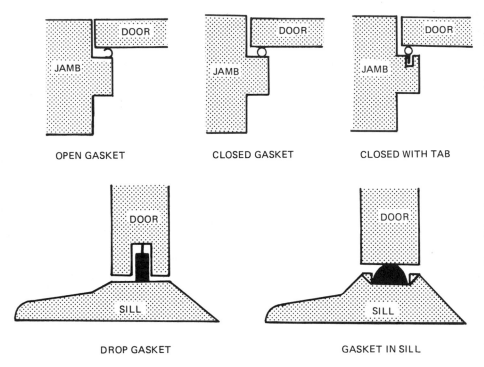

OPEN GASKET   CLOSED GASKET   CLOSED WITH TAB

DROP GASKET   GASKET IN SILL

Figure 15.9 Examples of acoustically effective doorjamb and sill constructions and devices.

doors. One practical method of raising the acoustical performance of a door is to include one or more layers of sheet lead on the inner surfaces. In addition, filling the cavity with absorbing material also improves the isolation performance. Generally, the sheet lead selected is 1/64 in. in thickness and weighs approximately 1 lb/ft^2. The absorbing material usually selected is rather dense, 3 to 4 lb/ft^3. With this laminate construction, sound isolation up to an STC of 42 is common.

As mentioned earlier, in most door installations, the area of primary concern for good sound isolation is the effectiveness of the door seal. Shown in Fig. 15.9 are some basic gasket configurations for sealing doorjambs, heads, and sills. It must be emphasized that the use of some form of gasket is absolutely essential to assure even a minimum level of privacy. In addition, these gaskets act as impact cushions during closure and hence reduce room-to-room or hallway disturbances.

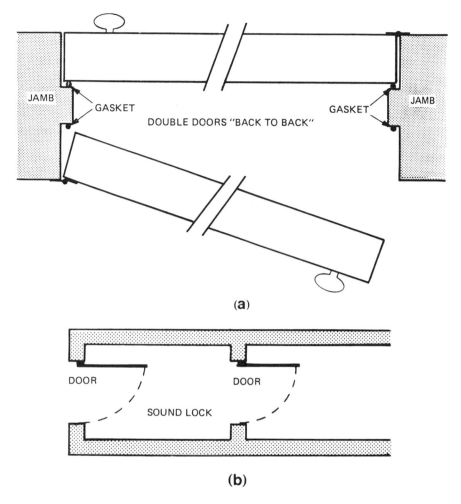

JAMB

GASKET

DOUBLE DOORS "BACK TO BACK"

GASKET

JAMB

(a)

DOOR

DOOR

SOUND LOCK

(b)

Figure 15.10 (a) Double doors and (b) acoustical lock for installations where extremely high noise levels are encountered.

Where extreme sound isolation is required, double doors back to back with gaskets should be considered, as illustrated in Fig. 15.10a. Another approach, the construction of a sound *lock*, as illustrated in Fig. 15.10b, should be considered where a high level of acoustical isolation is required.

It should be emphasized that sliding doors rarely provide effective isolation due to the difficulty in maintaining a tight acoustical seal at the jamb, head, and sill.

15.2.5 Windows

With respect to windows, here again acoustical seals are the area for
most concern. For single-glazed operating windows, wood frames,
in grooved guides, with weather stripping are best. Unfortunately,
for a typical 1/4-in.-thick pane, the average noise reduction will
rarely exceed 20 dB. However, with the addition of another pane and
a 2- to 3-in. spacing between panes, such as a storm window, the
attenuation will typically increase 6 to 8 dB.

   Where fixed plate glass is utilized, good acoustical isolation is
easier to achieve. Here the thickness of the glass can be selected up
to 1.5 in., which has a rather high surface density.

   The glass should be framed airtight in a resilient mount and be
caulked to assure elimination of acoustical leaks. For additional noise
reduction in the most demanding situations such as control rooms, an
additional pane separated 6 to 8 in. is recommended. Where two panes
are used, they should *not* be installed parallel but offset slightly to
minimize standing waves between the panes.

15.2.6 Summary

In summary, effective sound isolation in buildings does not just happen
but requires careful attention to detail from design through construction. In this brief treatment of the subject, only basic design guidelines have been presented. There is much more to the topic of sound
isolation; in fact, whole volumes have been published. However, there
is a *commonsense* aspect which must also be present for effective
isolation. For example, the back-to-back placement of electrical out-
lets or inset-type medicine cabinets can completely *scuttle* the acoustical
performance of a well-designed and well-constructed wall. Furthermore, room layouts that have the recreation room adjacent to a bed-
room or den only add to the problem of sound isolation.

15.3 MECHANICAL EQUIPMENT

Mechanical equipment associated with air conditioning, cooling, heating,
etc., can produce excessive airborne as well as structure-borne noise.
In most cases, the noise at the source is inherent and cannot be significantly or sufficiently lowered. Therefore, treatment of the noise
path is the only available approach to noise control.

   Mechanical room noise has as its origin motors, fans, pumps, etc.
Since we have treated noise control of this equipment in depth previously, only a few additional comments are required. Unquestionably,
the air-moving systems will require silencing, duct treatment, and
flow velocity control. Further, all rotating equipment, piping, con-

duits, etc., must have sufficient vibration isolation. For large installations such as apartments, hospitals, hotels, and office buildings,
the mechanical equipment is often housed in a single room. It usually
is prudent to give these rooms, if possible, a remote location relative
to noise-sensitive areas. The basement is the best location for obvious
reasons, and second best is the roof. When other locations must be
considered, locations along an external wall have advantages. Further,
if possible, select a location that has a natural *buffer* space adjacent
which is not extremely noise sensitive such as halls, utility rooms,
elevator shafts, stairways, etc.

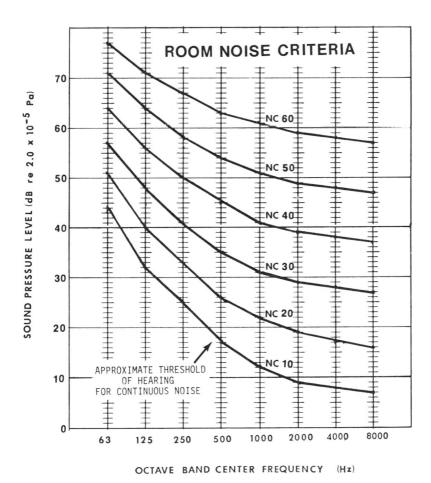

Figure 15.11 Spectral design criteria for room noise (NC curves).
Linear interpolation may be used to obtain NC 15, 25, 35, 45, 55, and
65.

Finally, resilient or floating floors, such as discussed in Sec. 15.2, should be seriously considered where mechanical equipment is installed on suspended flooring.

In treating mechanical noise associated with air conditioning, heating, ventilating, etc., a question frequently asked is "What do I design to? How quiet is quiet?" In short, what noise level design criterion is available for the many uses of rooms, space, etc. Shown in Fig. 15.11 are some spectral design guides (NC curves) for con-

Table 15.2 Recommended Noise Criteria of Acceptability for Rooms[a]

Type of Room or Space	NC, RC, or NCB Criteria
Recording or broadcast studios	10-20
Concert halls	15-20
Theatres, performance	20-25
Classrooms	25-30
Hospital rooms	25-35
Cinemas	30-35
Apartments, hotel rooms, residences	30-35
Offices, private	30-35
Corridors	30-35
Churches	30-35
Courtrooms	35-40
Libraries	35-40
Offices, open plan	35-40
Lobbies, waiting rooms, lounges	35-40
Restaurants, cafeterias	40-45
Washrooms	40-45
Stores (department stores, supermarkets)	40-50
Gymnasiums, pools	45-55
Tabulation/computer rooms	45-60
Manufacturing areas	50-70

[a] A range is given; the lower value is for the most noise-sensitive situation.
Source: Ref. 4.

tinuous broadband noise which have proved acceptable for a wide
variety of applications. These NC curves provide octave band level
design limits when used in conjunction with the noise criteria shown
in Table 15.2. For example, if one is designing the acoustical treatment of the air supply system for a library, the recommended noise
criterion is, from Table 15.2, NC 30 to 40. Now, referring to the
chart of Fig. 15.11 and selecting the curve NC 30, we obtain the
octave band sound levels that should not be exceeded to assure a negligible interference with the use of the library. In short, with these
band levels, the acoustical engineer has quantitative noise level limits
for selecting silencers, duct treatments, etc. The higher NC 40 might
be selected for less noise-sensitive areas such as the lobby, media
room, etc. If the use of the room were as a concert hall, the criterion
selected would be NC 15 to 20, which from Fig. 15.11 one sees clearly

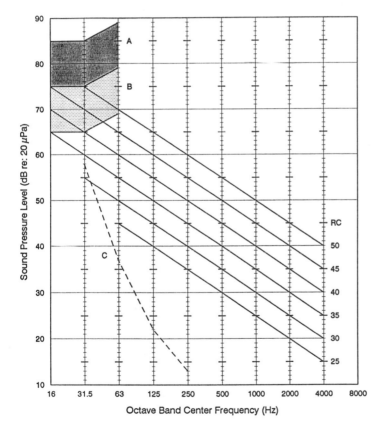

Figure 15.12   Room criterion (RC) curves. (From Ref. 5, with permission.)

is much more stringent and would require much more attention to design detail.

In addition to the NC curves, other criteria such as room criteria (RC) and balanced noise criteria (NCB) curves are gaining popularity with acoustical engineers.

For example, the RC curves shown in Fig. 15.12 are considered by the American Society of Heating, Refrigeration, and Air-Conditioning Engineers (ASHRAE) to be a preferred alternative to NC curves. This criterion accounts for the subjective response of individuals to spectral density and level. In addition, the RC curves include criteria at the octave bands with mean geometric center frequencies at 31.5 and 16 Hz. It should be noted that the extended low-frequency range is generally below the threshold of hearing (infrasound) and is sensed as a very low rumble or vibration.

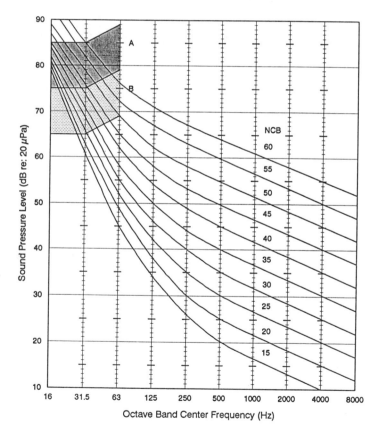

Figure 15.13 Balanced noise criterion (NCB) curves. (From Ref. 6, with permission.)

The methods for using the RC curves are tedious, very specialized, and beyond the scope of this text. The reader is therefore encouraged to pursue the references for details and applications.

The referenced NCB curves are a modified version of the NC curves. Here again, the criterion is extended to the 31.5 and 16 Hz octave bands (see Fig. 15.13). In addition, the curves are linear above 1000 Hz and the slope has been slightly increased. The NCB criterion can be applied in the same manner as the NC criterion, i.e., in conjunction with Table 15.2

The question of which criteria to use has been asked many times, and we shall leave the subject this way: the RC and NCB curves are more conservative than the NC curves. Thus, if the room or space usage is extremely noise sensitive, the RC or NCB curves can be applied. Rarely are there complaints of HVAC systems being too quiet!

15.4 PLUMBING NOISE

Plumbing noise has always been a troublesome source of annoyance, especially in apartment buildings. Presented in somewhat outline form are some design guidelines for its control.

1. The noise resulting from water rushing through pipes can be controlled by minimizing the number of bends, fittings, and valves. These items all induce turbulent flow in the fluid. Further, the flow velocity should be kept minimum by installing pressure-regulating and -reducing valves to maintain uniform pressure at each floor level.

2. Pipe *creaking* or *squeaking* noise generally associated with hot water pipes can be controlled by including flexible connections or

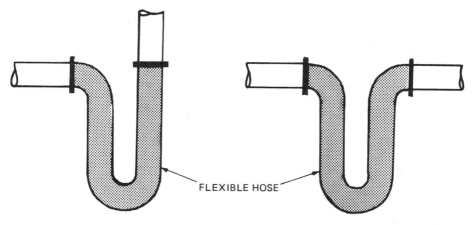

FLEXIBLE HOSE

Figure 15.14 Examples of flexible connections installed to allow pipe expansion and contraction.

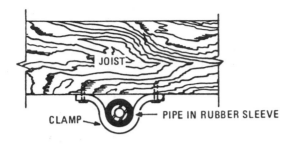

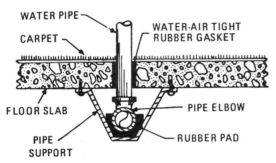

Figure 15.15 Methods of suspending and clamping pipes to reduce vibration and noise. (From Refs. 2 and 3.)

bends to allow the pipe to expand and contract. Shown in Fig. 15.14 are some examples of flexible connections.

3. Mounting or suspending pipes in elastomeric materials reduces pipe vibration and associated noise. Shown in Fig. 15.15 are some mounting and suspension methods.

4. Locating pipes in the walls of usually quiet or noise-sensitive areas should also be avoided.

Finally, the impulsive hammeringlike sound common to pipe systems is due to the sudden interruption, starting, or stopping of water flow. The flow discontinuity generates a pressure wave, which in a sense reflects back through the system, producing a series of impulsive sounds as if someone were striking the pipe with a hammer. Little can be done about this problem, except that including compressed air accumulators in long runs has been shown to be effective.

REFERENCES

1. P. H. Parkin and H. R. Humphreys. *Acoustics Noise and Buildings.* Praeger, New York, 1958.

2. R. D. Berendt and E. L. R. Corliss. *Quieting; a Practical Guide to Noise Control.* 1976.
3. R. Berendt, G. E. Winzer, and C. B. Burroughs. *A Guide to Airborne, Impact, and Structure Borne Noise-Control in Mutlifamily Dwellings.* U.S. Department of Housing and Urban Development, Washington, D.C., Sept. 1967.
4. *ASHRAE Guide and Data Book-Systems.* American Society of Heating, Refrigerating and Air Conditioning Engineers, Inc., New York, 1991.
5. W. E. Blazier, Jr. Revised noise criteria for application in the acoustical design and rating of HVAC systems. *Noise Control Eng. 16* (1981).
6. L. L. Beranek. Application of NCB noise criterion curves. *Noise Control Eng. 33* (1989).

BIBLIOGRAPHY

Acoustics in Plumbing systems. In *ASPE Databook.* American Society of Plumbing Engineers, Westlake, Calif., 1981–1982.
Beranek, L. L. *Noise and Vibration Control.* McGraw-Hill, New York, 1971.
Bishop, D. E., and R. D. Horonjeff. Procedures for Developing
Computing the Effective Perceived Noise Level for Flyover Aircraft Noise, Definitions and Procedures for. *American National Standards Institute S6.4-1073.* American National Standards Institute, New York, 1973.
Egan, D. M. *Concepts in Architectural Acoustics.* McGraw-Hill, New York, 1972.
Faulkner, L., *Handbook of Industrial Noise Control.* Industrial Press, New York, 1975.
Harris, C. M. *Handbook of Noise Control.* McGraw-Hill, New York, 1957.
Harris, C. M. *Handbook of Acoustical Measurements and Noise Control.* McGraw-Hill, New York, 1991.
Harris, C. M. (ed.). Specifications. In *Handbook of Utilities and Services for Buildings.* McGraw-Hill, New York, 1990.
Hedden, J. *Plumbing for Old and New Houses.* Creative Homeowner Press, Upper Saddle River, N.J., 1980.
Hirschorn, M. Noise Level Criteria. *Bulletin 6.0010.0.* Industrial Acoustics Company, Bronx, New York, 19
Information on Levels of Environmental Noise Requisite to Protect Health and Welfare with an Adequate Margin of Safety. *Report No. 550/9-74-004.* U.S. Environmental Protection Agency, Washington, D.C., 1974.
Irwin, J. D., and E. R. Graf. *Industrial Noise and Vibration Control.* Prentice-Hall, Englewood Cliffs, New Jersey, 1979.

Jones, R. S. *Noise and Vibration Control in Buildings.* McGraw-
    Hill, New York, 1984.
Noise Exposure Forecast Areas for Aircraft Flight Operations. *Re-
    port No. FAA-D5-67-10.* Federal Aviation Administration,
    Washington, D.C., 1967.
Public Health and Welfare Criteria for Noise. *Report No. 550/9-73-002.*
    U. S. Environmental Protection Agency, Washington, D.C., 1973.
Robinson, D. W. The Concept of Noise Pollution Level. *National
    Physical Laboratory Aero Report Ac 38.* National Physical Labora-
    tory, Teddington, Middlesex, England, 1969.
Robinson, D. W. *J. Sound Vib. 14* (1971).
Shultz, T. J. Noise Assessment Guidelines: Technical Background.
    Report No. TE/NA 172. Department of Housing and Urban
    Development, Washington, D.C., 1971.
Uniform Plumbing Code. International Association of Mechanical and
    Plumbing Officials, Walnut, Calif., 1988.
Water Hammer Arresters. The Plumbing and Drainage Institute,
    Indianapolis, September 1983.
Webster, J. C. SIL: past, present, future. *Sound Vib.* (Aug.
    1979.)
Yerges, L. F. *Sound, Noise and Vibration Control.* Van Nostrand
    Reinhold, New York, 1969.
Young, R. W. Sabine reverberation equation and sound power
    calculations. *J. Acoust. Soc. Am. 31* (1959).

EXERCISES

15.1 A small-theater designed for stage plays has a volume of 150,000
ft^3. What reverberation time at 500 Hz would assure good acoustics
if the theater were to be used dominantly for speech?

Answer: 1.3 seconds

15.2 Referring to Exerciese 15.1, what reverberation time would be
selected at
a.   125 Hz?
b.   250 Hz?
c.   1000 Hz?
d.   2000 Hz?
e.   4000 Hz?

Answer: (a) 2.0 s approx-
                   imately;
           (b) 1.5 s;
           (c) 1.3 s;
           (d) 1.3 s;
           (e) 1.3 s

15.3 If the theater of Exercise 15.1 was to be used for musical pro-
ductions, what reverberation time would be selected at
a. 500 Hz?
b. 250 Hz?
c. 125 Hz?

> *Answer:* (a) 1.7 s approx-
> imately;
> (b) 2.0 s;
> (c) 2.6 s

15.4 A symphony hall has a volume of 25,000 m^3. The estimated total
absorption in the hall with a full house is 2000 metric sabins at 500
Hz. Estimate the reverberation time at 500 Hz.

> *Answer:* 2.0 s

15.5 Referring to Exercise 15.4, how does this reverberation time
compare to the recommended reverberation time at 500 Hz?

> *Answer:* Approximately 0.25
> s below the recom-
> mended reverbera-
> tion time

15.6 If the relative humidity of the symphony hall in Exercise 15.4
is typically 40%, what is the reverberation time at 4000 Hz? (Hint:
To obtain the air absorption coefficient m in metric units, multiply
the values in Table 9.1 by 3.28.)

> *Answer:* 1.78 s

15.7 An architect is designing a library for a community college.
What noise criterion should be selected for the reading rooms?

> *Answer:* NC 30

15.8 Referring to Exercise 15.7, what are the octave band noise
level limits for the air-conditioning system?

> *Answer:* 57 dB at 63 Hz;
> 48 dB at 125 Hz;
> 41 dB at 250 Hz;
> 35 dB at 500 Hz;
> 31 dB at 1000 Hz;
> 29 dB at 2000 Hz;
> 28 dB at 4000 Hz;
> 27 dB at 8000 Hz;

16
COMMUNITY AND ENVIRONMENTAL NOISE CONTROL

The outdoor acoustical environment varies dynamically in magnitude
and character throughout most communities. The sound level variation can be temporal (depending on the time of day or season), spectral (depending on the source type), or spatial (depending on one's
location). Further, the number of noise sources is awesome, but recent studies have shown that transportation vehicles are the worst
offenders, with construction and industrial plants a solid second.
With these factors, it is clear that control of community noise is
neither simple nor straightforward. It must be emphasized, however,
that the number of basic noise mechanisms for community noise is not
large, and with a few minor exceptions such as tire noise, dogs, crick-
ets, etc., they have been discussed in detail in the preceding chapters. Further, systematic methods of noise control have also been
presented. We shall therefore focus attention in this chapter on
current methods of measuring and characterizing community noise
and, to some extent, estimating the community response.

Over the years, many methods have been developed for describing community noise, and we can anticipate an ongoing development
of new methods upon revision of the old. Therefore, in this chapter,
we shall review only the major methods and discuss the basic measurement and calculation procedures for the descriptors. It should be
emphasized that the methods selected for discussion should not be
considered complete but rather representative. With this background
and understanding, the noise control engineer can compare measured
data to regulatory noise criteria to determine compliance and the required level of noise reduction for design purposes.

16.1 DESCRIBING COMMUNITY NOISE

Perhaps the foremost function of describing community noise is to
relate the physical properties of the sound environment to human reaction or response. A noise descriptor is a physically measurable and/
or mathematically calculable unit by which the statistical response of
humans can be quantitatively or numerically scaled. In short, the
descriptor provides an objective scale for predicting the subjective
response of people to noise. The precise correlation between the
human reaction or response to noise has eluded researchers over the
years. However, four parameters appear to be first order in relation
to annoyance:

1. Intensity
2. Pitch
3. Time or duration
4. Intrusiveness

Intuitively, the intensity or loudness level of the noise must be considered foremost. Accordingly, if discrete or pure tones are present,
a strong negative response is generally evoked from the public, regardless of the level. Further, people tend to resent an extended
duration of noise and, to compound the matter, find intrusive or
intermittent duration extremely distracting and thus disturbing.
Despite the physical and phychological complexity, the current noise
rating schemes or descriptors show progress in statistically predicting
community response. As such, in adopting noise regulations, federal,
state, and municipal agencies are using the descriptors as criteria.

16.1.1 Audibility

Probably the oldest and simplest form of rating noise is in terms of
audibility or the capability of being heard. Being audible is not particularly subjective but depends strongly on the simultaneous presence of other sounds. As such the extraneous presence of background
noise may *mask* the sound of interest. Because of this masking factor
and the extreme severity of the parameter, no formal rating schemes
have been adopted using audibility as a descriptor. It is also pretty
well acknowledged that in modern living nobody is entitled to silence
as inferred by audibility. Further, quantitative levels in terms of
audibility appear to be, at best, nebulous in structure or order.

16.1.2 Weighted Sound Levels

It was obvious to early researchers that a weighted spectral sound
level, following closely the frequency response of the human ear, was

essential. Three frequency-weighted scales, A, B, and C, as mentioned in Chap. 3 and illustrated again in Fig. 16.1, have been used
extensively. These scales differ from each other in the amount each
discriminates against sound at lower frequencies. The A scale is most
discriminating and follows closely the response of the human ear to
low-level (below 60 dB) sounds. The B scale is less discriminating
and follows the response of the ear at moderately high levels (above
60 dB). One popular application of the B scale was in early research
regarding the noise impact of trucks. The C scale is essentially flat
or uniform over the range of human hearing of most interest. Therefore, for most practical purposes, the C scale represents absolute
sound level and provides a baseline for comparison with other scales.

Of these three, the A scale is used the most extensively, both as
an overall descriptor and as a parameter for calculating more sophisticated ratings. By itself, the A scale does not provide a measure
of pure tone content or time-varying intrusion, which relates heavily

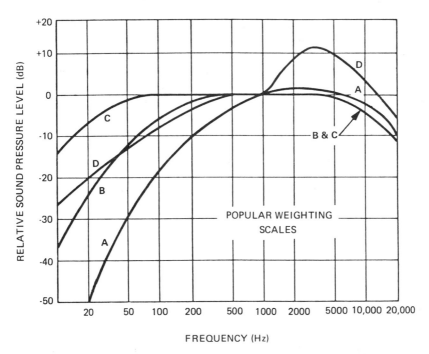

Figure 16.1 International standard A, B, and C weighting curves.
Also shown is the D weighting curve for jet aircraft noise. (From
Refs. 21 and 22.)

to annoyance. However, the simplicity of measurement is a very
positive advantage.

Two other weighting scales, the D and E scales, are gaining in
popularity, and the D scale is also shown in Fig. 16.1 [1]. Both,
however, place additional emphasis on higher frequencies (above 1000
Hz) and are applied most often to jet aircraft noise.

16.1.3 Loudness

Since loudness plays such a vital role in a listener's sensory response,
further elaboration to previous discussions is in order. In Chap. 3,
we discussed the calculation of loudness in terms of sones and loudness
levels in terms of phons. From these principal and extensive psychoacoustic studies, investigators have developed relationships for matching loudness. In particular, shown in Fig. 16.2 [1,2] are the standardized equal loudness contours showing subjectively the loudness of pure
tones compared to a reference 1000-Hz tone. These equal loudness
contours are the statistical results of observations of human subjects
in a free-field listening environment.

To see the importance of these contours, note that a 50-dB pure
tone at 1000 Hz has a loudness level of 50 phons. To be judged or
perceived as equally loud, a pure tone at 63 Hz would have to be at a
sound level of 65 dB, or 15 dB higher. Note also that at a level of
100 dB there is considerably less difference in the equal loudness contour at 63 and 1000 Hz. From this example, we see how both sound
level and spectral quality must always be considered. Stevens [23]
has extended the concept of loudness and loudness level to *perceived*
loudness level. The extensions include lower frequency ranges and
higher sound pressure levels (130 db).

Since sounds we are exposed to in our everyday living are generally more complex than pure tones at a steady level, the equal loudness descriptors cannot be applied directly for regulatory criteria,
noise impact assessment, and/or response. Extensive studies involving narrow bands of broadband noise are currently being conducted which should provide closer relationships. In any event,
these equal loudness contours form the basis for, or are present to
some extent in, almost all the sophisticated rating schemes or criteria.

16.1.4 Perceived Noise Level

To account for temporal variation in loudness such as aircraft flyovers, a rating called the perceived noise level (PNL), or symbolically L_{PN}, was developed [3–9]. Here again a jury of listeners was

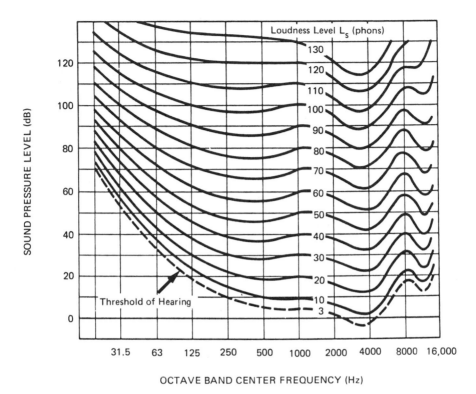

Figure 16.2 Curves of equal loudness level for pure tones in a frontal sound field, according to ISO Recommendation 226. These curves show how the frequency response of the human ear varies with the loudness level. (From Refs. 1 and 2.)

asked to compare sounds of equal duration for the subjective features of noisiness, annoyance, or unacceptability rather than just loudness. From the response of the listeners, contours of equal noisiness, in the units *noys* were developed analogously to the sone for loudness. Shown in Fig. 16.3 are the standardized equal perceived noisiness contours for octave bands [9,10]. With these contours or values in tabular form (see App. G), the total perceived noise PN can be calculated from [11]

$$PN = N_m(1 - K) + K\sum_{i=1}^{n} N_i \qquad [\text{noys}] \qquad (16.1)$$

where

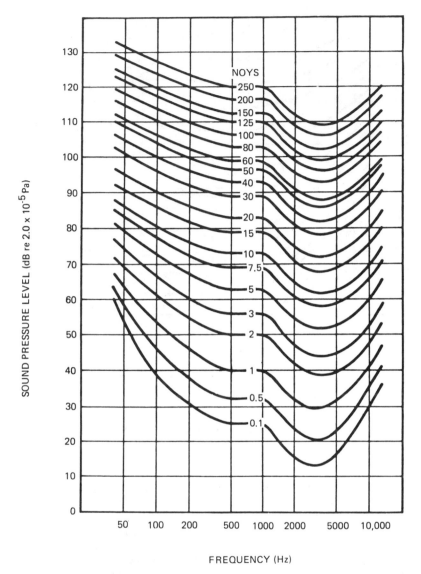

Figure 16.3 Curves for determining the noisiness of sound used in Kryter's method of calculating the perceived noise level (PNL). (From Refs. 1, 3, and 9.)

N_m = maximum perceived noisiness

N_i = noisiness in each band including N_m

K = factor: K = 0.3 for one octave bands, 0.2 for one-half
     octave bands, and 0.15 for one-third octave bands

n = number of bands

Now the total perceived noise PN is related to the perceived noise
level L_{PN} as follows:

$$L_{PN} = 40 + 33.3 \log(PN) \qquad [PNdB] \tag{16.2}$$

The unit for perceived noise level L_{PN} is PNdB.
Consider an example.

Example
Determine the perceived noise PN and perceived noise level L_{PN}
for the octave band analysis (instantaneous) of the following
aircraft flyover:

Center Frequency (Hz)	Sound Pressure Level (dB)
63	90
125	96
250	101
500	103
1000	101
2000	103
4000	98
8000	95

Solution
From the table in App. G, we get the noisiness of each octave
band, and for the sum we obtain

$$\sum_i N_i = 15 + 32 + 60 + 79 + 69 + 133 + 116 + 67$$

$$= 571 \text{ noys}$$

Using K = 0.3, since we have octave bands and since $N_m = 133$,
upon substitution in Eq. (16.1), we have

$$PN = 133(1 - 0.3) + 0.3 \times 571$$

$$= 93.1 + 171.3$$

$$= 264.4 \text{ noys}$$

From Eq. (16.3), the perceived noise level is

$$L_{PN} = 40 + 33.3 \log 264.4$$

$$= 40 + 80.7$$

$$= 120.7 \text{ PNdB}$$

A simpler method using the D weighted scale, available on some sound level meters, has been proposed and provides a good first-order estimate. Here the peak D weighted sound level is measured, and the perceived noise level L_{PN} is calculated from

$$L_{PN} = L_D + 7 \qquad [\text{PNdB}] \tag{16.3}$$

where L_D is the peak D weighted sound level.

Furthermore, for community noise, where aircraft noise is not dominant, the following relationship between A weighted sound level L_A and the perceived noise level L_{PN} also yields good first-order results:

$$L_{PN} = L_A + 13 \qquad [\text{PNdB}] \tag{16.4}$$

In summary, the perceived noise level can be used to predict or assess relative annoyance or noisiness due to broadband noise. It was designed specifically for jet aircraft noise but has been shown to be reasonably applicable to other forms of community noise.

16.1.5 Effective Perceived Noise Level

Further refinements to the perceived noise level to account more specifically for duration and the presence of clearly audible discrete tones have led to the development of the rating scheme called the effective perceived noise level (EPNL), symbolically L_{EPN}. Accordingly, EPNL is defined as

$$L_{EPL} = L_{PN} + C + D \qquad [\text{EPNdB}] \tag{16.5}$$

where

C = correction for pure tones (generally less than 3 dB)

D = time-integrated correction for noise duration

Now the actual calculation of EPNL as given in Eq. (16.5) is quite involved and in practice requires some form of computer-aided support.

However, a good first-order approximation for the effective perceived noise level can be obtained using

$$L_{EPN} \simeq L_{Dmax} + 7 + 10 \log(t_2 - t_1)/T_0 \qquad [EPNdB] \qquad (16.6)$$

where

L_{Dmax} = maximum D weighted sound level during an aircraft flyover (dB)

T_0 = normalizing constant having the dimensions of time, usually 15 s

$t_2 - t_1$ = time interval during which Lp is within 10 dB of its maximum value

It should be noted that the parameters to be measured in Eq. (16.6) require only basic laboratory instrumentation. This method is described in detail in ISO Recommendation 507 and its newest amendments (April 1968).

We shall not pursue this rating scheme further, as it applies almost exclusively to jet aircraft noise. The reader is, however, referred to the references for complete details, especially ANSI S6.4-1973. Computing the Effective Perceived Noise Levels for Flyover Aircraft Noise [12] and Federal Aviation Regulation Part 36, Aircraft Certification Procedure [7].

16.1.6 Noise Exposure Forecast

The noise exposure forecast NEF is a rating scheme usually used for land use planning in the vicinity of airports. The NEF is a single-number index relating community response to the following:

1. Number of aircraft operations (takeoffs and landings)
2. Types of aircraft (707, 747, etc.)
3. Time of day [daytime (7 a.m. to 10 p.m.) or nighttime (10 p.m. to 7 a.m.)]

The basic descriptor in the calculation is the day-night average sound level [(DNL) see section 16.1.11], and the final format is a set of equal exposure contours over the airport and its surroundings. Shown in Fig. 16.4 is an example of the NEF contours for Los Angeles International Airport.

For land-planning purposes, the NEF zones can be interpreted in terms of community reaction as follows:

1. NEF < 20: No complaints expected.
2. $20 \leq$ NEF < 30: May interfere with community activities in residential areas.

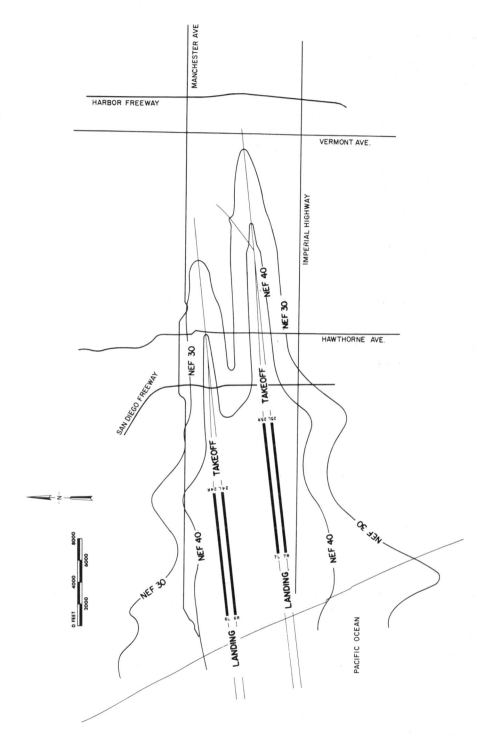

Figure 16.4 Example of noise exposure forecast for a large airport showing the contours of equal NEF values.

544

3. $30 \leq NEF < 40$: Individuals in residential areas may complain; group action possible.
4. $40 \leq NEF$: Repeated vigorous complaints expected; group action probable.

We shall not go into detail regarding the methods of calculation but refer the reader to Ref. 13. Here again the rating scheme is almost exclusively used for jet aircraft. However, some equivalent schemes for STOL, VTOL, helicopters, and private aircraft have been developed and are being evaluated.

16.1.7 Statistical Descriptors

Measured noise levels displayed as a function of time provide a useful scheme for describing the acoustical *climate* of a community. For example, shown in Fig. 16.5 is the time history or temporal pattern of the A weighted noise level for an 8-min sample of community noise [3]. This sample, obtained from a continuous graphic level recording, is useful for identifying the major sources of noise. However, in itself there is no quantitative means of comparing the acoustical environment to another. Illustrated in Fig. 16.6 is a statistical presentation of community noise showing the histogram and the cumulative distribution function. Here the percentage of time a given sound level was exceeded was measured on a commercially available instrument called a sound distribution analyzer. In this form, one can look at *key* intercepts of the cumulative distribution curve and draw some quantitative conclusions. For example, referring to Fig. 16.7, if one looks at the intercept of the cumulative distribution at 90%, symbolized L_{90}, we have the noise level exceeded 90% of the time. Since this represents "most" of the time, L_{90} generally provides a good measure of the ambient or background noise of the measurement site. Correspondingly, the intercept at 10%, L_{10}, the noise level exceeded only 10% of the time, provides a good measure of the intermittent or intrusive noise. Examples of intrusive noise include traffic, aircraft flyovers, barking dogs, etc. Another statistical parameter, L_{50}, or the level exceeded 50% of the time, yields a measure of the medium level. Naturally, any intercept may be used, and L_1 and L_{99} are used occasionally. It is, however, L_{90}, L_{50}, and L_{10} that are most often used, and with these, one can characterize environmental noise quantitatively. For example, if the value of L_{90} measured in a residential area were to increase from 45 to 50 dBA with the opening nearby of a new highway or industrial plant, one could reasonably assess the acoustical impact of the highway.

It must be emphasized that statistical parameters also vary with time. In Fig. 16.7a and 16.7b are charts showing the hourly variation of the L_{10}, L_{50}, and L_{90} parameters. Note in the case of the wooded

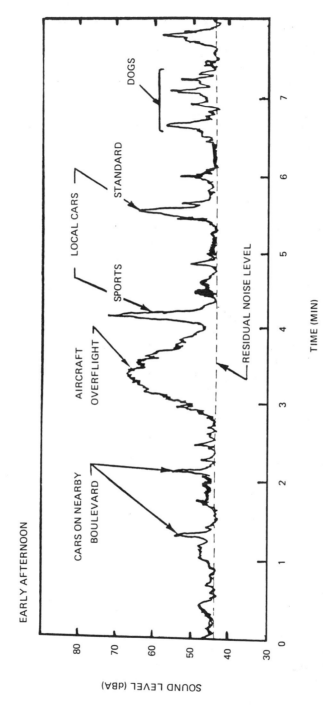

Figure 16.5 Time history of A weighted sound levels of some community noise. (From Ref. 10.)

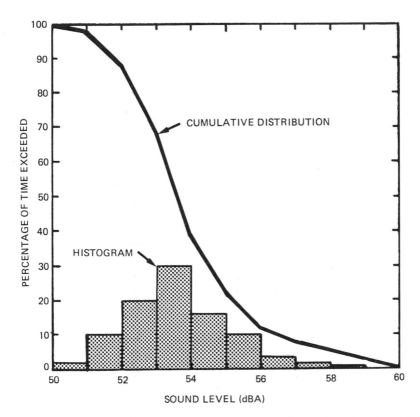

Figure 16.6 Statistically processed community noise showing histogram and cumulative distribution of A weighted sound levels.

area (16.7a) that there are dynamic changes in the ambient L_{90} due to natural uncontrollable noise sources such as insects and birds. Referring to Fig. 16.7b, observe that the level of intrusive noise L_{10} rises sharply in a residential area during the afternoon hours after school. The rather low ambient L_{90} in the early evening suggests the season of winter, as there is an absence of bird and insect noise.

In summary, these easy-to-measure statistical parameters provide a quantitative method of characterizing the time variation of community noise. Further, these parameters form the basis for other composite rating schemes.

16.1.8 Equivalent Sound Level

A single-number descriptor which is very powerful in describing time-varying environmental noise is the equivalent sound level (LEQ),

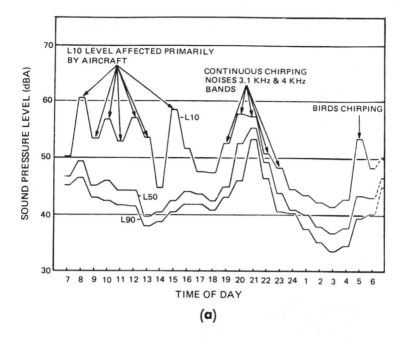

(a)

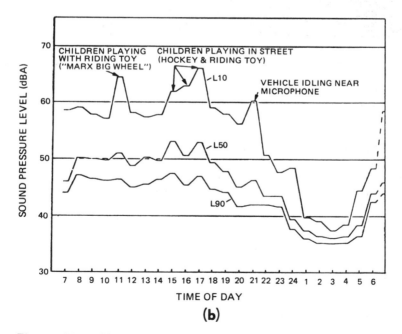

(b)

Figure 16.7 Charts showing the hourly variation of the statistical noise descriptors for (a) a wooded area and (b) a residential area. (From Ref. 10.)

symbolized L_{eq}. The equivalent sound level is defined mathematically as

$$L_{eq} = 10 \log_{10}\left(\frac{1}{T} \int_0^T 10^{L/10} \, dt\right) \qquad [dB] \qquad (16.7)$$

where

L = sound pressure level, a function of time (dB)

T = time interval of observation (s)

If A weighted sound levels are used, which is most often the case,
then LEQ is in terms of dBA. It is not easy to interpret Eq. (16.7);
however, one can consider LEQ as a constant sound level equivalent
on an energy basis of a time-varying sound level over the same time
period. An example will clarify the concept.

Shown in Fig. 16.8 is a chart, the solid line showing the time
variation of some community noise over a 24-h period. The dashed
line illustrates the measured LEQ for the same period and represents
a constant sound level that would have contained equivalent sound
energy.

Now, if the A weighted sound levels are obtained over discrete
time intervals, the A weighted equivalent sound level LEQ is computed
from the following equation, which is just a simplification of Eq. (16.7):

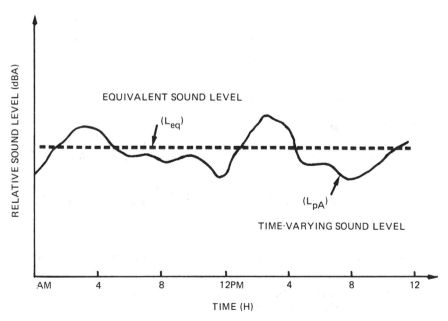

Figure 16.8 Chart illustrating the equivalent sound level for a 24-h
period.

$$L_{eq} = 10 \log \left[\sum_{j=1}^{N} (P_j) \, 10^{L_j/10} \right] \quad [dBA] \qquad (16.8)$$

where

  N = number of intervals (unitless)

  P_j = fraction of time in interval j (unitless)

  L_j = A weighted sound level of interval j (dBA)

Consider an example.

Example

By utilizing statistical measurement methods, the noise data shown in the following table were obtained over a 24-h period in a residential neighborhood. Calculate the daily equivalent sound level L_{eq}.

Interval j	A Weighted Sound Level L_j (dBA)	Fraction of Time in Interval P_j
1	50	0.01
2	55	0.10
3	60	0.30
4	65	0.40
5	70	0.10
6	75	0.09

Solution
From Eq. (16.9),

$$L_{eq} = 10 \log \left[0.01 \log^{-1}\left(\frac{50}{10}\right) + 0.1 \log^{-1}\left(\frac{55}{10}\right) + 0.30 \log^{-1}\left(\frac{60}{10}\right) \right.$$

$$\left. + 0.40 \log^{-1}\left(\frac{65}{10}\right) + 0.1 \log^{-1}\left(\frac{70}{10}\right) + 0.09 \log^{-1}\left(\frac{75}{10}\right) \right]$$

$$= 67.4 \text{ dBA}$$

Thus the equivalent sound level in the residential neighborhood is 67.4 dBA.

For the special case when a sound level is constant for a period of time and negligible thereafter, Eqs. (16.7) and (16.8) reduce to

$$L_{eq} = L_j + 10 \log_{10}(P_j) \quad [dB] \qquad (16.9)$$

Consider an example.

Example

The noise level from a compressor measured at the boundary line of a factory was 110 dBA. On Monday the compressor ran for 1 h, on Tuesday it ran for 2 h, and on Wednesday it ran for 10 h. Calculate the L_{eq} for each day.

Solution

Since we are dealing with days or 24 h as the total time duration, P_j for Monday is 1/24 h. Thus from Eq. (16.9), L_{eq} for Monday is

$$L_{eq} = 110 + 10 \log_{10}\left(\frac{1}{24}\right)$$

$$= 110 - 13.8$$

$$= 96.2 \text{ dBA}$$

For Tuesday, $P_j = 2/24$; thus,

$$L_{eq} = 110 + 10 \log_{10}\left(\frac{2}{24}\right)$$

$$= 110 - 10.8$$

$$= 99.2 \text{ dBA}$$

For Wednesday, $P_j = 10/24$, and

$$L_{eq} = 110 + 10 \log_{10}\left(\frac{10}{24}\right)$$

$$= 110 - 3.8$$

$$= 106.2 \text{ dBA}$$

From the results of this example, one sees the equal energy relationship clearly. That is, for the same level 110 dBA, doubling the duration from 1 to 2 h increased the LEQ by 3 dBA, consistent with previous considerations. Further, an increase in the duration by a factor of 10 raised the equivalent sound level 10 dBA. It follows then that the LEQ can be combined logarithmically in exactly the same manner as the sound power level L_W.

We shall leave the subject of equivalent sound level at this point, but it must be emphasized that the LEQ tends to correlate well with many psychoacoustical effects and is used extensively as the base descriptor for many other environmental noise rating schemes.

Another descriptor closely related to LEQ which deserves mentioning is the sound exposure level (SEL). The SEL is mathematically similar to LEQ except that the sound level is averaged over a reference duration of one second instead of over the entire measurement period. The SEL is used predominantly to compare transient noise level excursions or to calculate the corresponding LEQ for a given period using the individual SEL values for different events occurring within the period. The SEL descriptor, however, has not gained prominence in regulatory areas.

16.1.9 Noise Pollution Level

Related to the LEQ, but gaining widespread application to all types of environmental noise sources, is the descriptor noise pollution level NPL, or symbolically L_{NP} [14,15]. The NPL is defined mathematically as

$$L_{NP} = L_{eq} + k\sigma \qquad [dBA] \qquad (16.10)$$

where

L_{eq} = equivalent sound level over a given time period, usually A weighted (dBA)

σ = standard deviation of the instantaneous sound level considered as a statistical time series over the same time period

k = arbitrary coefficient, tentatively chosen as 2.56

In Eq. (16.10), the first term is largely a measure of sound level, and the second term is a measure of the variability in sound level. In short, the NPL combines the first-order effect of loudness with the disturbing quality of level fluctuation or the presence of intrusive events.

Current research has shown that for first-order approximations Eq. (16.10) can be rewritten in terms of the easy-to-measure statistical parameters as follows [13,16]:

$$L_{NP} = L_{eq} + (L_{10} - L_{90}) \qquad [dBA] \qquad (16.11)$$

and

$$L_{NP} = L_{50} + (L_{10} - L_{90}) + (L_{10} - L_{90})^2/60 \qquad [dBA] \qquad (16.12)$$

A note of caution: Where highway noise is dominant, especially where trucks form a large percentage of the mix, Eqs. (16.11) and (16.12) may give rather poor results.

16.1.10 Community Noise Equivalent Level

The community noise equivalent level CNEL is a single-number index designed to rate environmental noise on a daily or 24-h basis [17,18]. In this rating scheme, weightings are selected to account for increased sensitivity to noise in the evening and nighttime hours.

Mathematically, the CNEL is given by

$$CNEL = 10 \log \left[1/24 \left(12 \times 10^{L_d/10} + 3 \times 10^{(L_e+5)/10} \right. \right.$$

$$\left. \left. + 9 \times 10^{(L_n+10)/10} \right) \right] \quad [dBA] \qquad (16.13)$$

where

L_d = A weighted LEQ for daytime period (7 a.m. to 7 p.m.) (dBA)

L_e = A weighted LEQ for evening period (7 p.m. to 10 p.m.) (dBA)

L_n = A weighted LEQ for nighttime period (10 p.m. to 7 a.m.) (dBA)

Examining Eq. (16.13), one sees that the CNEL is really just a weighted modification of LEQ. In particular, the measured value for the LEQ for the evening period (the middle term in the argument of the logarithm) is increased or weighted by 5 dBA. Further, the LEQ for the nighttime period (the last term in the argument of the logarithm) is increased or weighted by 10 dBA. In short, without the weighting factors, the CNEL reduces to LEQ. Consider an example.

Example
Equivalent sound levels were measured for a 24-h period on the property line of a racetrack. Broken into daytime, evening, and nighttime periods, the LEQ levels were L_d = 70 dBA, L_e = 70 dBA, and L_n = 60 dBA. Calculate the CNEL for the 24-h survey.
Solution
From Eq. (16.13),

$$CNEL = 10 \log \left[\frac{1}{2} \log^{-1}(7) + \frac{1}{8} \log^{-1}(7.5) + \frac{3}{8} \log^{-1}(7) \right]$$

$$= 71.0 \text{ dBA}$$

Thus the CNEL is 71 dBA, even though the equivalent sound levels never exceeded 70 dBA.

16.1.11 Day–Night Average Sound Level

Another weighted single-number index is the day-night average sound level DNL, or symbolically L_{dn}. The DNL is identical to the CNEL except that the weighted evening term L_e is deleted and the daytime period is extended to 10 p.m., that is, 7 a.m. to 10 p.m. In equation form, we have

$$L_{dn} = 10 \log \left[1/24 \left(15 \times 10^{L_d/10} + 9 \times 10^{(L_n + 10)/10} \right) \right]$$

$$[dBA] \quad (16.14)$$

where

L_d = A weighted LEQ for daytime period (7 a.m. to 10 p.m.)

L_n = A weighted LEQ for nighttime period (10 p.m. to 7 a.m.)

It should be emphasized that most of the parameters and descriptors discussed herein can be measured directly. Community noise analyzers, as described in Chapter 5, are available and generally provide direct measurements of the statistical exceedance parameters and the descriptors LEQ, SEL, DNL, CNEL, etc. Shown in Appendix H are some state and municipal standards which illustrate how these desriptors are used in regulating community and environmental noise.

16.1.12 Speech Interference Level

Another descriptor which needs brief mention is the speech interference level SIL. Frequently in a work environment it is important to determine the effect of background noise on speech communication.

Table 16.1 Vocal Effort Required for Reliable Speech Communication at Given Speech Interference Levels[a]

Distance Between Speaker and Listener (ft)	Vocal Effort, SIL (dB)			
	Normal	Raised	Very Loud	Shout
1	69	75	81	87
2	63	69	75	81
4	57	63	69	75
8	51	57	63	69
16	45	51	57	63
32	39	45	51	57

[a]The table represents a condensation of data from several sources and assumes the voice is male. For female voices, subtract 5 dB from the SIL.
Source: Ref. 18, 19, and 20.

One method developed primarily by Beranek for steady continuous noise is the preferred speech interference level PSIL [13,19]. The PSIL is the arithmetic average of the sound pressure level in the octave bands whose center frequencies are 500, 1000, and 2000 Hz. In mathematical form, we have

$$\text{PSIL} = \frac{L_{p500} + L_{p1000} + L_{p2000}}{3} \quad \text{[dB]} \quad\quad (16.15)$$

where L_{p500}, L_{p1000}, and L_{p2000} are the octave band sound pressure levels at 500, 1000, and 2000 Hz, respectively.

With the speech interference level determined, one compares it to Table 16.1 [13] to assess the effect on communication at a given distance. For example, if the measured PSIL for an office were 62 dB, then from Table 16.1 one sees that reliable speech communication would be barely possible up to 2 ft at a normal voice level, or up to 4 ft at a raised voice level, or up to about 8 ft for a very loud voice level, and about 16 ft for shouting. The table assumes that the speaker and listener are facing each other and that there are no reflecting surfaces to enhance the voice sounds. Consider an example.

Example
  Noise levels near the filling machine in a bottling plant were 88 dB at 500 Hz, 85 dB at 1000 Hz, and 85 dB at 2000 Hz. (1) What is the preferred speech interference level, and (2) would one have to raise his voice to be understood?
Solution
From Eq. (16.15) we have

$$\text{PSIL} = \frac{88 + 85 + 85}{3}$$

$$= 86 \text{ dB}$$

From Table 16.1, we see that at 2 ft, shouting would not yield reliable conversation.

As a matter of interest, the overall A weighted noise level of the previous example is 90 dBA. As a rule of thumb, one typically has to raise his or her voice sharply to be understood at a distance of 2 ft when the level reaches 90 dBA.

16.1.13 Articulation Index

The articulation index (AI) is a single-number parameter used for predicting speech intelligibility. The index is determined by comparing the level of intruding speech to the level of the background noise at a given listening location. The measurement and calculation procedure used to determine the index is both involved and time consuming, and the index is rarely applicable to industrial noise control. As such, we shall say only that the general method involves determining the speech-to-noise ratio in 20 one-third octave bands (200 to 5000 Hz), weighting each, and summing. The net result is a number between 0 and 1.0. An AI of 0 represents no intelligibility, and an AI of 1.0 represents total intelligibility. It should be noted that when the articulation index exceeds 0.5, the relative understanding of complete sentences is generally good.

16.2 SUMMARY

With the number of descriptors, rating schemes, etc., the questions of which to use and when cannot be answered simply. The complexity of the human response along with the number of acoustical variables continues to confound the scientific community. However, strides of progress have been made, and many of the new schemes are merely refinements of the old and often are of second order or are applicable only to a particular type of environmental noise source. But regardless of the status, a basic understanding of the descriptors and the schemes is essential for the acoustical engineer to proceed systematically to assess noise impact for new plants or highways and/or to establish design criteria for controlling noise.

REFERENCES

1. *Acoustics Handbook.* Hewlett-Packard Company, Palo Alto, California, Nov. 1968.
2. D. W. Robinson and R. S. Dadson. *Br. J. Appl. Phys.* 7 (1956).
3. K. D. Kryter. *J. Acoust. Soc. Am.* 31 (1959).
4. K. D. Kryter. Possible Modifications to Procedures for the Calculation of Perceived Noisiness. *Report No. CR-1635.* National Aeronautics and Space Administration, Washington, D.C., 1969.
5. K. D. Kryter and K. S. Pearsons. *J. Acoust. Soc. Am.* 35 (1963).
6. K. D. Kryter and K. S. Pearsons. *J. Acoust. Soc. Am.* 36 (1964).
7. Aircraft Certification Procedure. *Federal Aviation Regulation, Part 36.* Federal Aviation Administration, Washington, D.C., 1969.

8. Procedure for Describing Aircraft Noise Around an Airport. *ISO R507-1970*. International Organization for Standardization, Geneva, Switzerland, 1970.

9. K. D. Kryter. *The Effects of Noise on Man*. Academic Press, New York, 1970.

10. Fundamentals of Noise: Measurement, Rating Schemes, and Standards. *Report No. NTID300.15*. U.S. Environmental Protection Agency, Washington, D.C., 1971.

11. J. D. Irwin and E. R. Graf. *Industrial Noise and Vibration Control*. Prentice-Hall, Englewood Cliffs, New Jersey, 1979.

12. Computing the Effective Perceived Noise Levels, for Flyover Aircraft Noise, Definitions and Procedures for. *American National Standards Institute S6.4-1973*. American National Standards Institute, New York, 1973.

13. Public Health and Welfare Criteria for Noise. *Report No. 550/9-73-002*. U.S. Environmental Protection Agency, Washington, D.C., 1973.

14. D. W. Robinson. *J. Sound Vib. 14* (1971).

15. T. J. Schultz. Noise Assessment Guidelines: Technical Background. *Report No. TE/NA 172*. Department of Housing and Urban Development, Washington, D.C., 1971.

16. L. L. Beranek (Ed.). *Noise and Vibration Control*. McGraw-Hill, New York, 1971.

17. Information on Levels of Environmental Noise Requisite to Protect Health and Welfare with an Adequate Margin of Safety. *Report No. 550/9-74-004*. U.S. Environmental Protection Agency, Washington, D.C., 1974.

18. L. L. Beranek. *Acoustics*. McGraw-Hill, New York, 1954.

19. J. C. Webster. SIL: past, present, future. *Sound Vibr.* (Aug. 1979).

20. M. Hirschorn. Noise Level Criteria. *Bulletin 6.0010.0*. Industrial Acoustics Company, Bronx, New York.

21. American National Standards Specification for Sound Level Meters. *American National Standards Institute S1.4-1971*. American National Standards Institute.

22. Frequency Weighting for the Measurement of Aircraft Noise (D-weighting). *IEC/537*. International Electrotechnical Commission, 1976.

23. S. S. Stevens. Perceived level of noise by Mark VIII and decibels (E). *J. Acoust. Soc. Am 51* (1972).

BIBLIOGRAPHY

Bishop, D. E., and R. D. Horonjeff. Procedures for Developing Noise Exposure Forecast Areas for Aircraft Flight Operations.

Report No. FAA-D5-67-10. Federal Aviation Administration,
Washington, D.C., 1967.

Faulkner, L. *Handbook of Industrial Noise Control.* Industrial
Press, New York, 1975.

Galloway, W. J., and D. E., Bishop. Noise Exposure Forecasts:
Evolution, Evaluation, Extensions and Land Use Interpretations.
Report No. FAA-NO-70-9. Federal Aviation Administration,
Washington, D.C., 1970.

Goldstein, J. Descriptors of Auditory Magnitude and Methods of
Rating Community Noise. *Community Noise, ASTM STP 692,*
R. J. Peppin and C. W. Rodman, (Eds.). American Society for
Testing and Materials, 1979.

Peppin, R. J., and Rodman, C. W. (Eds.). *Community Noise.*
American Society for Testing and Materials, Kansas City, Mo., 1979.

Shultz, T. J. Community noise ratings. *App. Acoust.* *2* (suppl.)
(1972).

EXERCISES

16.1 Following is an instantaneous spectral analysis obtained during
the landing approach of a jet aircraft (SST):

Octave Band Center Frequency (Hz)	Sound Pressure Level (dB)
63	70
125	73
250	75
500	80
1000	88
2000	86
4000	82
8000	78

Calculate
a. The perceived noise
b. The perceived noise level

Answer: (a) PN = 80 noys;
(b) PNL = 103 PNdB

16.2 During the flyover in Exercise 16.1, the D weighted sound level L_D reached a peak level of 95 dBD. Estimate the perceived noise level using the data.

Answer: 102 PNdB

16.3 An overall A weighted noise level of 92 dBA was measured at the property line of a foundary. Estimate the perceived noise level.

Answer: 105 PNdB

16.4 During an aircraft flyover, the peak D weighted sound level reached 102 dBD. By using a stopwatch and a sound level meter, it was observed that the duration at which the sound level was within 10 dB of its maximum value was 8 s. Estimate the effective perceived noise level for the flyover.

Answer: 106.3 EPNdB

16.5 Referring to Exercise 16.4, show that if the duration had been only 4 s the level would be 3 EPNdB lower.

16.6 A shrill whistle of 10-s duration sounds to alert motorists that a drawbridge is to be opened once each hour. If the sound level of the whistle at a nearby residential area is 110 dB, what is the daily equivalent sound level?

Answer: 84.4 dB

16.7 Following is a table of A weighted sound levels measured over a 24-h period:

i	L_{Ai} (dBA)	t_i (h)
1	60	8
2	65	4
3	70	6
4	75	6

Calculate the equivalent sound level for the period.

Answer: 70.5 dBA

16.8 By using a sound distribution analyzer near a construction site, the daily equivalent sound level LEQ, L_{10}, and L_{90} were measured and found to be 75, 80, and 71 dBA, respectively. Estimate the noise pollution level at the site.

Answer: L_{NP} = 84 dBA

16.9 Equivalent sound levels were measured continuously for a 24-h period near a highway. The levels for the daytime, evening, and nighttime periods were, respectively, 72, 68, and 65 dBA. Calculate the community noise equivalent level CNEL for the 24-h period.

Answer: 73.5 dBA

16.10 Referring to Exercise 16.9, calculate the day-night average sound level L_{dn} for the 24-h period, L_d = 71 dBA.

Answer: L_{dn} = 73.0 dBA

16.11 Noise levels in the cockpit of a two-engine aircraft were 80 dB at 500 Hz, 75 dB at 1000 Hz, and 70 dB at 2000 Hz. a. What is the preferred speech interference level? b. At what distance could clear communication be expected for a slightly raised voice?

Answer: (a) 75 dB;
(b) about 1 ft

17
COMMUNITY AND ENVIRONMENTAL NOISE
REGULATIONS

This chapter presents an overview of regulations regarding community and environmental noise. Specifically, typical formats and various technical approaches excerpted from actual regulations are discussed and compared. We shall limit our text to U.S. federal, state, and local jurisdictions.

17.1 FEDERAL REGULATIONS

With respect to environmental noise, there are no major federal regulations that apply directly to industrial noise emissions. However, as a cursory review, the Federal Aviation Administration (FAA) prescribes noise standards (FAR Part 36) [1] for various types of aircraft and associated flight takeoff and landing patterns. Further, the Department of Housing and Urban Development (HUD) has also established criteria [2] governing the approval of new construction or rehabilitation of housing where federal financial aid is involved, and the Environmental Protection Agency (EPA) has published guidelines [3] for the risk of complaint.

Briefly, with respect to aircraft noise, a maximum EPNL level of 108 dB is the performance standard to be met at three locations, i.e., takeoff, approach, and sideline. The HUD criteria of acceptability for proposed housing sites is a maximum DNL of 65 dBA, with other degrees of acceptability to 75 dBA, beyond which the site is deemed unacceptable. The guidelines of acceptability published by the EPA are for both indoor and outdoor spaces and specify a DNL of 45 dBA and 55 dBA, respectively. In meeting these design goals, activity interference is considered minimal.

Finally, the National Environmental Policy Act (NEPA) of 1969 empowers the federal government to require an environmental impact statement (EIS) when results of a proposed action might compromise the environment. Typically, these statements are mandated when new airport or highway construction is contemplated.

As mentioned before, these federal regulations have little direct application to industrial noise control; therefore we shall leave the discussion here and refer the reader to the references and bibliography for further details.

17.2   STATE REGULATIONS

Most states have enacted some form of regulation or statute regarding noise. The regulations or statutes often include performance standards for motor vehicles and the acoustic interaction among residences, commercial businesses, and heavy industries. Because there are no direct federal guidelines, each state that has enacted legislation has done so unilaterally. Thus there is little consistency in the form of the legislation or standards. One state that has taken a leadership role in the enactment and enforcement of noise control regulations is Connecticut. Shown in Fig. 17.1 are excerpts from Connecticut regulations regarding motor vehicles. Note that overall A weighted noise limits are defined for a variety of vehicles at a given measurement distance and vehicular speed. This format is common in many states and can be enforced readily. Note that a simple overall A weighted measurement is all that is required to establish the status of compliance.

Although it may be hard to see much connection between vehicular noise and industrial noise control, in many external industrial noise control programs, the pass-by or presence of idling trucks must be taken into account. Thus, an awareness of legal implications is essential and cannot be ignored.

With respect to external community and environmental noise, shown in Fig. 17.2 are excerpts from the State of Connecticut Department of Environmental Protection (DEP) administrative regulations. Note that the allowable noise levels and standards are based on emitter-to-receptor classification and distinguish between day and night operation. With respect to classification, Class C is typically industrial land usage, Class B is commercial land usage, and Class A is residential land usage. Daytime refers to the hours of 7:00 a.m. to 10:00 p.m., while nighttime is 10:00 p.m. to 7:00 a.m. In all cases the most stringent standard is for residential reception at night. This emphasis is typical of most performance standards. It should be emphasized that the Connecticut regulations also define noise level limits for impulse noise, ultrasonic and infrasonic noise, prominent pure tones, and the presence of high ambient background levels. See Appendix H for complete details.

ESTABLISHMENT OF MAXIMUM PERMISSIBLE
NOISE LEVELS FOR VEHICLES

The following regulation with the exception of section 14-80a-3 became effective on March 22, 1973. Section 14-80a-3 became effective on January 10, 1974.

Sec. 14-80a-1. Vehicles not subject to registration exempted. This section shall apply only to vehicles subject to registration by the Department of Motor Vehicles.

Sec. 14-80a-2. Definitions. For the purposes of these regulations, the term "motor vehicle" means any vehicle as defined in subdivision (26) of section 14-1, and "vehicle" means any vehicle as defined in subdivision (56) of section 14-1, which vehicle is subject to registration by the Department of Motor Vehicles.

Sec. 14-80a-3. General noise limits. No person shall operate either a vehicle or combination of vehicles at any time or under any condition of grade, load, acceleration or deceleration in such a manner as to exceed the following noise limit for the category of vehicle, within the highway speed limits, where applicable, specified in this section:

	Speed Limit of 35 MPH or Less	Speed Limit of More Than 35 MPH
(1) Any motor vehicle with a manufacturer's gross vehicle weight rating of ten thousand pounds or more and any combination of vehicles towed by such motor vehicle:		
(A) Before January 1, 1975	86dB(A)	90dB(A)
(B) On or after January 1, 1975	84dB(A)	88dB(A)
(2) Any motorcycle:		
(A) Before January 1, 1975	82dB(A)	86dB(A)
(B) On or after January 1, 1975	80dB(A)	84dB(A)
(3) Any other motor vehicle and any combination of vehicles towed by such motor vehicle:	76dB(A)	82dB(A)

Sec. 14-80a-4. Distance when measuring noise. The noise limits established by section 14-80a-3 and section 14-80a-7 for categories (1), (2) and (3) shall be based on a distance of fifty feet from the center of the lane of travel within the speed limit specified in that section. Law enforcement authorities may provide for measuring at a distance closer than fifty feet from the center of the lane of travel. In such a case, the measuring devices shall be so calibrated as to provide for measurements equivalent to the noise limit established by section 14-80a-3 at fifty feet. In addition, measurements made at distances greater than fifty feet, such as in adjacent lanes of travel will be based on a distance of fifty feet. In all such cases, a tolerance of 2dB(A) in sound level readings will be allowed.

Figure 17.1 Excerpts from state of Connecticut Department of Motor Vehicles regulations regarding noise.

Department of Environmental Protection
Control of Noise

Classification of Land According to Use

Sec. 22a-69-2.1. Basis
Noisy Zone classifications shall be based on the actual use of any
parcel or tract under single ownership as detailed by the Standard Land Use Classification Manual of Connecticut (SLUCONN).

Sec. 22a-69-2.3. Class A noise zone
Lands designated Class A shall generally be residential areas where
human beings sleep or areas where serenity and tranquility are essential to the intended use of the land.

Sec. 22a-69-2.4. Class B noise zone
Lands designated Class B shall generally be commercial in nature,
areas where human beings converse and such conversation is essential to the intended use of the land.

Sec. 22a-69-2.5. Class C noise zone
Lands designated Class C shall generally be industrial where protection against damage to hearing is essential, and the necessity
for conversation is limited.

Sec. 22a-69-3. Allowable noise levels

Sec. 22a-69-3.5. Noise zone standards

(a) No person in a Class C Noise Zone shall emit noise exceeding
the levels stated herein and applicable to adjacent Noise Zones:

	Receptor			
	C	B	A/Day	A/Night
Class C Emitter to	70 dBA	66 dBA	61 dBA	51 dBA

(b) No person in a Class B Noise Zone shall emit noise exceeding
the levels stated herein and applicable to adjacent Noise Zones:

	Receptor			
	C	B	A/Day	A/Night
Class B Emitter to	62 dBA	62 dBA	55 dBA	45 dBA

(c) No person in a Class A Noise Zone shall emit noise exceeding
the levels stated herein and applicable to adjacent Noise Zones:

	Receptor			
	C	B	A/Day	A/Night
Class A Emitter to	62 dBA	55 dBA	55 dBA	45 dBA

Levels emitted in excess of the values listed above shall be considered excessive noise.

Figure 17.2 Excerpts from state of Connecticut regulations regarding noise.

In summary, the regulatory format standards and measurement procedures for most states are generally very different. For example, the regulations of the Commonwealth of Massachusetts stipulate that the ambient background noise levels cannot be increased by more than 10 dBA and, further, no prominent discrete tones are permitted. Very simply, no other factors are included. The state of New Jersey has regulations similar to Connecticut but, in addition, requires meeting an octave band criterion. The states of New York and California, at this writing, have no well-defined regulations for stationary noise sources.

Finally, it should be reemphasized that those states with performance standards generally consider land usage, rather than local zoning, for structuring emitter-to-receptor limits.

17.3 LOCAL AND MUNICIPAL REGULATIONS

Generally, all local or municipal governments have legislation in the form of codes, ordinances, and so forth that regulate noise. The regulations are often in the form of "nuisance laws," where disturbing the peace, disorderly conduct, or other subjective descriptors are mentioned. These subjective regulations are confounding to the acoustical engineer as well as to the enforcement officer in that objective standards are not often present to assess the status of compliance or design criteria for noise abatement. The only practical approach in these situations is to utilize other or nearby local or state performance standards as guidelines for limits of acceptability.

Fortunately, many cities, townships, and towns have enacted noise regulations with well-defined performance standards. Shown in Fig. 17.3 are excerpts from the administrative codes of New York City. Note that the noise emission limits or standards are in terms of LEQ measured over a duration of one hour. Here again, considerable relief is given to commercial and manufacturing land use zones, while the most stringent requirements are low-density residential zones during the nighttime hours of 10:00 p.m. to 7:00 a.m.

It should be noted that when measuring ambient sound levels, contributions from natural sounds, vehicular traffic, and aircraft noise are not to be included in the measurements. This exclusion creates a dilemma for the acoustical engineer. How does one measure the LEQ level for an hour and not include birds, insects, traffic, aircraft, etc., in an urban environment? To be sure, integrating sound level meters are not selective with respect to the source of noise. This example leads us to an even more basic problem that

Local Laws of the City of New York

Ambient noise quality criteria and standards.

Ambient noise quality criteria and standards are herein established and tabulated below for each of the three ambient noise quality zones. . . . Not included in the standard are contributions to
the sound level from natural sources·such as birds and thunder
and sound sources outside the boundaries of the noise source such
as public highways, vehicular traffic and overflying aircraft.

Ambient noise quality zone	Day-time standards (7 am–10 am)	Night-time standards (10 pm–7 am)
Noise quality zone N-1 (Low Density residential R?; land use zones R-1 to R-3)	L_{eq} = 60 dB(A) measured for any one hour	L_{eq} = 50 dB(A) measured for any one hour
Noise quality zone N-2 (High density residential R?; land-use zones R-4 to R-10)	L_{eq} = 65 dB(A) measured for any one hour	L_{eq} = 55 dB(A) measured for any one hour
Noise quality zone N-3 (All commercial and manufac-turing land-use zones)	L_{eq} = 70 dB(A) measured for any one hour	L_{eq} = 70 dB(A) measured for any one hour

These criteria and standards as set forth in this section shall
apply to all stationary activities and to all mobile activities whenever they may be stationary, with the following exceptions:

(a) Construction activities conforming with section 1403.3.4.11
of article IV of part III of chapter fifty-seven of the administrative
code of the city of New York.

(b) Devices, vehicles, equipment and other noise producing
items or circumstances for which provisions are set forth elsewhere
in the administrative code of the city of New York.

Figure 17.3 Excerpts from the administrative code of New York City
regarding noise.

is almost always present in determining the noise emitted by industrial facilities. Typically, the applicable regulations will require
that, to demonstrate compliance, noise levels from the emitter must
be measured at the receptor's property line. Often other sources
of noise are present, and the problem is to quantitatively assess

the contribution of the emitter in question. Unless the various
sources have strong spectral or temporal identifying characteristics or can be shot down, some uncertainty is inherent.

   In short, precise apportionment of noise emitted from industrial
facilities is difficult and often the subject of controversy in establishing the status of compliance.

   Shown in Fig. 17.4 are excerpts from the noise regulations of
the city of Boston, Massachusetts. Note here that octave band
performance standards or emission limits are given for residential
business, and industrial zoning areas. Again, the standards are
severely reduced for noise emissions during nighttime hours. Note
that the octave band with a center frequency of 31.5 Hz is also
included.

   With respect to meeting community noise regulations, several
options are generally available to the acoustical engineer. Consider the case when a project is still at the design phase (pre-
construction). At this time, the major sources of noise can be
identified by a thorough acoustical design review. In particular,
sources at grade outside the building or on the roof, along with
all penetrations such as fan inlets or exhausts, must be listed.
If potentially noisy equipment has been already specified, the
acoustical engineer can obtain from the manufacturer the acoustical power levels of the sources or, more often, sound pressure
levels at a given distance. This is especially relevant for HVAC
equipment such as fans, blowers, cooling towers, air cooled con-
densers, etc. With the radiation characteristics of each source
determined, the divergence to the property line or noise-sensitive
locations can easily be calculated as shown in Chapter 4, Sound
Propagation. The total or cumulative effect of each noise source
at the property can then be calculated as shown in Chapter 2,
Combining Decibels.

   Finally, the results of the calculations can then be compared
to the regulations, and any levels exceeding the standards can be
noted. The excessive level differences are just the acoustical performance or noise reduction required for the applied noise control
measure. Consider an example:

Example
   An air-cooled condenser with four fans is being proposed for
   the HVAC system on a new office building. The unit will be
   located at grade level near the rear of the building. The manufacturer states that noise levels are 75 dBA at 5' from any
   side of the unit. What will the noise level be 100' from the
   unit at the nearest residential property line?
   Solution
   With respect to noise levels at the property line, Eq. (4.6)
   can be used directly.

$$L_{p,100} = L_{p,5} - 20 \log (100/5)$$
$$= 75 - 20 \log 20$$
$$= 75 - 26$$
$$= 49 \text{ dBA}$$

Now, if the measured overall A weighted level at the residential property line is 49 dBA and the nighttime standard is 45 dBA, the difference (4 dBA) is the level exceeding the criteria. It follows that the acoustical performance of the selected noise reduction measure or measures must be at least 4 dBA. Experience has shown that prudent design practice would include a design margin of at least 5 dBA, or a total design goal of 9 dBA.

If, in the early stages of design, specific equipment has not already been selected, the acoustical engineer can compile a list of performance specifications for each source. These specifications can then be applied to purchase orders for the potentially noisy equipment. In this way, the onus of meeting acoustical perform-ance can be transferred to the equipment manufacturer or contrac-tor.

Consider now the case where the equipment is already installed or in situ. A sound level survey at the property line or noise-sensitive locations is now essential to establish the status of com-pliance. Naturally, the measurement methods and measured parame-ters must be consistent with the regulations. The measured levels can be compared to the applicable criteria, and any levels exceed-ing the standards noted. Here again, the difference between the measured levels and the standard is just the required acoustical performance of the noise reduction measures to be selected.

The potential for anomalies associated with meteorological dif-fraction must also be considered if divergent distances to sensitive locations exceed 200 meters. As noted in Chapter 4, variation in sound level of up to 20 dB can be anticipated from these effects.

Finally, sound transmission through building walls, windows, and doors must also be considered in a thorough design review. These considerations are especially important for large facilities containing equipment with high noise levels, such as electricity generating stations, cogeneration plants, stand-by power facilities, petrochemical processes, and so on. Here again, for a systematic approach, the sound levels (octave band levels) in the interior of the building must be determined. The transmission loss or noise reduction of the walls, windows, doors, or roof elements are then estimated. The resultant level outside the building is the differ-ence between the interior levels and the wall attenuation. At this point, the divergence to the critical noise-sensitive location can be calculated, as illustrated in the example. It should be emphasized

2.4   Noise in Industrial Zoning Districts

No person shall create or cause to be emitted from or by any source subject to Regulation 2, any noise which causes or results in a maximum noise level, measured at any lot line of any lot in recreational or business use in any Industrial Zoning District in conformance with the Boston Zoning Code, in excess of any level of the "Industrial District Noise Standard," Regulation 2.5.

2.5   Zoning District Noise Standards

Noise standards referred to in these Regulations for the several zoning districts of the City of Boston, as defined in and established persuant to the Boston Zoning Code, are as established by the following table:

Table of Zoning District Noise Standards

Maximum Allowable Octave Band Sound Pressure Levels

Octave band center frequency of measurement (Hz)	Residential		Residential/ industrial		Business	Industrial
	Daytime	All other times	Daytime	All other times	Anytime	Anytime
31.5	76	68	79	72	79	83
63	75	67	78	71	78	82
125	69	61	73	65	73	77
250	62	52	68	57	68	73
500	56	46	62	51	62	67
1000	50	40	56	45	56	61
2000	45	33	51	39	51	57
4000	40	28	47	34	47	53
8000	38	26	44	32	44	50
Single number equivalent	60 dBA	50 dBA	65 dBA	55 dBA	65 dBA	70 dBA

Figure 17.4   Excerpts from the noise regulations of the city of Boston, Massachusetts.

that walls, doors, and windows may have to be treated as plane radiating surfaces, not simple point sources, if the noise-sensitive receiving locations are in the near field. See Chapter 4 for calculation methods.

From these discussions and examples, it is clear that meeting community or environmental regulatory criteria can be approached both systematically and quantitatively. A note of caution: because of uncontrolled construction and equipment manufacturing variations, outdoor reflections, estimates and simplifications in the analysis itself, final results should be considered at best as first-order approximations. As such, an adequate design margin should always be included to assure meeting design goals.

In final summary, some form of community noise or environmental regulation applies to all land usage throughout the United States and most industrial nations of the world. The regulations vary from subjctive codes to well-defined and easy-to-measure performance standards. It is the author's experience that where environmental issues such as air, water, or noise pollution are present, compliance with noise standards is the most difficult to assess. As such, when legal actions are involved between industrial facilities and their neighbors, noise is often the last issue to be settled.

REFERENCES

1. Federal Aviation Regulations, Part 36: Noise Standards, Aircraft Type Certification. *Federal Register 34*, no. 226, pp. 18,815–18,878, Nov. 25, 1969.
2. Environmental Criteria and Standards. *Federal Register 44* (135 PT51) 40,860–40,866, 1979.
3. EPA "Levels" document, 1974.

BIBLIOGRAPHY

Harris, G. M. *Handbook of Acoustical Measurements and Noise Control.* McGraw-Hill, New York, 1991.
Shultz, T. J. *Community Noise Rating.* ASP, New York, 1982.
Peppin, Rodman Community Noise ASTM. Kansas City, MO, 1978.

18
PERSONAL HEARING PROTECTION

Despite the progress in noise control engineering, there are many noisy environments where noise reduction measures are neither technically nor environmentally feasible. In these environments, personal hearing protection is often the only measure available to significantly reduce daily noise exposure. We shall discuss in this chapter the various types of hearing protection devices or defenders (HPD), the acoustical performance, methods of estimating resultant levels, and selection guidelines.

18.1 HEARING PROTECTION DEVICES

There are two basic types of hearing protection devices: earplugs and earmuffs. Earplugs are inserted into the ear canal. Earmuffs cover the outer ear as an acoustical shield or barrier. We shall elaborate on each type.

18.1.1 Earplugs

Earplugs can be broken down into reusable types and disposable types. Reusable earplugs are generally made of soft plastic or silicone rubber. Shown in Fig. 18.1 are examples of commercially available reusable earplugs. These plugs, when properly fitted to the ear canal, provide a high degree of sound attenuation. The performance of these plugs depends dynamically on the fit, maintenance, and installation. Because of their solid construction, a feeling of pressure or other discomfort is unfortunately often experienced by the wearer.

In addition to the preformed plugs just described, custom molded earplugs are also available. Here, a soft rubber material

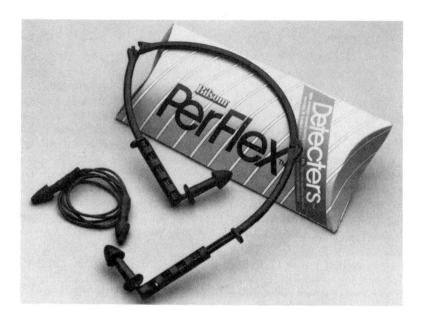

Figure 18.1   Reusable earplugs.   (Courtesy of Bilsom International Inc., Sterling, Va.)

is formed or molded in the individual's outer ear canal, providing a custom fit.   As such, many of the problems associated with fit and comfort are avoided. Here again, a high degree of attenuation is achieved but performance variability from wearer to wearer can be anticipated.

Disposable earplugs are generally made of soft mineral or organic fibers or closed cell forms.   Shown in Fig. 18.2 are examples of disposable earplugs.   These plugs may be discarded after a single use, and since they are porous and soft, most of the discomfort associated with solid earplugs is avoided.

18.1.2 Earmuffs

Earmuffs are basically hard plastic cups which cover the outer ear. A cushion seal to the head is maintained by a spring-loaded headband.   Shown in Fig. 18.3 are examples of earmuffs.

The effectiveness of earmuffs depends mainly on the cushion-to-head seal.   The headband must exert sufficient uniform pressure to

Figure 18.2 Disposable earplugs. (Courtesy of Bilson International Inc., Sterling, Va.)

Figure 18.3 Examples of earmuffs. (Courtesy of Willson Safety
Products, Reading, Pa.)

conform the cushions to the shape of the head. In this way, an
airtight seal is maintained. Eyeglass rims often inhibit a good seal
in front of the outer ear and thus sharply compromise the protection
potential of the muff. With respect to comfort, added weight and
annoying pressure to the head are complaints often lodged against
earmuffs. It should be noted that earmuffs can be attached to
hard hats, as shown in Fig. 18.4.

18.2 ACOUSTICAL PERFORMANCE OF
HEARING PROTECTION DEVICES

The acoustical performance of hearing protection devices is determined by measurements in accordance with industry standards.
There are two basic methods: the real ear method [1,2,3] and
the physical method [4]. In the real ear method, the subjective
response of the subject is determined while wearing and not wearing the device. This difference then represents the attenuation of
a particular device under laboratory conditions.

   The physical method utilizes a fixture resembling a human head
and ear. The insertion loss of the device is then determined by

Figure 18.4   Earmuffs attached to hard hat or safety cap.   (Courtesy of Bilsom International, Inc., Sterling, Va.)

measuring the difference in sound levels between the outside and inside of the head fixture.

Of the two methods, the real ear is more popular.   Nearly all performance data provided by the manufacturer is derived from the real ear method.   Shown in Fig. 18.5 is an excerpt from a catalog of the measured attenuation for a line of hearing protectors.   Note that the attenuation is obtained in conventional octave bands, but data is also presented for octave bands whose center frequency is 3150 Hz and 6300 Hz.   It should be emphasized that real ear testing is subjective and variation can be expected from subject to subject.   The variation is due mainly to factors such as fit, leakage, etc.   As such, a mean or average value is given, along with the standard deviation about the mean.   With these statistical parameters, the spectral attenuation of a device can be estimated to a high level of confidence.

Shown in line 1 of Table 18.1 are actual measured octave band sound levels in a packaging area in a food processing plant.   Subtracting the octave band A scale correction from the actual measured

	Frequency HZ	125	250	500	1000	2000	3150	4000	6300	8000	NRR as tested in most common position
AllFit™ ear plugs	Mean Value	29	30	32	34	37	40	40	41	42	28dB
	1 Std. Deviation	4.4	3.5	2.8	2.5	2.9	3.2	4.1	4.5	5.9	
ULTRA SOFT™ ear plugs	Mean Value	25	27	29	32	35	38	38	39	42	26dB
	1 Std. Deviation	3.9	3.6	2.0	2.2	3.5	4.0	2.8	4.2	4.3	
SOFT™ ear plugs	Mean Value	22	25	29	34	37	45	46	41	38	26dB
	1 Std. Deviation	3.5	2.3	2.5	3.1	2.4	3.4	4.2	4.4	3.6	
P.O.P™ ear plugs	Mean Value	23	25	26	26	34	39	41	41	38	22dB
	1 Std. Deviation	3.6	2.8	2.5	3.0	3.0	2.2	2.8	3.2	3.6	
Per-Fit® ear plugs	Mean Value	29	29	30	32	37	43	40	39	35	26dB
	1 Std. Deviation	3.7	3.8	3.2	2.1	2.5	3.0	5.1	4.6	3.1	
PerFlex™ banded ear plugs	Mean Value	31	29	32	30	35	36	38	40	41	22dB
	1 Std. Deviation	4.8	4.6	5.5	4.6	4.7	5.7	5.8	3.6	4.4	
quietzone™ ear plugs	Mean Value	20	23	24	28	34	36	37	37	37	24dB
	1 Std. Deviation	2.0	2.0	2.0	2.0	2.0	3.0	3.0	3.0	3.0	
1-Fit™ ear plugs	Mean Value	33	32	33	28	34	42	45	46	45	23dB
	1 Std. Deviation	3.8	4.0	4.3	3.8	4.1	5.3	5.0	3.3	4.5	
Whisper™ ear plugs	Mean Value	23	24	29	33	36	38	41	40	41	26dB
	1 Std. Deviation	3.0	2.0	3.0	2.0	2.0	2.0	2.0	3.0	3.0	
Viking 29™ Muff foam seals	Mean Value	23	25	31	36	40	42	42	39	37	29dB
	1 Std. Deviation	2.2	2.0	1.9	2.2	2.0	2.3	3.2	3.2	3.0	
Bilsom Comfort™ Muff foam seals	Mean Value	17	21	24	36	38	41	42	39	36	25dB
	1 Std. Deviation	2.1	1.9	2.0	1.8	1.8	2.4	1.6	2.5	1.9	
Warrior™ Muff foam seals	Mean Value	13	17	26	33	37	38	40	43	42	23dB
	1 Std. Deviation	1.9	2.0	2.4	2.4	2.8	3.1	3.7	3.0	4.1	
Bilsom Blue™ Muff foam seals	Mean Value	17	20	26	33	40	45	47	45	44	25dB
	1 Std. Deviation	1.9	1.3	2.4	1.7	1.5	1.9	1.4	1.9	2.8	
Bilsom Special™ Muff liquid seals	Mean Value	15	19	24	33	39	40	39	36	34	22dB
	1 Std. Deviation	2.7	2.9	3.0	2.2	2.3	1.5	3.3	2.7	1.4	
Bilsom Economy™ and Light™ Muff foam seals	Mean Value	12	15	25	35	37	39	37	37	34	22dB
	1 Std. Deviation	2.1	1.4	1.6	2.2	1.9	2.3	1.8	2.3	1.6	

Figure 18.5 Measured attenuation for a variety of earplugs and earmuffs. (Courtesy of Bilsom International, Inc., Sterling, Va.)

Table 18.1 Method for Calculating Resultant A Weighted Level at Eardrum

	Octave band center frequency (Hz)							dBA
	125	250	500	1K	2K	4K	8K	
1. Measured octave band noise levels	75	84	87	90	91	91	84	
2. A scale corrections	−16.1	−8.6	−3.2	0.0	+1.2	+1.0	−1.1	
3. A weighted noise levels[a]	59	75	84	90	92	92	83	97
4. Attenuation of P. O. P. (TM) earplug	23	25	26	26	34	41	38	
5. Resultant level at eardrum	36	50	58	64	58	51	45	66
6. 2X standard deviation	7.2	5.6	5.0	6.0	6.0	5.6	7.2	
7. Resultant level at eardrum to a high level of confidence	43	56	63	70	64	57	52	72

[a]Rounded to nearest decibel level.

octave band levels in line 1, one obtains the A weighted noise levels near the packaging line. Now, if we subtract the mean octave band attenuation for P.O.P (TM) earplugs (Fig. 18.4) from the A weighted noise levels, we obtain the resultant mean A weighted noise level at the eardrum of an employee. Note that the resultant level of 67 dBA is well below 85 dBA, the action level specified in the OSHA hearing conservation amendment [5]. Unfortunately, the mean attenuation is rarely achieved in the workplace or other noisy environments. In fact, studies have shown [6] that resultant levels can be 10 dB or more higher. As such, two standard deviations of the measured variation are often added to the resultant levels, as shown in line 6 of Table 18.1, as an added measure of safety. Note that the estimated resultant level at the eardrum is now 10 dBA higher, or 77 dBA.

The calculations just presented are somewhat tedious, but another method to estimate the acoustical performance of hearing protection devices is available. Shown also in Figure 18.5 is a column labeled Noise Reduction Rating (NRR). The noise reduction rating is a single number index [7] also determined by the hearing protection device manufacturer in addition to the spectral attenuation. The measurement procedure is similar to the octave band procedure except that the subjects listen to a "pink noise" spectrum, i.e., a spectrum of equal amplitude sound pressure level in each octave band.

To use the NRR to estimate the effectiveness of a device in the workplace, the following procedure is utilized: (1) Subtract 7 dB from the NRR, then (2) subtract the remainder from the overall A weighted noise level measured in the workplace. Consider the following example.

Example

The measured noise levels in a weaving room are 94 to 96 dBA. Earplugs whose NRR rating is 22 dB are being considered for use. What is the anticipated A weighted resultant noise level at the ear canal of an employee?

Solution

Step 1. Subtract 7 dB from the NRR, i.e., 22 −7 = 15 dB.

Step 2. As a worst case, 96 dBA is selected and the difference yields 96 −15 = 81 dBA.

Thus the resultant level is 81 dBA.

It should be noted that the procedure just outlined utilizing the NRR can be used to establish compliance with OSHA regulations and the hearing conservation amendment. However, it is acknowledged that the calculated effectiveness, just as in the octave band method, is rarely achieved in the workplace. As such, OSHA has recommended [8] that, as a margin of safety, the NRR be derated by a factor of 1/2, or 50%, for estimating actual workplace exposure. That is, if a hearing protection device with an NRR of 22 dB is being considered, derating as recommended by OSHA by 50% would yield an NRR of 11 dB.

It must be noted that both earplugs and earmuffs can be worn concurrently. The attenuation is not cumulative, i.e., it is not the algebraic sum of both. However, some manufacturers do provide octave band performance data for the combination with generally sharply improved performance in the frequency range below 500 Hz.

It should also be mentioned that earmuffs utilizing the active noise reduction approach have been developed. These devices are extremely effective in the spectral range below 1000 Hz. Because of the required ancillary electronics, these devices have been limited to military or aerospace applications.

In summary, estimating the actual attenuation of hearing protection devices from laboratory test data has inherent risks. Variables associated with fit, vibration, bone conduction, actual spectral noise levels, and so forth, limit the estimate accuracy to, at best, the first order, i.e., plus or minus 10 dB. Therefore, a margin of safety should always be included in any selection.

REFERENCES

1. Method for the Measurement of Real-Ear Attenuation of Hearing Protectors, ANSI S12.6-1984, American National Standards Institute, New York, 1984.
2. Method for the Measurement of Real-Ear Protection of Hearing Protectors and Physical Attenuation of Earmuffs, ASA STD.1-1975 (ANSI S3.19-1974). Acoustical Society of America, New York, 1975.
3. Measurement of Sound Attenuation of Hearing Protectors—Subjective Method, ISO 4869:1981. International Organization for Standardization, Geneva, Switzerland, 1975.
4. Simplified Method for a Measurement of Insertion Loss of Hearing Protectors of Ear-Muff Type for Quality Inspection Purposes, ISO/TR 6290:1989. International Organization for Standardization, Geneva, Switzerland, 1989.
5. Occupational Safety and Health Administration. Occupational Noise Exposure, Hearing Conservation Amendment. *Federal Register 48*, no. 46, pp. 9738—9785, 1983.
6. Industrial Hygiene Technical Manual, Change 2. Occupational Safety and Health Administration, March 1, 1987, pp. VI-13—VI-19, Washington, D.C.
7. Environmental Protection Agency, Noise Labeling Requirements for Hearing Protectors. *Federal Register 42*, no. 190, 40 CFR Part 211, pp. 56120—56147, 1979.

8. Industrial Hygiene Technical Manual, Change 2. Occupational
    Safety and Health Administration, March 1, 1987, pp. VI-13–
    VI-19, Washignton, D.C.

EXERCISES

18.1 The following octave band noise levels were measured in a
foundry: 125 Hz, 92 dB; 250 Hz, 94 dB; 500 Hz, 93 dB; 1000 Hz,
92 dB, 2000 Hz, 92 dB; 4000 Hz, 88 dB; 8000 Hz, 85 dB. If Quiet-
zone (TM) earplugs (Figure 18.5) were selected for use in the area,
what would the mean resultant octave noise levels be at an employee's
eardrum?

Answer: 72, 71, 69, 64, 58,
51, and 48 dB

18.2 Referring to Exercise 18.1, what are the resultant levels if 2
standard deviations are taken as a margen of safety?

Answer: 76, 75, 73, 68, 62,
57, and 54 dB

18.3 Referring to Exercise 18.2, what is the resultant overall A
weighted level?

Answer: 74 dBA

18.4 The overall A weighted noise level at a stone crushing facility
is 101 dBA. With Warrior (TM) earmuffs, what would the resultant
level be if used in the most common position? Use the NNR method.

Answer: Use the NRR method.
85 dBA

APPENDIXES

A

THE INTERNATIONAL SYSTEM OF UNITS*

The International System of Units (SI) consists of base units, supplementary units, derived units, and a series of approved prefixes for the formation of multiples and submultiples of the units. Listed in this appendix are those units relevant to the study of noise control.

Quantity	Unit	SI Symbol	Formula
Base units			
Length	Meter	m	—
Mass	Kilogram	kg	—
Time	Second	s	—
Electric current	Ampere	A	—
Thermodynamic temperature	Kelvin	°K	—
Amount of substance	Mole	mol	—
Supplementary Units			
Plane angle	Radian	rad	—
Solid angle	Steradian	sr	—
Derived units			
Acceleration	Meter per second squared	—	m/s^2
Angular velocity	Radian per second	—	rad/s
Area	Square meter	—	m^2
Density	Kilogram per cubic meter	—	kg/m^3
Electric capacitance	Farad	F	$A\text{-}s/V$
Electric resistance	Ohm	Ω	V/A
Energy	Joule	J	$N\text{-}m$

*Système International d'Unités.

582

Appendix A (Continued)

Quantity	Unit	SI Symbol	Formula
Force	Newton	N	$kg\text{-}m/s^2$
Frequency	Hertz	Hz	$1/s$
Magnetomotive force	Ampere	A	—
Power	Watt	W	J/s
Pressure	Pascal	Pa	N/m^2
Quantity of electricity	Coulomb	C	$A\text{-}s$
Quantity of heat	Joule	J	$N\text{-}m$
Radiant intensity	Watt per steradian	—	W/sr
Specific heat	Joule per kilogram-Kelvin	—	$J/kg\text{-}°K$
Velocity	Meter per second	—	m/s
Viscosity, dynamic	Pascal-second	—	$Pa\text{-}s$
Voltage	Volt	V	W/A
Volume	Cubic meter	—	m^3
Wave number	Reciprocal meter	—	$Wave/m$
Work	Joule	J	$N\text{-}m$

B CONVERSION FACTORS

To Convert	Into	Multiply by	Conversely, Multiply by
atm (atmosphere)	mm Hg at 0°C	760	1.316×10^{-3}
	lb/in.^{2}	14.70	6.805×10^{-2}
	N/m^2 (Pa)	1.0132×10^{5}	9.872×10^{-6}
	kg/m^2	1.033×10^{4}	9.681×10^{-5}
Btu	Joules	1054.8	9.480×10^{-4}
°C (Celsius)	°F (fahrenheit)	[°C × 9/5] + 32	(°F − 32) × 5/9
cm (centimeter)	in. (inch)	0.3937	2.540
	ft (foot)	3.281×10^{-2}	30.48
Degrees (angle)	Radians	1.745×10^{-2}	57.30
Dyne	lb (force)	2.248×10^{-6}	4.448×10^{5}
	N (newton)	10^{-5}	10^{5}
Dynes/cm^2	lb/ft^2	2.090×10^{-3}	478.5
	N/m^2 (Pa)	10^{-1}	10
Ergs	Foot-pounds	7.376×10^{-8}	1.356×10^{7}
Ergs	Joules	10^{-7}	10^{7}
Ergs per second	Watts	10^{-7}	10^{7}
Fathoms	Feet	6	0.16667

ft (foot)	in. (inch)	12	0.08333
	cm (centimeter)	30.48	3.281×10^{-2}
	m (meter)	0.3048	3.281
ft^2	in.^{2}	144	6.945×10^{-3}
ft^3	in.^{3}	1728	5.787×10^{-4}
	cm^3	2.832×10^4	3.531×10^{-5}
	m^3	2.832×10^{-2}	35.31
Gallons	Cubic meters	3.785×10^{-3}	264.2
Gallons (liquid United State)	Gallons (liquid British Imperial)	0.8327	1.201
Horsepower (550 ft-lb/sec)	Foot-pounds (force) per minute	3.3×10^4	3.030×10^{-6}
Horsepower (550 ft/lb/sec)	Kilowatts	0.745	1.342
hp (horsepower)	W (watt)	745.7	1.341×10^{-3}
in. (inch)	ft (foot)	0.0833	12
	cm (centimeter)	2.540	0.3937
	m (meter)	0.0254	39.37
in.^{2}	ft^2	6.945×10^{-3}	144
	cm^2	6.452	0.1550
	m^2	6.452×10^{-4}	1550

Appendix B (Continued)

To Convert	Into	Multiply by	Conversely, Multiply by
in.^{3}	ft^3	5.787×10^{-4}	1.728×10^3
	cm^3	16.387	6.102×10^{-2}
	m^3	1.639×10^{-5}	6.102×10^4
Joules	Foot-pounds	0.7376	1.356
Joules	Ergs	10^7	10^{-7}
kg (kilogram)	lb (weight)	2.2046	0.4536
	slug	0.06852	14.594
Kilowatt-hours	Joules	3.6×10^8	2.778×10^{-7}
Liters	Cubic meters	0.001	1000
Liters	Cubic inches	61.02	1.639×10^{-2}
Liters	Gallons (liquid United States)	0.2642	3.785
$\log_e$ N, or ln N	$\log_{10}$N	0.4343	2.303
m (meter)	in. (inch	39.371	2.540×10^{-2}
	ft (foot)	3.2808	0.30481
	cm (centimeter)	10^2	10^{-2}
Microbar (dynes/cm^2)	lb/in.^{2}	1.4513×10^{-5}	6.890×10^4

Source	Target		
Microbars	lb/ft^2	2.090×10^{-3}	478.5
	N/m^2 (Pa)	10^{-1}	10
	Newtons per square meter	10^{-1}	10
Miles (nautical)	Feet	6080.20	1.645×10^{-4}
Miles (statute)	Feet	5280	1.894×10^{-4}
Miles per hour	Kilometers per hour	1.609	0.6214
Np (neper)	dB (decibel)	8.686	0.1151
N (newton)	lb (force)	0.2248	4.448
	Dynes	10^5	10^{-5}
N/m^2 (pascal, Pa)	lb/in.^{2} (force)	1.4513×10^{-2}	6.890×10^3
	lb/ft^2 (force)	2.090×10^{-2}	47.85
	Dynes/cm^2	10	10^{-1}
lb (pound)	N (newton)	4.448	0.2248
lb (pound)	Slug	0.03108	32.17
	kg (kilogram)	0.4536	2.2046
Pounds of water (distilled)	Cubic feet	1.603×10^{-2}	62.38
Pounds of water (distilled)	Gallons	0.1198	8.347
Pounds per square inch	Kilograms per square meter	703.1	1.422×10^{-3}
Poundals (force)	Dynes	1.383×10^4	7.233×10^{-5}

Appendix B (Continued)

To Convert	Into	Multiply by	Conversely, Multiply by
Rayls	mks rayls	10	10^{-1}
Slugs (mass)	Pounds (force)	32.174	3.108×10^{-2}
Watts	Ergs per second	10^7	10^{-7}
W (watt)	hp (horsepower)	1.341×10^{-3}	745.7

C

STANDARDS AND PROCEDURES

RELEVANT NATIONAL AND INTERNATIONAL ORGANIZATIONS

International Organizations

International Organization for Standardization (ISO): 1, rue de Varembe; 1211 Geneva 20, Switzerland.

International Electrotechnical Commission (IEC): 1, rue de Varembe; 1211 Geneva 20, Switzerland.

National Organizations

Acoustical Society of America (ASA): Standards Secretariat, 335 East 45th Street, New York, N.Y. 10017.

Air Conditioning and Refrigeration Institute (ARI); 1815 North Fort Meyer Drive, Arlington, Va. 22209.

Air Moving and Conditioning Association (AMCA): 205 West Touby Avenue, Park Ridge, Ill. 60068.

American Gear Manufacturers Association (AGMA): 1 Thomas Circle, Washington, D.C. 20005.

American National Standards Institute, Inc. (ANSI): 1430 Broadway, New York, N.Y. 10018.

American Society of Heating, Refrigerating and Air-Conditioning Engineers (ASHRAE): 345 47th Street, New York, N.Y. 10017.

American Society for Testing and Materials (ASTM): 1916 Race Street, Philadelphia, Pa. 19103.

Compressed Air and Gas Institute (CAGI): 122 East 42nd Street, New York, N.Y. 10017.

Institute of Electrical and Electronic Engineers (IEEE): 345 East 47th Street, New York, N.Y. 10017.

Institute of Noise Control Engineering (INCE): Box 3206, Poughkeepsie, N.Y. 12603.

Instrument Society of America (ISA): 400 Stanwix Street, Pittsburgh,
    Pa. 15222.
National Electrical Manufacturers Association (NEMA): 2101 L Street N.
    W., Washington, D.C. 20037.
National Fluid Power Association (NFPA): P.O. Box 49, Thiensville,
    Wisc. 53092.
National Machine Tool Builders Association (NMTBA): 7901 West Park
    Drive, McLean, Virginia 22102
Society of Automotive Engineers, (SAE): 400 Commonwealth Drive,
    Warrendale, Pa. 15096.
Society of the Plastics Industry (SPI): New York, N.Y. 10617.

LIST OF BASIC STANDARDS

International Standards

ISO International Standard 3740. Determination of Sound Power Levels
    of Noise Sources—Guidelines for Use of Basic Standards and
    Preparation of Noise Test Codes (1978).
ISO International Standard 3741. Determination of Sound Power Levels
    of Noise Sources—Precision Methods for Broad-Band Sound Sources
    Operating in Reverberation Rooms (1975).
ISO International Standard 3742. Determination of Sound Power Levels
    of Noise Sources—Precision Methods for Discrete-Frequency and
    Narrow-Band Sound Sources Operating in Reverberation Rooms
    (1975).
ISO International Standard 3744. Determination of Sound Power Levels
    of Noise Sources—Engineering Methods for Free-Field Conditions
    over a Reflecting Plane.
ISO International Standard 3745. Determination of Sound Power Levels
    of Noise Sources—Precision Methods for Sources Operating in
    Anechoic Rooms.
ISO International Standard 3746. Determination of Sound Power Levels
    of Noise Sources—Survey Method.
ISO International Standard 3747. Determination of Sound Power Levels
    of Noise Sources—Methods Using a Reference Sound Source.
IEC Recommendation, Publication 179. Precision Sound Level Meters
    (1973).
IEC Recommendation, Publication 179A. Additional Characteristics
    for the Measurement of Impulsive Sounds (1973).
IEC Recommendation, Publication 225. Octave, Half-Octave and Third-
    Octave Band Filters Intended for the Analysis of Sounds and Vi-
    brations (1966).
IEC Recommendation, Publication 327. Precision Method for the Pres-
    sure Calibration of One-Inch Standard Condenser Microphones by
    the Reciprocity Technique (1971).

IEC Recommendation, Publication 402. Simplified Methods for Pressure Calibration of One-Inch Condenser Microphones by the Reciprocity Technique (1972).

Test Codes and Procedures

ANSI S1.4-1983 (ASA 47). American National Standard Specification for Sound Level Meters. This Standard includes ANSI S1.4A-1985 Amendment to ANSI S1.4-1983.

ANSI S1.6-1984 (R 1990) (ASA 53). American National Standard Preferred Frequencies, Frequency Levels, and Band Numbers for Acoustical Measurements.

ANSI S1.8-1989 (ASA 84). American National Standard Reference Quantities for Acoustical Levels.

ANSI S1.10-1966 (R 1986). American National Standard Method for the Calibration of Microphones.

ANSI S1.11-1986 (ASA 65). American National Standard Specification for Octave-Band and Fractional-Octave-Band Analog and Digital Filters.

ANSI S1.12-1967 (R 1986). American National Standard Specifications for Laboratory Standard Microphones.

ANSI S1.13-1971 (R 1986). American National Standard Methods for the Measurement of Sound Pressure Levels.

ANSI S1.25-1991 (ASA 98). Specification for Personal Noise Dosimeters.

ANSI S1.26-1978 (R 1989) (ASA 23). American National Standard Method for the Calculation of the Absorption of Sound by the Atmosphere.

ANSI S1.40-1984 (R 1990) (ASA 40). American National Standard Specifications for Acoustical Calibrators.

ANSI S1.42-1986 (ASA 64). American National Standard Design Response of Weighting Networks for Acoustical Measurements.

ANSI S12.1-1983 (R 1990) (ASA 49). American National Standard Guidelines for the Preparation of Standard Procedures for the Determination of Noise Emission from Sources.

ANSI S12.3-1985 (R 1990) (ASA 57). American National Standard Statistical Methods for Determining and Verifying Stated Noise Emission Values of Machinery and Equipment.

ANSI S12.4-1986 (ASA 63). American National Standard Method for Assessment of High-Energy Impulsive Sounds with Respect to Residential Communities.

ANSI S12.5-1990 (ASA 87). American National Standard Requirements for the Performance and Calibration of Reference Sound Sources.

ANSI S12.6-1984 (R 1990) (ASA 55). American National Standard Method for the Measurement of the Real-Ear Attenuation of Hearing Protectors. (This standard is a revision of the real-ear measurement section of ANSI S3.19-1974.)

ANSI S12.7-1986 (ASA 62). American National Standard Methods
for Measurements of Impulse Noise

ANSI S12.8-1986 (ASA 73). American National Standard Methods
and Procedures for Description and Measurement of Environmental Sound, Part 1.

ANSI S12.10-1985 (R 1990) (ASA 61). American National Standard
Methods for the Measurement and Designation of Noise Emitted by
Computer and Business Equipment. (Revision of ANSI S1.29-1979).

ANSI S12.23-1989 (ASA 83). American National Standard Method
for the Designation of Sound Power Emitted by Machinery and
Equipment.

ANSI S12.34-1988 (ASA 77). American National Standard Engineering Methods for the Determination of Sound Power Levels of
Noise Sources for Essentially Free-Field Conditions over a Reflecting Plane.

ANSI S12.36-1990 (ASA 89). American National Standard Survey
Methods for the Determination of Sound Power Levels of Noise
Sources.

ASA STD. 3-1975. Test-Site Measurement of Noise Emitted by Engine Powered Equipment.

ASHRAE Standard 36-72. Methods of Testing for Sound Rating
Heating, Refrigerating, and Air-Conditioning Equipment (super-
sedes ASHRAE Standards 36-72, 36A-63, and 36B-63).

ASTM §E 90-90. Test Method for Laboratory Measurement of Airborne Sound Transmission Loss of Building Partitions.

ASTM §E 336-90. Test Method for Measurement of Airborne Sound
Insulation in Buildings.

ASTM §E 413-87. Classification for Rating Sound Insulation.

ASTM E 477-90. Test Method for Measuring Acoustical and Airflow
Performance of Duct Liner Materials and Prefabricated Silencers.

ASTM E 492-90. Test Method for Laboratory Measurement of Impact
Sound Transmission Through Floor-Ceiling Assemblies Using the
Tapping Machine.

ASTM §E 596-90. Method for Laboratory Measurement of the Noise
Reduction of Sound-Isolating Enclosures.

ASTM §E 756-83. Method for Measuring Vibration-Damping Properties of Materials.

ASTM E 795-91. Practices for Mounting Test Specimens During
Sound Absorption Tests.

ASTM E 989-89. Classification for Determination of Impact Insulation Class (IIC).

ASTM E 1014-84 (1990). Guide for Measurement of Outdoor A-Weighted
Sound Levels.

ASTM E 1050-90. Test Method for Impedance and Absorption of Acoustical Materials Using a Tube, Two Microphones, and a Digital Frequency Analysis System

ASTM E 1124-86. Test Method for Field Measurement of Sound Power Level by the Two-Surface Method.

ASTM E 1130-90. Test Method for Objective Measurement of Speech Privacy in Open Offices Using Articulation Index.

ASTM E 1222-90. Test Method for Laboratory Measurement of the Insertion Loss of Pipe Lagging Systems.

ASTM E 1265-90. Test Method for Measuring Insertion Loss of Pneumatic Exhaust Silencers.

ASTM E 1332-90. Classification for Determination of Outdoor-Indoor Transmission Class.

ASTM E 1408-91. Test Method for Laboratory Measurement of the Sound Transmission Loss of Door Panels and Door Systems.

IEEE 85. Test Procedure for Airborne Sound Measurements on Rotating Electric Machinery (1973).

SAE Standard J336a. Sound Level for Truck Cab Interior (1973).

SAE Standard J672a. Exterior Loudness Evaluation of Heavy Trucks and Buses (1970).

SAE Standard J952b. Sound levels for Engine Powered Equipment (1969).

SAE Standard J986a. Sound level for Passenger Cars and Light Trucks (1973) (ANSI S6.3).

SAE Standard J88a. Exterior Sound Level Measurement Procedure for Power Mobile Construction Machinery (1973).

AMCA Standard 300-67. Test Code for Sound Rating.

AGMA Standard 293.03. Specification for Measurement of Sound on High Speed Helical and Herringbone Gear Units (1968).

AFBMA Standard No. 13. Rolling Bearing Vibration and Noise (1968).

AHAM Standard No. RAC-2SR. Room Air Conditioner Sound Rating (1971).

DEMA Test Cose for the Measurement of Sound from Heavy-Duty Reciprocating Engine (1972).

NEMA Standard MG1-12.49. Motors and Generators. Methods of Measuring Machine Noise (1972).

NEMA Standard TR1-1972. Transformers, Regulators and Reactors (Sec. 9-04, Audible Sound Level Tests).

NFPA TS.9.70.12. Method of Measuring Sound Generated by Hydraulic Fluid Power Pumps (1970).

NFPA T3.9.14. Method of Measuring Sound Generated by Hydraulic Fluid Power Motors (1971).

NMBTA Technique. Noise Measurement Techniques (1970).

D RECOMMENDED DESCRIPTORS AND ABBREVIATIONS

Term	Unweighted	A Weighting	Alternative[a] A Weighting	Other Weighting[b]
1. Sound (pressure) level[c]	L_p	L_A	L_{pA}	L_B, L_{pB}
2. Sound power level	L_W	L_{WA}		L_{WB}
3. Maximum sound level	L_{pmax}	L_{max}	L_{Amax}	L_{Bmax}
4. Peak sound (pressure) level	L_{pk}	L_{Apk}		L_{Bpk}
5. Level exceeded x% of the time	L_{px}	L_x	L_{Ax}	L_{Bx}
6. Equivalent sound level	L_{peq}	L_{eq}	L_{Aeq}	L_{Beq}
7. Equivalent sound level over time (T)[d]	$L_{peq(T)}$	$L_{eq(T)}$	$L_{Aeq(T)}$	$L_{Beq(T)}$
8. Day sound level	L_{pd}	L_d	L_{Ad}	L_{Bd}
9. Night sound level	L_{pn}	L_n	L_{An}	L_{Bn}
10. Day-night sound level	L_{pdn}	L_{dn}	L_{Adn}	L_{Bdn}
11. Yearly day-night sound level	$L_{pdn(Y)}$	$L_{dn(Y)}$	$L_{Adn(Y)}$	$L_{Bdn(Y)}$
12. Sound exposure level	L_{Sp}	L_S	L_{SA}	L_{SB}

[a] *Alternative* symbols may be used to assure clarity or consistency.
[b] Only B weighting shown; applies also to C, D, E, weighting.
[c] The term *pressure* is used only for the unweighted level.
[d] Unless otherwise specified, time is in hours (e.g., the hourly equivalent level is $L_{eq}(1)$). Time may be specified in nonquantitative terms (e.g., could be specified as L_{eq}(WASH) to mean the washing cycle noise for a washing machine).

Abbreviation	Expansion	Abbreviation	Expansion
AI	Articulation index	ISO	International Organization for Standardization
ANSI	American National Standards Institute	NC	Noise Criterion Level
ASA	Acoustical Society of America	NEF	Noise exposure forecast
ASHRAE	American Society of Heating, Refrigerating and Air-Conditioning Engineers	NEL	Noise exposure level
ASME	American Society of Mechanical Engineers	NEMA	National Electrical Manufacturers Association
ASTM	American Society for Testing and Materials	NIOSH	National Institute for Occupational Safety and Health
CHABA	Committee on Hearing and Bio-Acoustics	NIPTS	Noise-induced permanent threshold shift
CNEL	Community noise equivalent level	NPL	Noise pollution level
DL	Day average sound level	NRC	Noise reduction coefficient
DNAL	Day-night A-weighted sound level	OSHA	Occupational Safety and Health Administration
DNL	Day-night average sound level	PNL	Perceived noise level
DOD	Department of Defense	PWL	Sound power level
EPNL	Effective perceived noise level	RC	Room criterion level
FAA	Federal Aviation Administration	SAE	Society of Automotive Engineers
FAR	Federal Aviation Regulations	SEL	Sound exposure level
HVAC	Heating, ventilating, and air conditioning	STC	Sound transmission class
IEC	International Electrochemical Commission	TL	Transmission loss
IIC	Impact isolation class	TTS	Temporary threshold shift
INCE	Institute of Noise Control Engineering	TWA	8-Hour time-weighted average sound level

E

DEPARTMENT OF LABOR OCCUPATIONAL NOISE EXPOSURE STANDARD AND PERMISSIBLE ULTRASONIC THRESHOLD LEVELS

Since noise levels are rarely constant in an industrial environment, it is the footnote associated with the exposure table in the following extract that is most relevant. Here one can calculate the daily exposure when two or more periods of exposure are known. An example follows the extract to illustrate the method.

---

*Code of Federal Regulations, Title 29, Chapter XVII, Part 1910, Subpart G, 36 FR 10466, May 19, 1971. Ammended by 46 FR 4161, Jan. 16, 1981; 46 FR 42632, Aug. 21, 1981.

(*Editor's note:* The Department of Labor's noise exposure standard first was promulgated under the Walsh-Healey Public Contracts Act. It was adopted under the Occupational Safety and Health Act on May 29, 2971, and is applicable under the general industry, construction, and longshore standards.)

§ 1910.95 Occupational noise exposure.

(a) Protection against the effects of noise exposure shall be provided when the sound levels exceed those shown in Table G-16 when measured on the A scale of a standard sound level meter at slow response. When noise levels are determined by octave band analysis, the equivalent A-weighted sound level may be determined as follows:

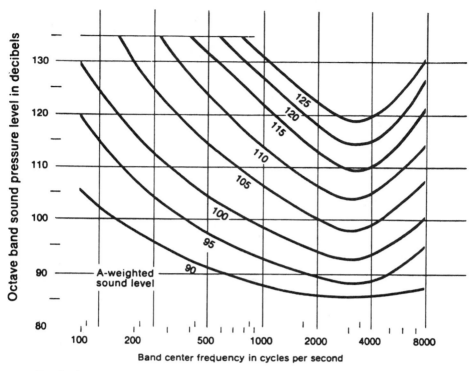

Equivalent sound level contours. Octave band sound pressure levels may be converted to the equivalent A-weighted sound level by plotting them on this graph and noting the A-weighted sound level corresponding to the point of highest penetration into the sound level contours. This equivalent A-weighted sound level, which may differ from the actual A-weighted sound level of the noise, is used to determine exposure limits from Table I.G-16.

(b) (1) When employees are subjected to sound exceeding those listed in Table G-16, feasible administrative or engineering controls shall be utilized. If such controls fail to reduce sound levels within the levels of Table G-16, personal protective equipment shall be provided and used to reduce sound levels within the levels of the table.

(2) If the variations in noise level involve maxima at intervals of 1 second or less, it is to be considered continuous.

Table G-16   Permissible Noise Exposures

Duration per day (hours)	Sound level dBA slow response
8	90
6	92
4	95
3	97
2	100
$1\frac{1}{2}$	102
1	105
$\frac{1}{2}$	110
$\frac{1}{4}$ or less	115

Note: When the daily noise exposure is composed of two or more periods of noise exposure of different levels, their combined effect should be considered, rather than the individual effect of each. If the sum of the following fractions: $C1/T1 + C2/T2$ Cn/Tn exceeds unity, then, the mixed exposure should be considered to exceed the limit value. Cn indicates the total time of exposure at a specified noise level, and Tn indicates the total time of exposure permitted at that level.

Exposure to impulsive or impact noise should not exceed 140 dB peak sound pressure level.

HEARING CONSERVATION AMENDMENT

(c) Hearing conservation program. (1) The employer shall administer a continuing, effective hearing conservation program, as described in paragraphs (c) through (o) of this section, whenever employee noise exposures equal or exceed an 8-hour time-weighted average sound level (TWA) of 85 decibels measured on the A scale (slow response) or, equivalently, a dose of fifty percent. For purposes of the hearing conservation program, employee noise exposures shall be computed in accordance with Appendix A and Table G-16a, and without regard to any attenuation provided by the use of personal protective equipment.

(2) For purposes of paragraphs (c) through (n) of this section, an 8-hour time-weighted average of 85 decibels or a dose of fifty percent shall also be referred to as the action level.

(d) Monitoring. (1) When information indicates that any employee's exposure may equal or exceed an 8-hour time-weighted average of 85 decibels, the employer shall develop and implement a monitoring program. (i) The sampling strategy shall be designed to identify employees for inclusion in the hearing conservation program and to enable the proper selection of hearing protectors.

(ii) Where circumstances such as high worker mobility, significant variations in sound level, or a significant component of impulse noise make area monitoring generally inappropriate, the employer shall use representative personal sampling to comply with the monitoring requirements of this paragraph unless the employer can show that area sampling produces equivalent results.

(2)(i) All continuous, intermittent and impulsive sound levels from 80 decibels to 130 decibels shall be integrated into the noise measurements.

(ii) Instruments used to measure employee noise exposure shall be calibrated to ensure measurement accuracy.

(3) Monitoring shall be repeated whenever a change in production process, equipment or controls increases noise exposures to the extent that:

(i) Additional employees may be exposued at or above the action level; or

(ii) The attenuation provided by hearing protectors being used by employees may be rendered inadequate to meet the requirements of paragraph (j) of this section.

(e) Employee notification. The employer shall notify each employee exposed at or above an 8-hour time-weighted average of 85 decibels of the results of the monitoring.

(f) Observation of monitoring. The employer shall provide affected employees or their representatives with an opportunity to observe any noise measurements conducted pursuant to this section.

(g) Audiometric testing program. (1) The employer shall establish and maintain an audiometric testing program as provided in this paragraph by making audiometric testing available to all employees whose exposures equal or exceed an 8-hour time-weighted average of 85 decibels.

Now it is interesting to note that the exposures listed in Table G-16 in the extract are really selected values of the formula and the continuous chart shown in Fig. E.1. It is from this relationship that electronic dosimeters are designed and programmed. It also enables one to measure noise exposure on a continuous basis.

It must be emphasized that since in most industrial noise environments the daily noise exposure of employees varies dynamically, the exposure variation can be attributed to fluctuating noise levels at a specific work location, or to temporal (time) changes of employee work locations, or a combination of both. As such, in most cases, no simple number adequately describes the employee's daily

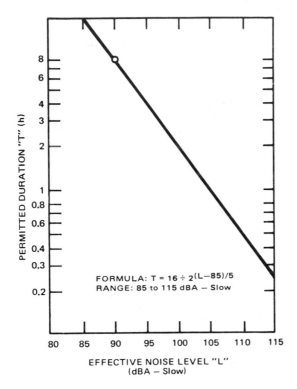

Figure E.1

noise exposure. Presented here are several methods for sampling and calculating to a high level of confidence the average daily noise exposure of employees in accordance with current OSHA regulations.

BACKGROUND

Federal (OSHA) regulations define the daily noise exposure or dose D from 1; footnote to Table G-16 as:

$$D = \frac{C_1}{T_1} + \frac{C_2}{T_2} + \frac{C_3}{T_3} \cdots\cdots \frac{C_n}{T_n} \tag{1}$$

where

C_n = time of exposure at a specified noise level (hours) and
T_n = total time of exposure permitted at the specified level
    (hours)

The hours of permissible exposure T_n is given in Table G-16.

Generally, D is expressed in percentage hence the partial sums in Eq. (1) are multiplied by 100. An example will best illustrate the calculation.

Example

An employee works most of the day in an area in which the sound level is 90 dBA, but for 15 minutes out of each of 7 hours, he is in an area of 100 dBA, and for one 15 minute period each day, he is in an area of 105 dBA. What is the employee's daily noise exposure?

Solution

From Table G-16, we get 6 hours at 90 dBA: permissible duration of exposure 8 hours; 1-3/4 hours at 100 dBA: permissible exposure, 2 hours; and 1/4 hour at 105 dBA: permissible exposure, 1 hour. In tabular form, we have:

dBA	Actual time C (hours)	Permissible time, T (hours)
90	6	8
100	1-3/4	2
105	1/4	1

Substitution into Eq. (1) yields:

$$D = \frac{6}{8} + \frac{1.75}{2} + \frac{0.25}{1} = 1.87$$

Or, on a percentage basis, $1.87 \times 100 = 187\%$.

No calculation method could be more straightforward or simpler. Unfortunately, in many industrial environments, the noise levels at the specified durations as given in the example are not constant on a daily or even on an hourly basis. This variation in exposure is due to fluctuating noise levels at a given work station, or to different noise levels at several work stations, or often to a combination of both. With such random variability, it is not possible to apply the method as outlined in the example.

To resolve the dilemma, noise dose meters or dosimeters have been developed which, when worn by an employee, will continuously calculate the time weighted dose in accordance with Eq. (1).

In short, if worn for a day, these dosimeters will provide a digital readout of the daily dose, as a percentage, regardless of the variation in noise level.

The problems associated with this electronic approach are simply that: (1) obtaining an 8-hour dose for dozens or hundreds of employees is prohibitively time consuming; and (2) one still has only a single daily noise dose, which may not be representative due to day-to-day variation.

The solution to the dilemma is naturally to apply a statistical approach. In this way, the actual *range* of daily exposure can be estimated to a high level of confidence. In addition, with some clever sampling techniques, the grueling, time-consuming effort can usually be reduced to reasonable duration.

STATISTICS

For simplicity, suppose we have measured the daily exposure of an employee for a week and found the *daily* exposure values to be, respectively: 150%, 130%, 170%, 200%, and 150%. Let us now calculate the mean, or average, daily exposure of the employee.

The average daily dose $\overline{D}$ is given by Eq. (2);

$$\overline{D} = \sum_{i=1}^{N} \frac{D_i}{N} \ \% \tag{2}$$

where

D_i = daily exposure samples %
N = number of daily exposure samples

From the example, the average daily exposure is then from Eq. (2):

$$\overline{D} = \frac{150 + 130 + 170 + 200 + 150}{5} = 160\%$$

We could now say that based on the sample, the *average* daily exposure $\overline{D}$ of the employee is 160%.

Further, from the small sampling theory, it can be shown that the true mean lies in the interval

$$\overline{D} \pm t_p \frac{s}{\sqrt{N - 1}} \ \% \tag{3}$$

at a level of confidence depending on t_p. Here, s = standard deviation of sample

$$s = \sqrt{\frac{\displaystyle\sum_{i=1}^{N} (\overline{D} - D_i)^2}{N - 1}} \qquad (4)$$

and

D_i = daily dose measurements %
N = number of samples
$\overline{D}$ = average daily dose %

The values of t_p are obtained from Table E.1 and are just the percentile confidence coefficients for the "Student" t distribution. For a 95% level of confidence, one takes $t_{.975}$ and the table values according to the sample size $\nu = N - 1$.

Returning to the example and Table E.1, $t_{.975} = 2.78$, (N - 1 = 4), s = 26.45, and the confidence interval for the mean daily exposure is:

$$160\% \pm \frac{2.78 \times 26.46}{\sqrt{5 - 1}}$$

Or 160% ± 36.8%, or symbolically,

$$123.2\% < \overline{D} < 196.8\%$$

Our conclusion is that based on the sample of 5 daily dose measurements the average daily exposure $\overline{D}$ of the employee lies between 123.2% and 196.8%, or expressed in terms of equivalent sound pressure levels as given in Table E.2, the average daily exposure $\overline{D}$ of the employee lies in the interval 91.4 dBA to 94.8 dBA equivalent, 19 days out of 20. The 19 days out of 20 represents the 95% confidence limits.

In summary, we now have a meaningful estimate of the daily exposure that accounts for the inherent variability found in most industrial environments.

It should be emphasized that the calculations for this example can be done on a relatively inexpensive hand-held programmable calculator; that is, large computing capabilities are not required. Let us now consider another example to further illustrate the approach.

Example

Daily noise exposure at 10 punch press operator stations varies dynamically due to fluctuating noise levels. One-hour-long dosimeter samples were taken at each station and each one-hour dose was considered representative of a daily dose. Therefore, each one-hour dose was extrapolated to a daily dose. In tabular form below are the daily dose samples for 10 stations.

Station:	1	2	3	4	5	6	7	8	9	10
Daily dose %	102	130	90	110	85	85	90	92	80	85

Table E.1 Percentile Values (t_p) for Student's t
Distribution with ν Degrees of Freedom

ν	$t_{.995}$	$t_{.99}$	$t_{.975}$	$t_{.95}$	$t_{.90}$
1	63.66	31.82	12.71	6.31	3.08
2	9.92	6.96	4.30	2.92	1.89
3	5.84	4.54	3.18	2.35	1.64
4	4.60	3.75	2.78	2.13	1.53
5	4.03	3.36	2.57	2.02	1.48
6	3.71	3.14	2.45	1.94	1.44
7	3.50	3.00	2.36	1.90	1.42
8	3.36	2.90	2.31	1.86	1.40
9	3.25	2.82	2.26	1.83	1.38
10	3.17	2.76	2.23	1.81	1.37
11	3.11	2.72	2.20	1.80	1.36
12	3.06	2.68	2.18	1.78	1.36
13	3.01	2.65	2.16	1.77	1.35
14	2.98	2.62	2.14	1.76	1.34
15	2.95	2.60	2.13	1.75	1.34
16	2.92	2.58	2.12	1.75	1.34
17	2.90	2.57	2.11	1.74	1.33
18	2.88	2.55	2.10	1.73	1.33
19	2.86	2.54	2.09	1.73	1.33
20	2.84	2.53	2.09	1.72	1.32
21	2.83	2.52	2.08	1.72	1.32
22	2.82	2.51	2.07	1.72	1.32
23	2.81	2.50	2.07	1.71	1.32
24	2.80	2.49	2.06	1.71	1.32
25	2.79	2.48	2.06	1.71	1.32
26	2.78	2.48	2.06	1.71	1.32
27	2.77	2.47	2.05	1.70	1.31
28	2.76	2.47	2.05	1.70	1.31
29	2.76	2.46	2.04	1.70	1.31
30	2.75	2.46	2.04	1.70	1.31
40	2.70	2.42	2.02	1.68	1.30
60	2.66	2.39	2.00	1.67	1.30
120	2.62	2.36	1.98	1.66	1.29
∞	2.58	2.33	1.96	1.645	1.28

Source: R. A. Fisher and F. Yates, *Statistical Tables for Biological, Agricultural and Medical Research* (5th edition), Table III, Oliver and Boyd Ltd., Edinburgh.

What is the average daily noise exposure of the operators at a 95% level of confidence?

Solution

From Eq. (2) the average daily dose is:

$$\overline{D} = \frac{102 + 130 + 90 + 110 + 85 + 85 + 90 + 92 + 80 + 85}{10} \quad \%$$

$$= 94.9\%$$

and from Eq. (4), s = 15.2%. The 95% confidence interval is obtainable from Eq. (3):

$$94.9 \pm \frac{t_{.975} \times 15.2}{\sqrt{10 - 1}} \quad \%$$

Now, $t_{.975}$ = 2.26 for N = 10 or ν = 10 - 1 = 9 from Table E.1, and the interval is

$$94.9 \pm 11.5 \quad \text{or} \quad 83.4 < \overline{D} < 106.4\%$$

Hence, the average daily exposure of the operators lies between 83.4% and 106.4% at a 95% level of confidence.

In terms of equivalent dBA, the average daily exposure, from Table E.2, in the range of 88.7 dBA and 90.5 dBA, 19 days out of 20. More important, the exposure can be considered only marginally excessive with respect to current federal OSHA regulations.

Table E.2 Conversion from Percent Noise Exposure to Equivalent Sound Pressure Level (L_p)

Percent noise exposure	Equivalent L_p (dBA)	Percent noise exposure	Equivalent L_p (dBA)	Percent noise exposure	Equivalent L_p (dBA)
10	73.4	60	86.3	86	88.9
15	73.6	65	86.9	87	89.0
20	78.4	70	87.4	88	89.1
25	80.0	75	87.9	89	89.2
30	81.3	80	88.4	90	89.2
35	82.4	81	88.5	91	89.3
40	83.4	82	88.6	92	89.4
45	84.2	83	88.7	93	89.5
50	85.0	84	88.7	94	89.6
55	85.7	85	88.8	95	89.6

(continued)

Table E.2   (Cont.)

Percent noise exposure	Equivalent L_p (dBA)	Percent noise exposure	Equivalent L_p (dBA)	Percent noise exposure	Equivalent L_p (dBA)
96	89.7	210	95.4	620	103.2
97	89.8	220	95.7	630	103.3
98	89.9	230	96.0	640	103.4
99	89.9	240	96.3	650	103.5
100	90.0	250	96.6	660	103.6
101	90.1	260	96.9	670	103.7
102	90.2	270	97.2	680	103.8
103	90.3	280	97.4	690	103.9
104	90.4	290	97.7	700	104.0
105	90.4	300	97.9	710	104.1
106	90.5	310	98.2	720	104.2
107	90.6	320	98.4	730	104.3
108	90.6	330	98.6	740	104.4
109	90.7	340	98.9	750	104.5
110	90.8	350	99.0	760	104.6
111	90.8	360	99.2	770	104.6
112	90.8	370	99.4	780	104.8
113	90.9	380	99.6	790	104.9
114	91.0	390	99.8	800	105.0
115	91.1	400	100.0	810	105.1
116	91.1	410	100.2	820	105.2
117	91.2	420	100.4	830	105.3
118	91.3	430	100.5	840	105.4
119	91.3	440	100.7	850	105.4
120	91.3	450	100.8	860	105.5
125	91.6	460	101.0	870	105.6
130	91.6	470	101.2	880	105.7
135	92.2	480	101.3	890	105.8
140	92.4	490	101.5	900	105.8
145	92.7	500	101.6	910	105.9
150	92.9	510	101.8	920	106.0
155	93.2	520	101.9	930	106.1
160	93.4	530	102.0	940	106.2
165	93.6	540	102.2	950	106.5
170	93.8	550	102.3	960	106.3
175	94.0	560	102.4	970	106.4
180	94.2	570	102.6	980	106.5
185	94.4	580	102.7	990	106.5
190	94.6	590	102.8	999	106.6
195	94.8	600	102.9		
200	95.0	610	103.0		

It must be emphasized that the accuracy and dependability of
the statistical approach depends on the quality of the sample data.
If the dose samples are not representative, the values will not be
representative, and here is where good judgment is required. Let
us now consider some guidelines for obtaining representative samples.

SAMPLING TECHNIQUES

As mentioned before, the variation of daily noise exposure is usually
due to fluctuation in noise levels at a work station or changes in
levels as an employee moves from station to station. Let us consider first the case of fluctuating noise levels and guidelines for
obtaining representative noise dose samples.

Case 1: Fluctuating Noise Levels

Fluctuating noise levels at a given work station will be either cyclical, such as commonly found around machine tools, assembly lines,
molding equipment, etc., or random, such as usually associated
with construction equipment, packaging lines, woodworking machines, metal forming areas, etc.

Cyclical

Consider the cyclical variation first. Shown in Fig. E.2 is an illustration of the time history of the noise at an operator's station
of a chucking machine. Note that the cyclical nature of the level
variation has a period of about 20 seconds. If one were to obtain
a noise dose for 5 minutes, the daily or 8-hour dose of the operator could be extrapolated with a high degree of accuracy.
    Naturally, it would be better to accumulate several five-minute
measurements spaced over the day. The downtime of the machine
can be factored into the daily estimate by reviewing production
records.

Random

Shown in Fig. E.3 is the time history of random type daily exposure of a saw operator. For random level variations there is no
choice but to extend the sample duration. From the author's experience, samples of duration — 20 to 30 minutes — taken 4 times
a day generally provide a representative sample. This is especially
true for assembly areas, packaging lines, and in most metal forming industrial environments. The four samples should be spaced
over the work day, say, early in the (day) shift, before noon, after

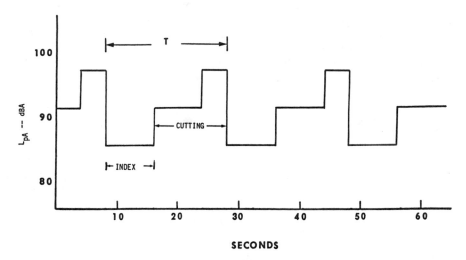

Figure E.2   Cyclical fluctuating noise levels from a machine tool.

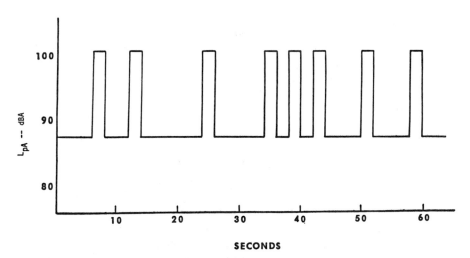

Figure E.3   Random fluctuating noise levels at a saw operator's station.

lunch and near the end of the shift.   Here again, production records should be reviewed to account for significant downtime.

In summary, for fluctuating levels at a given work station,
short dose measurements spaced over the daily work shift will
yield reliable extrapolated daily noise exposures.   The daily exposures will be representative only if the short samples are indeed
representative.

Case II: Change in Work Station (Movement)

When an employee's exposure changes as he or she moves from one
work station to another, generally the level changes dynamically.
Shown in Fig. E.4 is an example of the time history of a forklift
operator who works in three different noise environments.   To get
a representative sample of a daily dose here, short dosimeter samples could be taken in each environment, but good records or careful observation over several days would be required to access accurately the time spent in each environment.   In short, a combination
of dosimeter samples at each work station with careful estimates of
time spent at each station is required to estimate the daily exposure.
Given the daily estimates, the average daily exposure can be computed and the interval of confidence determined as outlined in the
previous section.

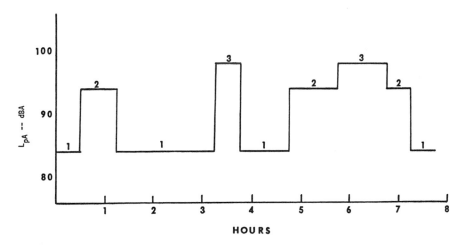

Figure E.4   Daily exposure variation where forklift operator works
in three different noise level environments.

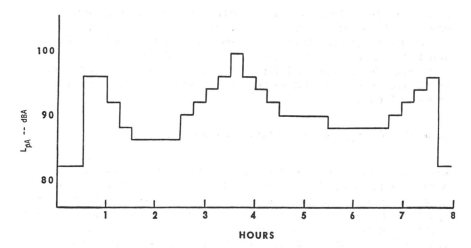

Figure E.5 Illustration of random fluctuation of utility (clean-up) worker as he moves around noisy machines.

        Shown in Fig. E.5 is the time history of a utility man. Here, the variation in levels is due to his constant motion as he comes close to noisy machines in his clean-up capacity. This variation is obviously random in character and will vary from day to day. However, 4 dose samples of 30 minute duration, taken daily over a week, will yield representative daily dose data for further processing. If the daily variation is large, the confidence intervals will be correspondingly wide.

        In summary, from dosimeter samples taken at the various work stations and larger samples for employees who roam in the course of their work, representative daily dose estimates can be obtained. These daily doses can be statistically processed as outlined in the previous section to give accurate estimates of the range of average daily exposure.

FINAL SUMMARY

With careful attention and judgment to obtaining representative daily exposure estimates, the range of employee exposure can be accurately measured to a high degree of confidence. The calculations are straightforward and the measurement sample times can be short, relative to an 8-hour shift.

It must be emphasized, however, that if the range of daily variations is large and random, the method outlined will not save time. In short, there is no mathematical method that will take small samples of dynamically changing noise levels and provide to a high level of confidence a narrow estimate of the mean.

There are no federal exposure limits for uppersonic or ultrasonic acoustic radiation. However, shown in Table E.3 are threshold limit values published by the American Conference of Governmental Industrial Hygienists. These limits should be considered as guidelines only.

Table E.3 Permissible Airborne Uppersonic and Ultrasonic Acoustic Radiation Exposure Levels

Mid-frequency of third-octave band kHz	One-third octave—band level in dB re 20 µPa
10	80
12.5	80
16	80
20	105
25	110
31.5	115
40	115
50	115

Note: Subjective annoyance may occur in some sensitive individuals at levels between 75 and 105 dB at 20 kHz 1/3 octave band, and hearing protection or engineering controls may be needed to minimize or prevent the annoyance.
Source: Threshold Limit Values and Biological Exposure Indices. American Conference of Governmental Industrial Hygienists.

F
SOUND ABSORPTION COEFFICIENTS OF COMMON BUILDING MATERIALS, AUDIENCES, SEATS, MUSICIANS, ETC.

Material	Absorption Coefficients (Hz)					
	125	250	500	1000	2000	4000
Brick						
Unpainted	0.03	0.03	0.03	0.04	0.05	0.07
Painted	0.01	0.01	0.02	0.02	0.02	0.03
Concrete						
Unpainted	0.01	0.01	0.02	0.02	0.02	0.02
Painted	0.01	0.01	0.01	0.02	0.02	0.02
Concrete block						
Porous, unpainted	0.36	0.44	0.31	0.29	0.39	0.25
Painted	0.10	0.05	0.06	0.07	0.09	0.08
Floors						
Concrete or terrazzo	0.01	0.01	0.01	0.02	0.02	0.02
Resilient tile on concrete	0.02	0.03	0.03	0.03	0.03	0.02
Parquet flooring	0.05	0.03	0.06	0.09	0.10	0.20
Varnished wood joist floor	0.15	0.11	0.10	0.07	0.06	0.07
Glass						
Large panes of heavy plate glass	0.18	0.06	0.04	0.03	0.02	0.02
Standard window	0.35	0.25	0.18	0.12	0.07	0.04
Gypsum board, 1/2 in., nailed to 2 × 4 in. studs, 16 in. o.c.	0.29	0.10	0.05	0.04	0.07	0.09
Marble or glazed tile	0.01	0.01	0.01	0.01	0.02	0.02
Plasters						
Gypsum or lime, smooth	0.01	0.01	0.02	0.03	0.04	0.05
On lath, over airspace or on joists or studs	0.30	0.15	0.10	0.05	0.04	0.05

Appendix F (Continued)

Material	Absorption Coefficients (Hz)					
	125	250	500	1000	2000	4000
(Plasters, continued)						
Space unit (typical) 12 × 12 in., 2 in. thick cellular glass tile, installed 32 in. o.c., per unit	0.13	0.74	2.35	2.53	2.03	1.73
Mineral or glass wool blanket 1 in. thick, acoustical quality, mounted against solid backing, covered with open weave fabric	0.15	0.35	0.70	0.85	0.90	0.90
Same, covered with 5% open perforated sheet metal	0.10	0.35	0.85	0.85	0.35	0.15
Same, covered with 10% open perforated sheet metal	0.15	0.30	0.75	0.85	0.75	0.40
Mineral or glass wool blanket 2 in. thick, acoustical quality, mounted over 1-in. airspace, covered with open weave fabric	0.35	0.70	0.90	0.90	0.95	0.90
Same, covered with 20% open perforated sheet metal	0.40	0.80	0.90	0.85	0.75	0.40
Plywood panels						
1/4 in., mounted over 3-in. airspace, with 1-in. glass fiber blanket behind the panel	0.60	0.30	0.10	0.09	0.09	0.09
Audience and seats						
Audience, seated in upholstered seats, per ft^2 of floor area	0.60	0.74	0.88	0.96	0.93	0.85
Unoccupied cloth-covered upholstered seats, per ft^2 of floor area	0.49	0.66	0.80	0.88	0.82	0.70
Wooden pews, occupied, per ft^2 of floor area	0.57	0.61	0.75	0.86	0.91	0.86
Unoccupied leather covered upholstered seats, per ft^2 of floor area	0.44	0.54	0.60	0.62	0.58	0.50

Appendix F (Continued)

Material	Absorption Coefficients (Hz)					
	125	250	500	1000	2000	4000
Carpets						
Heavy, on concrete	0.02	0.06	0.14	0.37	0.60	0.65
Same, on 40-oz hairfelt or foam rubber	0.08	0.24	0.57	0.69	0.71	0.73
Same, with impermeable latex backing, on 40-oz hairfelt or foam rubber	0.08	0.27	0.39	0.34	0.48	0.63
Curtains						
Light velour, 10 oz/yd^2 hung straight, in contact with wall	0.03	0.04	0.11	0.17	0.24	0.35
Medium velour, 14 oz/yd^2, draped to half area	0.07	0.31	0.49	0.75	0.70	0.60
Heavy velour, 18 oz/yd^2, draped to half area	0.14	0.35	0.55	0.72	0.70	0.65
Fibrous glass curtain, 6.1 oz/yd^2, draped to half area, 5 in. from rigid backing	0.08	0.13	0.21	0.29	0.23	0.29
Same, 8.4 oz/yd^2	0.09	0.32	0.68	0.83	0.76	0.76
Musician, with seat and instrument, per person	4.0	8.5	11.5	14.0	13.0	12.0
Water surface Swimming pool	0.01	0.01	0.01	0.02	0.02	0.03

G NOYS AS A FUNCTION OF SOUND PRESSURE LEVEL, L_p

	(1) 50	(2) 63	(3) 80	(4) 100	(5) 125	(6) 160	(7) 200	(8) 250	(9) 315	(10) 400	(11) 500	(12) 630	(13) 800
L_p							Band Center Frequency (Hz)						
29													
30													
31													
32													
33													
34													
35													
36													
37													
38													
39													
40										1.0	1.0	1.0	1.0
41										1.1	1.1	1.1	1.1
42									1.0	1.1	1.1	1.1	1.1
43									1.1	1.2	1.2	1.2	1.2
44								1.0	1.1	1.3	1.3	1.3	1.3
45								1.1	1.2	1.4	1.4	1.4	1.4
46							1.0	1.2	1.3	1.5	1.5	1.5	1.5
47							1.1	1.3	1.4	1.6	1.6	1.6	1.6
48						1.0	1.2	1.4	1.5	1.7	1.7	1.7	1.7
49						1.1	1.3	1.5	1.6	1.9	1.9	1.9	1.9
50						1.2	1.4	1.6	1.7	2.0	2.0	2.0	2.0
51					1.0	1.3	1.5	1.7	1.9	2.1	2.1	2.1	2.1

SPL													
52	2.3	2.3	2.3	2.3	2.0	1.9	1.6	1.4	1.1				
53	2.5	2.5	2.5	2.5	2.1	2.0	1.7	1.5	1.2				
54	2.6	2.6	2.6	2.6	2.3	2.1	1.9	1.6	1.3	1.0			
55	2.8	2.8	2.8	2.8	2.4	2.3	2.0	1.7	1.4	1.1			
56	3.0	3.0	3.0	3.0	2.6	2.4	2.2	1.9	1.5	1.2	1.0		
57	3.2	3.2	3.2	3.2	2.8	2.6	2.4	2.0	1.7	1.3	1.1		
58	3.5	3.5	3.5	3.5	3.0	2.8	2.6	2.2	1.8	1.4	1.2		
59	3.7	3.7	3.7	3.7	3.2	3.0	2.8	2.4	2.0	1.5	1.3		
60	4.0	4.0	4.0	4.0	3.5	3.2	3.0	2.6	2.2	1.8	1.4	1.0	
61	4.3	4.3	4.3	4.3	3.7	3.5	3.2	2.8	2.4	2.0	1.5	1.1	
62	4.6	4.6	4.6	4.6	4.0	3.7	3.5	3.0	2.6	2.2	1.7	1.2	
63	4.9	4.9	4.9	4.9	4.3	4.0	3.7	3.2	2.8	2.4	1.8	1.3	
64	5.3	5.3	5.3	5.3	4.6	4.3	4.0	3.5	3.0	2.6	2.0	1.5	1.0
65	5.7	5.7	5.7	5.7	5.0	4.6	4.3	3.7	3.2	2.8	2.2	1.6	1.1
66	6.1	6.1	6.1	6.1	5.4	5.0	4.6	4.0	3.5	3.0	2.4	1.8	1.2
67	6.5	6.5	6.5	6.5	5.9	5.4	5.0	4.3	3.7	3.3	2.6	2.0	1.4
68	7.0	7.0	7.0	7.0	6.4	5.9	5.4	4.6	4.0	3.6	2.8	2.2	1.6
69	7.5	7.5	7.5	7.5	6.9	6.4	5.9	5.0	4.3	3.9	3.0	2.3	1.8
70	8.0	8.0	8.0	8.0	7.5	6.9	6.4	5.4	4.6	4.2	3.3	2.5	2.0
71	8.6	8.6	8.6	8.6	8.0	7.5	6.9	5.9	5.0	4.6	3.6	2.8	2.2
72	9.2	9.2	9.2	9.2	8.7	8.0	7.5	6.4	5.4	5.0	3.9	3.0	2.3
73	9.8	9.8	9.8	9.8	9.3	8.7	8.0	6.9	5.9	5.4	4.2	3.3	2.5
74	10.6	10.6	10.6	10.6	10	9.3	8.7	7.5	6.4	5.9	4.6	3.7	2.8
75	11.3	11.3	11.3	11.3	11	10	9.3	8.0	6.9	6.4	5.0	4.1	3.0
76	12	12	12	12	11	11	10	8.7	7.5	6.9	5.4	4.5	3.3
77	13	13	13	13	12	11	11	9.3	8.3	7.5	5.9	5.0	3.7
78	14	14	14	14	13	12	11	10	9.1	8.3	6.4	5.4	4.1
79	15	15	15	15	14	13	12	11	10	9.1	6.9	5.9	4.5

Preferred frequencies for octave band filters marked by vertical lines.

Appendix G (Continued)

					Band Center Frequency (Hz)						
(14) 1000	(15) 1250	(16) 1600	(17) 2000	(18) 2500	(19) 3150	(20) 4000	(21) 5000	(22) 6300	(23) 8000	(24) 10,000	(25) 12,500
					1.0	1.0					
				1.0	1.0	1.1					
				1.1	1.1	1.1	1.0				
			1.0	1.1	1.2	1.2	1.1				
			1.1	1.2	1.3	1.3	1.2	1.0			
		1.0	1.2	1.3	1.4	1.4	1.3	1.1			
		1.1	1.3	1.4	1.5	1.5	1.4	1.2			
		1.2	1.3	1.5	1.6	1.6	1.5	1.4			
		1.3	1.4	1.6	1.8	1.8	1.6	1.5	1.0		
	1.0	1.3	1.5	1.8	1.9	1.9	1.7	1.6	1.1		
	1.1	1.4	1.6	1.9	2.0	2.0	1.9	1.8	1.2		
1.0	1.2	1.5	1.7	2.0	2.2	2.2	2.0	1.9	1.4		
1.1	1.3	1.6	1.8	2.2	2.4	2.4	2.2	2.0	1.5	1.0	
1.1	1.3	1.7	2.0	2.4	2.6	2.6	2.4	2.2	1.7	1.1	
1.2	1.4	1.8	2.2	2.6	2.8	2.8	2.6	2.4	1.8	1.2	
1.3	1.5	2.0	2.4	2.8	3.0	3.0	2.8	2.6	2.0	1.4	
1.4	1.6	2.1	2.6	3.0	3.2	3.2	3.0	2.8	2.2	1.5	
1.5	1.7	2.3	2.8	3.2	3.4	3.4	3.2	3.0	2.4	1.7	1.0
1.6	1.8	2.4	3.0	3.4	3.6	3.6	3.4	3.2	2.6	1.8	1.1
1.7	2.0	2.6	3.2	3.6	3.9	3.9	3.6	3.4	2.8	2.0	1.2
1.9	2.1	2.8	3.4	3.9	4.1	4.1	3.9	3.6	3.0	2.2	1.4
2.0	2.3	3.0	3.6	4.1	4.4	4.4	4.1	3.9	3.2	2.4	1.5
2.1	2.4	3.2	3.9	4.4	4.7	4.7	4.4	4.1	3.4	2.6	1.7

1.8	2.8	3.6	4.4	4.7	5.0	5.0	4.7	4.1	3.5	2.6	2.3
2.0	3.0	3.9	4.7	5.0	5.3	5.3	5.0	4.4	3.7	2.8	2.5
2.2	3.2	4.1	5.0	5.3	5.7	5.7	5.3	4.7	4.0	3.0	2.6
2.3	3.5	4.4	5.3	5.7	6.1	6.1	5.7	5.0	4.3	3.2	2.8
2.5	3.7	4.7	5.7	6.1	6.5	6.5	6.1	5.3	4.6	3.5	3.0
2.8	4.0	5.0	6.1	6.5	7.0	7.0	6.5	5.7	5.0	3.7	3.2
3.0	4.3	5.3	6.5	7.0	7.5	7.5	7.0	6.1	5.3	4.0	3.5
3.3	4.6	5.7	7.0	7.5	8.0	8.0	7.5	6.5	5.7	4.3	3.7
3.6	5.0	6.1	7.5	8.0	8.7	8.7	8.0	7.0	6.1	4.6	4.0
3.9	5.3	6.5	8.0	8.7	9.3	9.3	8.7	7.5	6.5	5.0	4.3
4.2	5.7	7.0	8.7	9.3	10	10	9.3	8.0	7.0	5.3	4.6
4.6	6.1	7.5	9.3	10	11	11	10	8.7	7.5	5.7	4.9
5.0	6.5	8.0	10	11	11	11	11	9.3	8.0	6.1	5.3
5.3	7.0	8.7	11	11	12	12	11	10	8.7	6.5	5.7
5.7	7.5	9.3	11	12	13	13	12	11	9.3	7.0	6.1
6.1	8.0	10	12	13	14	14	13	11	10	7.5	6.5
6.5	8.7	11	13	14	15	15	14	12	11	8.0	7.0
7.0	9.3	11	14	15	16	16	15	13	71	8.7	7.5
7.5	10	12	15	16	17	17	16	14	12	9.3	8.0
8.0	11	13	16	17	19	19	17	15	13	10	8.6
8.7	11	14	17	19	20	20	19	16	14	11	9.2
9.3	12	15	19	20	21	21	20	17	15	11	9.8
10	13	16	20	32	23	23	21	19	16	12	10.6
11	14	17	21	23	24	24	23	20	17	13	11.3
11	15	19	23	24	26	26	24	21	19	14	12
12	16	20	24	26	28	28	26	23	20	15	13
13	17	21	26	28	30	30	28	24	21	16	14
14	19	23	28	30	32	32	30	26	23	17	15

Appendix G (Continued)

L_p	(1) 50	(2) 63	(3) 80	(4) 100	(5) 125	(6) 160	(7) 200	(8) 250	(9) 315	(10) 400	(11) 500	(12) 630	(13) 800
80	5.0	6.4	7.5	10	11	11	13	14	15	16	16	16	16
81	5.5	6.9	8.3	11	11	12	14	15	16	17	17	17	17
82	6.1	7.5	9.1	11	12	13	15	16	17	18	18	18	18
83	6.8	8.3	10	12	13	14	16	17	19	20	20	20	20
84	7.5	9.1	12	13	14	15	17	19	20	21	21	21	21
85	8.3	10	13	14	15	16	19	20	21	23	23	23	23
86	9.1	12	13	15	16	17	20	21	23	24	24	24	24
87	10	13	14	16	17	19	21	23	24	26	26	26	26
88	11	13	15	17	19	20	23	24	26	28	28	28	28
89	12	14	16	19	20	21	24	26	28	30	30	30	30
90	14	15	17	20	21	23	26	28	30	32	32	32	32
91	15	16	19	21	23	24	28	30	32	34	34	34	34
92	16	17	20	23	24	26	30	32	35	37	37	37	37
93	17	19	21	24	26	28	32	35	37	39	39	39	39
94	19	20	23	26	28	30	35	37	40	42	42	42	42
95	20	21	24	28	30	32	37	40	42	45	45	45	45
96	21	23	26	30	32	35	40	42	45	49	49	49	49
97	23	24	28	32	35	37	42	45	47	52	52	52	52
98	24	26	30	35	37	40	45	47	50	56	56	56	56
99	26	28	32	37	40	42	47	50	55	60	60	60	60
100	28	30	35	40	42	45	50	55	60	64	64	64	64

Band Center Frequency (Hz)

101	30	32	37	42	45	47	55	60	64	69	69	69	69
102	32	35	40	45	47	50	60	64	69	74	74	74	74
103	35	37	42	47	50	55	64	69	74	79	79	79	79
104	37	40	45	50	55	60	69	74	79	84	84	84	84
105	40	42	47	55	60	64	74	79	84	91	91	91	91
106	42	45	50	60	64	69	79	84	91	97	97	97	97
107	45	47	55	64	69	74	84	91	97	104	104	104	104
108	47	50	60	69	74	79	91	97	104	111	111	111	111
109	50	55	64	74	79	84	97	104	111	119	119	119	119
110	55	60	69	79	84	91	104	111	119	128	128	128	128
111	60	64	74	84	91	97	111	119	128	137	137	137	137
112	64	69	79	91	97	104	119	128	137	147	147	147	147
113	69	74	84	97	104	111	128	137	147	158	158	158	158
114	74	79	91	104	111	119	137	147	158	169	169	169	169
115	79	84	97	111	119	128	147	158	169	181	181	181	181
116	84	91	104	119	128	137	158	169	181	194	194	194	194
117	91	97	111	128	137	147	169	181	194	208	208	208	208
118	97	104	119	137	147	158	181	194	208	223	223	223	223
119	104	111	128	147	158	169	194	208	223	239	239	239	239
120	111	119	137	158	169	181	208	223	239	256	256	256	256
121	119	128	147	169	181	194	223	239	256	274	274	274	274
122	128	137	158	181	194	208	239	256	274	294	294	294	294
123	137	147	169	194	208	223	256	274	294	315	315	315	315
124	147	158	181	208	223	239	274	294	315	338	338	338	338
125	158	169	194	223	239	256	294	315	338	362	362	362	362
126	169	181	208	239	256	274	315	338	362	388	388	388	388
127	181	194	223	256	274	294	338	362	388	416	416	416	416
128	194	208	239	274	294	315	362	388	416	446	446	446	446
129	208	223	256	294	315	338	388	416	446	478	478	478	478

Appendix G (Continued)

					Band Center Frequency (Hz)							
(14) 1000	(15) 1250	(16) 1600	(17) 2000	(18) 2500	(19) 3150	(20) 4000	(21) 5000	(22) 6300	(23) 8000	(24) 10,000	(25) 12,500
16	19	24	28	32	35	35	32	30	24	20	15
17	20	26	30	35	37	37	35	32	26	21	16
18	21	28	32	37	40	40	37	35	28	23	17
20	23	30	35	40	42	42	40	37	30	24	19
21	24	32	37	42	45	45	42	40	32	26	20
23	26	35	40	45	47	47	45	42	35	28	21
24	28	37	42	47	50	50	47	45	37	30	23
26	30	40	45	50	55	55	50	47	40	32	24
28	32	42	47	55	60	60	55	50	42	35	26
30	35	45	50	60	63	63	60	55	45	37	28
32	37	47	55	63	67	67	63	60	47	40	30
34	40	50	60	67	71	71	67	63	50	42	32
37	42	55	63	71	75	75	71	67	55	45	35
39	45	60	67	75	80	80	75	71	60	47	37
42	47	63	71	80	86	86	80	75	63	50	40
45	50	67	75	86	93	93	86	80	67	55	43
49	55	71	80	93	100	100	93	86	71	60	46
52	60	75	86	100	108	108	100	93	75	63	50
56	64	80	93	108	116	116	108	100	80	67	55
60	69	86	100	116	125	125	116	108	86	71	60
64	74	93	108	125	133	133	125	116	93	75	63
69	79	100	116	133	142	142	133	125	100	80	67
74	84	108	125	142	150	150	142	133	108	86	71

75	93	116	142	150	162	162	150	133	116	91	79
80	100	125	150	162	173	173	162	142	125	97	84
86	108	133	162	173	186	186	173	150	133	104	91
93	116	142	173	186	200	200	186	162	142	111	97
100	125	150	186	200	215	215	200	173	150	119	104
108	133	162	200	215	232	232	215	186	162	128	111
116	142	173	215	232	250	250	232	200	173	137	119
125	150	186	232	250	266	266	250	215	186	147	128
133	162	200	250	266	284	284	266	232	200	158	137
142	173	215	266	284	300	300	284	250	215	169	147
150	186	232	284	300	324	324	300	266	232	181	158
162	200	250	300	324	346	346	324	284	250	194	169
173	215	266	324	346	372	372	346	300	266	208	181
186	232	284	346	372	400	400	372	324	284	223	194
200	250	300	372	400	430	430	400	346	300	239	208
215	266	324	400	430	454	454	430	372	324	256	223
232	284	346	430	454	500	500	464	400	346	274	239
250	300	372	464	500	532	532	500	430	372	294	256
266	324	400	500	532	568	568	532	464	400	315	274
284	346	430	532	568	600	600	568	500	430	338	294
300	372	464	568	600	648	648	600	532	464	362	315
324	400	500	600	643	692	692	648	568	500	388	338
346	430	532	648	692	744	744	692	600	532	416	362
372	464	568	692	744	800	800	744	648	568	446	388
400	500	600	744	800	860	860	800	692	600	478	416
430	532	648	800	860	928	928	860	744	648	512	446
464	568	692	860	928	1000	1000	928	800	692	549	478

Appendix G (Continued)

				Band Center Frequency (Hz)							
(14) 1000	(15) 1250	(16) 1600	(17) 2000	(18) 2500	(19) 3150	(20) 4000	(21) 5000	(22) 6300	(23) 8000	(24) 10,000	(25) 12,000
512	588	744	860	1000	1064	1064	1000	928	744	600	500
549	630	800	928	1064	1136	1136	1064	1000	800	648	532
588	676	860	1000	1136	1200	1200	1136	1064	860	692	568
630	724	928	1064	1200	1296	1296	1200	1136	928	744	600
676	776	1000	1136	1296	1384	1384	1296	1200	1000	800	648
724	832	1064	1200	1384	1488	1488	1384	1296	1064	860	692
776	891	1136	1296	1488	1600	1600	1488	1384	1136	928	744
832	955	1200	1384	1600	1720	1720	1600	1488	1200	1000	800
891	1024	1296	1488	1720	1856	1856	1720	1600	1296	1064	860
955	1098	1384	1600	1856	2000	2000	1856	1720	1384	1136	928
1024	1176	1488	1720	2000			2000	1856	1488	1200	1000
1098	1261	1600	1856					2000	1600	1296	1064
1176	1351	1720	2000						1720	1384	1136
1261	1448	1856							1856	1488	1200
1351	1552	2000							2000	1600	1296
1448	1663									1720	1384
1552	1783									1856	1488
1663	1911									2000	1600
1783											1720
1911											2000

Band Center Frequency (Hz)

L_p	(1) 50	(2) 63	(3) 80	(4) 100	(5) 125	(6) 160	(7) 200	(8) 250	(9) 315	(10) 400	(11) 500	(12) 630	(13) 800
130	223	239	274	315	338	362	416	446	478	512	512	512	512
131	239	256	294	338	362	388	446	478	512	549	549	549	549
132	256	274	315	362	388	416	478	512	549	588	588	588	588
133	274	294	338	388	416	446	512	549	588	630	630	630	630
134	294	315	362	416	446	478	549	588	630	676	676	676	676
135	315	338	388	446	478	512	588	630	676	724	724	724	724
136	338	362	416	478	512	549	630	676	724	776	776	776	776
137	362	388	446	512	549	588	676	724	776	832	832	832	832
138	388	416	478	549	588	630	724	776	832	891	891	891	891
139	416	446	512	588	630	676	776	832	891	955	955	955	955
140	446	478	549	630	676	724	832	891	955	1024	1024	1024	1024
141	478	512	588	676	724	776	891	955	1024	1098	1098	1098	1098
142	512	549	630	724	776	832	955	1024	1098	1176	1176	1176	1176
143	549	588	676	776	832	891	1024	1098	1176	1261	1261	1261	1261
144	588	630	724	832	891	955	1098	1176	1261	1351	1351	1351	1351
145	630	676	776	891	955	1024	1176	1261	1351	1448	1448	1448	1448
146	676	724	832	955	1024	1098	1261	1351	1448	1552	1552	1552	1552
147	724	776	891	1024	1098	1176	1351	1448	1552	1663	1663	1663	1663
148	776	832	955	1098	1176	1261	1448	1552	1663	1783	1783	1783	1783
149	832	891	1024	1176	1261	1351	1552	1663	1783	1911	1911	1911	1911
150	891	955	1098	1261	1351	1448	1663	1783	1911				

H

REGULATIONS OF CONNECTICUT STATE AGENCIES

TITLE 22a

ENVIRONMENTAL PROTECTION

SECTION 22a-69-1 TO 22a-69-7.4

CONTROL OF NOISE

1518.8   10-78

Department of Environmental Protection

TABLE OF CONTENTS

Control of Noise

Control of Noise

Sec. 22a-69-1. Definitions

Sec. 22a-69-1.1. General

(a) **adaptive reuse** means remodeling and conversion of an obsolete or unused building or other structure for alternate uses. For example, older industrial buildings, warehouses, offices, hotels, garages, etc., could be improved and converted for reuse in terms of industrial processes, commercial activities, educational purposes, residential use as apartments, or other purposes.

(b) **aircraft** means any engine-powered device that is used or intended to be used for flight in the air and capable of carrying humans. Aircraft shall include civil, military, general aviation and VTOL/STOL aircraft.

(i) **aircraft, STOL** means any aircraft designed for, and capable of, short take-off and landing operations.

(ii) **aircraft, VTOL** means any aircraft designed for, and capable of, vertical take-off and landing operations such as, but not limited to, helicopters.

(c) **airport** means an area of land or water that is used, or intended to be used, for the landing and takeoff of aircraft and is licensed by the State of Connecticut Bureau of Aeronautics for such use. "Airport" shall include all buildings and facilities if any. "Airport" shall include any facility used, or intended for use, as a landing and take-off area for VTOL/STOL aircraft, including, but not limited to, heliports.

(d) **ANSI** means the American National Standards Institute or its successor body.

(e) **best practical noise control measures** means noise control devices, technology and procedures which are determined by the Commissioner to be the best practical, taking into consideration the age of the equipment and facilities involved, the process employed, capital expenditures, maintenance cost, technical feasibility, and the engineering aspects of the applicable noise control techniques in relation to the control achieved and the non-noise control environmental impact.

(f) **commissioner** means the Commissioner of the Department of Environmental Protection or his/her designated representative.

(g) **construction** means any, and all, physical activity at a site necessary or incidental to the erection, placement, demolition, assembling, altering, blasting, cleaning, repairing, installing, or equipping of buildings or other structures, public or private highways, roads, premises, parks, utility lines, or other property, and shall include, but not be limited to, land clearing, grading, excavating, filling and paving.

(h) **daytime** means 7:00 a.m. to 10:00 p.m. local time.

(i) **director** means the Director of the Office of Noise Control in the Department of Environmental Protection.

(j) **emergency** means any occurrence involving actual or imminent danger to persons or damage to property which demands immediate action.

(k) **intrusion alarm** means a device with an audible signal which, when activated, indicates intrusion by an unauthorized person. Such alarm may be attached to, or within, any building, structure, property or vehicle.

(l) **ISO** means the International Organization for Standardization, or its successor body.

(m) **lawn care and maintenance equipment** means all engine or motor-powered garden or maintenance tools intended for repetitive use in residential areas, typically capable of being used by a homeowner, and including, but not limited to, lawn mowers, riding tractors, snow-blowers, and including equipment intended for infrequent service work in inhabited areas, typically requiring skilled operators, including, but not limited to, chain saws, log chippers or paving rollers.

(n) **nighttime** means 10:00 p.m. to 7:00 a.m. local time.

(o) **noise zone** means an individual unit of land or a group of contiguous parcels under the same ownership as indicated by public land records and, as relates to noise emitters, includes contiguous publicly dedicated street and highway rights-of-way, railroad rights-of-way and waters of the State.

(p) **office of noise control** means the office within the Department of Environmental Protection designated by the Commissioner to develop, administer and enforce the provisions of Chapter 442 of the Connecticut General Statutes.

(q) **OSHA** means the Occupational Safety and Health Act and any amendments thereto or successor regulations administered by the U.S. and Connecticut Departments of Labor, or successor bodies.

(r) **person** means any individual, firm, partnership, association, syndicate, company, trust, corporation, municipality, agency, or political or administrative subdivision of the State or other legal entity of any kind.

(s) **public emergency sound signal** means an audible electronic or mechanical siren or signal device attached to an authorized emergency vehicle or within or attached to a building for the purpose of sounding an alarm relating to fire or civil preparedness. Such signal may also be attached to a pole or other structure.

(t) **SAE** means the Society of Automotive Engineers, Inc., or its successor body.

(u) **safety and protective devices** means devices that are designed to be used, and are actually used, for the prevention of the exposure of any person or property to imminent danger, including, but not limited to, unregulated safety relief valves, circuit breakers, protective

fuses, back-up alarms required by OSHA or other state
or federal safety regulations, horns, whistles or other
warning devices associated with pressure buildup.

(v) **site** means the area bounded by the property line
on or in which a source of noise exists.

(Effective June 15, 1978)

Sec. 22a-69-1.2. Acoustic terminology and definitions

(a) All acoustical terminology used in these Regulations shall be in conformance with the American National
Standards Institute (ANSI), "Acoustical Terminology,"
contained in publication S1.1 as now exists and as may
be hereafter modified. The definitions below shall apply
if the particular term is not defined in the aforesaid
ANSI publication.

(b) **audible range of frequency** means the frequency
range 20 Hz to 20,000 Hz which is generally considered
to be the normal range of human hearing.

(c) **background noise** means noise which exists at a
point as a result of the combination of many distant
sources, individually indistinguishable. In statistical
terms, it is the level which is exceeded 90% of the time
(L_{90}) in which the measurement is taken.

(d) **continuous noise** means ongoing noise, the intensity of which remains at a measurable level (which may
vary) without interruption over an indefinite period or a
specified period of time.

(e) **decibel** (dB) means a unit of measurement of the
sound level.

(f) **excessive noise** means emitter Noise Zone levels
from stationary noise sources exceeding the Standards
set forth in Section 3 of these Regulations beyond the
boundary of adjacent Noise Zones.

(g) **existing noise source** means any noise source(s)
within a given Noise Zone, the construction of which
commenced prior to the effective date of these Regulations.

(h) **fluctuating noise** means a continuous noise whose
level varies with time by more than 5 dB.

(i) **frequency** means the number of vibrations or alterations of sound pressure per second and is expressed in
Hertz.

(j) **hertz (Hz)** means a unit of measurement of frequency formerly stated as, and numerically equal to,
cycles per second.

(k) **impulse noise** means noise of short duration (generally less than one second), especially of high intensity,
abrupt onset and rapid decay, and often rapidly changing spectral composition.

(l) **infrasonic sound** means sound pressure variations
having frequencies below the audible range for humans,
generally below 20 Hz; subaudible.

(m) L_{10} means the A-weighted sound level exceeded 10% of the time period during which measurement was made.

(n) L_{50} means the A-weighted sound level exceeded 50% of the time period during which measurement was made.

(o) L_{90} means the A-weighted sound level exceeded 90% of the time period during which measurement was made.

(p) **octave band sound pressure level** means the sound pressure level for the sound contained within the specified preferred octave band, stated in dB, as described in ANSI S1.6-1967: Preferred Frequencies and Band Numbers for Acoustical Measurements.

(q) **peak sound pressure level** means the absolute maximum value of the instantaneous sound pressure level occurring in a specified period of time.

(r) **prominent discrete tone** means the presence of acoustic energy concentrated in a narrow frequency range, including, but not limited to, an audible tone, which produces a one-third octave sound pressure level greater than that of either adjacent one-third octave and which exceeds the arithmetic average of the two adjacent one-third octave band levels by an amount greater than shown below opposite the center of frequency for the one-third octave band containing the concentration of acoustical energy.

1/3 Octave Band Center Frequency (Hz)	dB
100	16
125	14
160	12
200	11
250	9
315	8
400	7
500	6
630	6
800	5
1000	4
1250	4
1600	4
2000	3
2500	3
3150	3
4000	3
5000	4
6300	4
8000	5
10000	6

(s) **reference pressure** is 0.00002 Newtons per square meter (N/M^2), or 20 microPascals, for the purposes of these Regulations.

(t) **sound** means a transmission of energy through solid, liquid, or gaseous media in the form of vibrations which constitute alterations in pressure or position of the particles in the medium and which, in air, evoke physiological sensations, including, but not limited to, an auditory response when impinging on the ear.

(u) **sound analyzer** means a device, generally used in conjunction with a sound level meter, for measuring the sound pressure level of a noise as a function of frequency in octave bands, one-third octave bands or other standard ranges.   The sound analyzer shall conform to Type E, Class II, as specified in ANSI S1.11-1971 or latest revision.

(v) **sound level** means a frequency weighted sound pressure level, obtained by the use of metering characteristics and the weighting A, B, or C as specified in ANSI, "Specifications for Sound Level Meters," S1.4-1971 or latest revision.   The unit of measurement is the decibel.   The weighting employed must always be stated as dBA, dBB, or dBC.

(w) **sound level meter** means an instrument, including a microphone, an amplifier, an output meter, and frequency weighting networks for the measurement of sound levels.   The sound level meter shall conform to ANSI Specifications for Sound Level Meters S1.4-1971.

(x) **sound pressure level (SPL)** means twenty times the logarithm to the base ten of the ratio of the sound pressure in question to the standard reference pressure of 0.00002 N/M^2.   It is expressed in decibel units.

(y) **ultrasonic sound** means sound pressure variations having frequencies above the audible sound spectrum for humans, generally higher than 20,000 Hz; superaudible.

(z) **vibration** means an ascillatory motion of solid bodies of deterministic or random nature described by displacement, velocity, or acceleration with respect to a given reference point.
(Effective June 15, 1978)

Sec. 22a-69-1.3.   Coordination with other laws

(a) Nothing in these Regulations shall authorize the construction or operation of a stationary noise source in violation of the requirements of any other applicable State law or regulation.

(b) Nothing in these Regulations shall authorize the sale, use or operation of a noise source in violation of the laws and regulations of the Connecticut Department of Motor Vehicles, the Federal Aviation Administration, the U.S. Environmental Protection Agency, or any amendments thereto.
(Effective June 15, 1978)

Sec. 22a-69-1.4.   Incorporation by reference

(a) The specifications, standards and codes of agencies of the U.S. Government and organizations which are not agencies of the U.S. Government, to the extent that they are legally incorporated by reference in these Regulations, have the same force and effect as other standards in these Regulations.

(b) These specifications, standards and codes may be examined at the Office of Noise Control, Department of Environmental Protection, State of Connecticut.

(c) Any changes in the specifications, standards and codes incorporated in these Regulations are available at the Office listed in (b) above. All questions as to the applicability of such changes should also be referred to this Office.

(Effective June 15, 1978)

Sec. 22a-69-1.5. Compliance with regulations no defense to nuisance claim

Nothing in any portion of these Regulations shall in any manner be construed as authorizing or legalizing the creation or maintenance of a nuisance, and compliance of a source with these Regulations is not a bar to a claim of nuisance by any person. A violation of any portion of these Regulations shall not be deemed to create a nuisance per se.

(Effective June 15, 1978)

Sec. 22a-69-1.6. Severability

If any provision of these Regulations or the application thereof to any person or circumstances is held to be invalid, such invalidity shall not' affect other provisions or applications of any other part of these Regulations which can be given effect without the invalid provisions or application; and to this end, the provisions of these Regulations and the various applications thereof are declared to be severable.

(Effective June 15, 1978)

Sec. 22a-69-1.7. Exclusions

These Regulations shall not apply to:

(a) Sound generated by natural phenomena, including, but not limited to, wind, storms, insects, amphibious creatures, birds, and water flowing in its natural course.

(b) The unamplified sounding of the human voice.

(c) The unamplified sound made by any wild or domestic animal.

(d) Sound created by bells, carillons, or chimes associated with specific religious observances.

(e) Sound created by a public emergency sound signal attached to an authorized emergency vehicle in the immediate act of responding to an emergency, as authorized by subsection (d) of Section 14.80 and Section 14-1a of Chapter 246 of the General Statutes and all amendments thereto, or located within or attached to a building, pole or other structure for the purpose of sounding an alarm relating to fire or civil preparedness.

(f) Sound created by safety and protective devices.

(g) Farming equipment or farming activity.

(h) Back-up alarms required by OSHA or other State or Federal safety regulations.

(i) Sound created by any mobile source of noise. Mobile sources of noise shall include, but are not limited to, such sources as aircraft, automobiles, trucks, and boats. This exclusion shall cease to apply when a mobile source of noise has maneuvered into position at the loading dock, or similar facility, has turned off its engine and ancillary equipment, and has begun the physical process of removing the contents of the vehicle.
(Effective June 15, 1978)

Sec. 22a-69-1.8. Exemptions

Exempted from these Regulations are:

(a) Conditions caused by natural phenomena, strike, riot, catastrophe, or other condition over which the apparent violator has no control.

(b) Noise generated by engine-powered or motor-driven lawn care or maintenance equipment shall be exempted between the hours of 7:00 a.m. and 9:00 p.m. provided that noise discharged from exhausts is adequately muffled to prevent loud and/or explosive noises therefrom.

(c) Noises created by snow removal equipment at any time shall be exempted provided that such equipment shall be maintained in good repair so as to minimize noise, and noise discharged from exhausts shall be adequately muffled to prevent loud and/or explosive noises therefrom.

(d) Noise that originates at airports that is directly caused by aircraft flight operations specifically preempted by the Federal Aviation Administration.

(e) Noise created by the use of property for purposes of conducting speed or endurance events involving motor vehicles shall be exempted but such exemption is effective only during the specific period(s) of time within which such use is authorized by the political subdivision or governmental entity having lawful jurisdiction to sanction such use.

(f) Noise created as a result of, or relating to, an emergency.

(g) Construction noise.

(h) Noise created by blasting other than that conducted in connection with construction activities shall be exempted provided that the blasting is conducted between 8:00 a.m. and 5:00 p.m. local time at specified hours previously announced to the local public, or provided that a permit for such blasting has been obtained from local authorities.

(i) Noise created by on-site recreational or sporting activity which is sanctioned by the state or local government provided that noise discharged from exhausts is adequately muffled to prevent loud and/or explosive noises therefrom.

(j) Patriotic or public celebrations not extending longer than one calendar day.

(k) Noise created by aircraft, or aircraft propulsion components designed for or utilized in the development of aircraft, under test conditions.

(l) Noise created by products undergoing test, where one of the primary purposes of the test is evaluation of product noise characteristics and where practical noise control measures have been taken.

(m) Noise generated by transmission facilities, distribution facilities and substations of public utilities providing electrical powers, telephone, cable television or other similar services and located on property which is not owned by the public utility and which may or may not be within utility easements.

(Effective June 15, 1978)

Sec. 22a-69-1.9. Burden of persuasion regarding exclusions and exemptions

In any proceeding pursuant to these Regulations, the burden of persuasion shall rest with the party attempting to enforce the Regulations. Notwithstanding the foregoing, if an exclusion or exemption stated in these Regulations would limit an obligation, limit a liability, or eliminate either an obligation or a liability, the person who would benefit from the application of the exclusion or exemption shall have the burden of persuasion that the exclusion or exemption applies and that the terms of the exclusion or exemption have been met. The Department shall cooperate with and assist persons in determining the application of the provisions of these Regulations.

(Effective June 15, 1978)

Sec. 22a-69-2. Classification of land according to use

Sec. 22a-69-2.1. Basis

Noisy Zone classifications shall be based on the actual use of any parcel or tract under single ownership as detailed by the Standard Land Use Classification Manual of Connecticut (SLUCONN).

(Effective June 15, 1978)

Sec. 22a-69-2.2. Multiple uses

Where multiple uses exist within a given Noise Zone, the least restrictive land use category for the Emitter and Receptor shall apply regarding the noise standards specified in Section 3 of these Regulations.

(Effective June 15, 1978)

Sec. 22a-69-2.3. Class A noise zone

Lands designated Class A shall generally be residential areas where human beings sleep or areas where serenity and tranquility are essential to the intended use of the land.

Class A Land Use Category. The land uses in this category shall include, but not be limited to, single and multiple family homes, hotels, prisons, hospitals, religious facilities, cultural activities, forest preserves, and land intended for residential or special uses requiring such protection.

The specific SLUCONN categories in Class A shall include:

1. Residential
11 Household Units*
12 Group Quarters
13 Mobile Home Parks and Courts
19 Other Residential
5. Trade
583 Residential Hotels
584 Hotels, Tourist Courts and Motels
585 Transient Lodgings
6. Services
651 Medical and Other Health Services; Hospitals
674 Correctional Institutions
691 Religious Activities
7. Cultural, Entertainment and Recreational
711 Cultural Activities
712 Nature Exhibitions
713 Historic and Monument Sites,
*Mobile homes are included if on foundations
9. Undeveloped, Unused and Reserved Lands and Water Areas
92 Reserved Lands
941 Vacant Floor Area—Residential
(Effective June 15, 1978)

Sec. 22a-69-2.4.   Class B noise zone

Lands designated Class B shall generally be commercial in nature, areas where human beings converse and such conversation is essential to the intended use of the land.

Class B Land Use Category. The land uses in this category shall include, but not be limited to, retail trade, personal, business and legal services, educational institutions, government services, amusements, agricultural activities, and lands intended for such commercial or institutional uses.

The specific SLUCONN categories in Class B shall include:

4. Transportation, Communication and Utilities
46 Automobile Parking
47 Communication
5. Trade
51 Wholesale Trade
52 Retail Trade – Building Materials
53 Retail Trade – General Merchandise
54 Retail Trade – Food

55 Retail Trade – Automotive Dealers and Gasoline Service Stations
56 Retail Trade – Apparel and Accessories
57 Retail Trade – Furniture, Home Furnishings and Equipment
58 Retail Trade – Eating, Drinking and Lodging— Except 583, 584, and 585
59 Retail Trade – N.E.C.*
6. Services
61 Finance, Insurance and Real Estate Services
62 Personal Services
63 Business Services—Except 637
64 Repair Services
65 Professional Services—Except 651
67 Government Services—Except 672, 674, and 675
68 Educational Services
69 Miscellaneous Services—Except 691
7. Cultural, Entertainment and Recreational
71 Cultural Activities and Nature Exhibitions—Except 711, 712, and 713
72 Public Assembly
73 Amusements
74 Recreational Activities
75 Resorts and Group Camps
76 Parks
79 Other, N.E.C.*
*Not Elsewhere Classified
8. Agriculture
81 Agriculture
82 Agricultural Related Activities
9. Undeveloped, Unused, and Reserved Lands and Water Area
91 Undeveloped and Unused Land Area
93 Water Areas
94 Vacant Floor Area—Except 941
99 Other Undeveloped Land and Water Areas, N.E.C.*
*Not Elsewhere Classified
(Effective June 15, 1978)

Sec. 22a-69-2.5. Class C noise zone

Lands designated Class C shall generally be industrial where protection against damage to hearing is essential, and the necessity for conversation is limited.

Class C Land Use Category. The land uses in this category shall include, but not be limited to, manufacturing activities, transportation facilities, warehousing, military bases, mining, and other lands intended for such uses.
The specific SLUCONN categories in Class C shall include:
2. Manufacturing – Secondary Raw Materials
3. Manufacturing – Primary Raw Materials

4. Transportation, Communications and Utilities—
Except 46 and 47
6. Services
637 Warehousing and Storage Services
66 Contract Construction Services
672 Protective Functions and Related Activities
675 Military Bases and Reservations
8. Agriculture
83 Forestry Activities and Related Services
84 Commercial Fishing Activities and Related Services
85 Mining Activities and Related Services
89 Other Resource Production and Extraction, N.E.C.*
*Not Elsewhere Classified
(Effective June 15, 1978)

Sec. 22a-69-3.   Allowable noise levels

Sec. 22a-69-3.1.   General prohibition

No person shall cause or allow the emission of excessive
noise beyond the boundaries of his/her Noise Zone so as
to violate any provisions of these Regulations.
(Effective June 15, 1978)

Sec. 22a-69-3.2.   Impulse noise

(a) No person shall cause or allow the emission of
impulse noise in excess of 80 dB' peak sound pressure
level during the nighttime to any Class A Noise Zone.

(b) No person shall cause or allow the emission of
impulse noise in excess of 100 dB peak sound pressure at
any time to any Noise Zone.
(Effective June 15, 1978)

Sec. 22a-69-3.3.   Prominent discrete tones

Continuous noise measured beyond the boundary of
the Noise Zone of the noise emitter in any other Noise
Zone which possesses one or more audible discrete tones
shall be considered excessive noise when a level of 5 dBA
below the levels specified in Section 3 of these Regulations is exceeded.
(Effective June 15, 1978)

Sec. 22a-69-3.4.   Infrasonic and ultrasonic

No person shall emit beyond his/her property infrasonic or ultrasonic sound in excess of 100 dB at any time.
(Effective June 15, 1978)

Sec. 22a-69-3.5.   Noise zone standards

(a) No person in a Class C Noise Zone shall emit noise
exceeding the levels stated herein and applicable to adjacent Noise Zones:

	Receptor			
	C	B	A/Day	A/Night
Class C Emitter to	70 dBA	66 dBA	61 dBA	51 dBA

Levels emitted in excess of the values listed above shall be considered excessive noise.

(b) No person in a Class B Noise Zone shall emit noise exceeding the levels stated herein and applicable to adjacent Noise Zones:

		Receptor		
	C	B	A/Day	A/Night
Class B Emitter to	62 dBA	62 dBA	55 dBA	45 dBA

Levels emitted in excess of the values listed above shall be considered excessive noise.

(c) No person in a Class A Noise Zone shall emit noise exceeding the levels stated herein and applicable to adjacent Noise Zones:

		Receptor		
	C	B	A/Day	A/Night
Class A Emitter to	62 dBA	55 dBA	55 dBA	45 dBA

Levels emitted in excess of the values listed above shall be considered excessive noise.

(Effective June 15, 1978)

Sec. 22a-69-3.6. High background noise areas

In those individual cases where the background noise levels caused by sources not subject to these Regulations exceed the standards contained herein, a source shall be considered to cause excessive noise if the noise emitted by such source exceeds the background noise level by 5 dBA, provided that no source subject to the provisions of Section 3 shall emit noise in excess of 80 dBA at any time, and provided that this Section does not decrease the permissible levels of the other Sections of this Regulation.

(Effective June 15, 1978)

Sec. 22a-69-3.7. Existing noise sources

Existing noise sources constructed between the effective date of these Regulations and January 1, 1960 shall be provided a permanent five (5) dBA maximum noise level allowance over levels otherwise herein required regardless of subsequent changes in ownership or facility utilization processes at the location of the existing noise source. Existing noise sources constructed prior to 1960 shall be provided a permanent ten (10) dBA maximum noise level allowance over levels otherwise herein required regardless of subsequent changes in ownership or facility utilization processes at the location of the existing noise source. Additionally, all existing noise sources shall be provided twenty-four (24) months in order to achieve compliance with these Regulations if a notice of violation has been, or may be, issued to the source. This time period begins with the effective date of these Regulations, not with the date of the notice of violation.

(Effective June 15, 1978)

Sec. 22a-69-3.8.   Adaptive reuse of existing buildings

Buildings and other structures that exist as of the effective date of these Regulations which have been remodeled or converted for adaptive reuse or which may be remodeled or converted at a future date shall be provided a permanent five (5) dBA maximum noise level allowance above the Emitter Class of the new use of the building over levels otherwise herein required.

(Effective June 15, 1978)

Sec. 22a-69-4.   Measurement procedures

Acoustic measurements to ascertain compliance with these Regulations shall be in substantial conformity with standards and Recommended Practices established by professional organizations such as ANSI and SAE.

(a) Personnel conducting sound measurements shall be trained and experienced in the current techniques and principles of sound measuring equipment and instrumentation. The Commissioner shall establish sufficiently detailed measurement procedure guidelines specifying, but not necessarily being limited to, the following: The appropriate utilization of fast or slow sound level meter dampening when making sound level measurements, the rise time specified in microseconds for measuring impulse noise, the need for a whole circuit in such measurements, and the proper weighting to be used in measuring impulse noise.

(b) Instruments shall conform to the following standards of their latest revisions:

(i) ANSI S1.4-1971, "Specifications for Sound Level Meters," Type 1 or 2.

(ii) ANSI S1.11-1966, "Specifications for Octave, One-Half Octave and One-Third Octave Band Filter Sets," Type E, Class II.

(iii) If a magnetic tape recorder or a graphic level recorder or other indicating device is used, the system shall meet the applicable requirements of SAE Recommended Practice J184, "Qualifying a Sound Data Acquisition System."

(c) Instruments shall be set up to conform to ANSI S1.13-1971, "Methods for the Measurement of Sound Pressure Levels."

(d) Instrument manufacturer's instructions for use of the instruments shall be followed, including acoustical calibration of equipment used.

(e) The determination of L_{90} to ascertain background levels requires a statistical analysis. A graphic level recording and visual interpretation of the chart recording to determine the levels is an acceptable method. Instruments designed to determine the cumulative distribution of noise levels are also acceptable used either in the field or in the labortaory to analyze a tape recording. Dynamic visual estimations from a sound level meter are not an acceptable method for determining such levels. Sound level sampling techniques are acceptable and will often be the most practical to employ. Such a technique using Connecticut Noise Survey Data Form #101 with accompanying instructions is acceptable.

(f) In measuring compliance with Noise Zone Standards, the following short-term noise level excursions over the noise level standards established by these Regulations shall be allowed, and measurements within these ranges of established standards shall constitute compliance therewith:

Allowable levels above standards (dBA)	Time period of such levels (minutes/hour)
3	15
6	7½
8	5

(g) Measurements taken to determine compliance with Section 3 shall be taken at about one foot beyond the boundary of the Emitter Noise Zone within the receptor's Noise Zone. The Emitter's Noise Zone includes his/her individual unit of land or group of contiguous parcels under the same ownership as indicated by public land records. The Emitter's Noise Zone also includes contiguous publicly dedicated street and highway rights-of-way, railroads rights-of-way and waters of the State.
(Effective June 15, 1978)

Sec. 22a-69-5. Other provisions

Sec. 22a-69-5.1. Intrusion alarms

No person shall cause, suffer, allow or permit the operation of any intrusion alarm which, from time of activation of audible signal, emits noise for a period of time exceeding ten minutes when attached to any vehicle or thirty minutes when attached to any building or structure.
The repetition of activation of the audible signal of an intrusion alarm due to malfunction, lack of proper maintenance, or lack of reasonable care shall be considered excessive noise.
(Effective June 15, 1978)

Sec. 22a-69-6. Airport facilities

Sec. 22a-69-6.1. Extent of regulation

Airport facilities are subject to Section 3 to the extent not preempted by state or federal law or regulation.
(Effective June 15, 1978)

Sec. 22a-69-6.2. Reserved

(This subsection is reserved for possible future regulations regarding the assessment of, and long-range plans for, the reduction of airport facility noise impacts to the extent not preempted by state or federal law or regulation.)

(Effective June 15, 1978)

Sec. 22a-69-7.   Variances and enforcement procedures

Sec. 22a-69-7.1.   Variances

(a) Any person who owns or operates any stationary noise source may apply to the Commissioner for a variance or a partial variance from one or more of the provisions of these Regulations. Applications for a variance shall be submitted on forms furnished by the Commissioner and shall supply such information as he/she requires, including, but not limited to:

(i) Information on the nature and location of the facility or process for which such application is made.

(ii) The reason for which the variance is required, including the economic and technical justifications.

(iii) The nature and intensity of noise that will occur during the period of the variance.

(iv) A description of interim noise control measures to be taken by the applicant to minimize noise and the impacts occurring therefrom.

(v) A specific schedule of the best practical noise control measures, if any, which might be taken to bring the source into compliance with those Regulations from which a variance is sought, or a statement of the length of time during which it is estimated that it will be necessary for the variance to continue.

(vi) Any other relevant information the Commissioner may require in order to make a determination regarding the application.

(b) Failure to supply the information required by the form furnished by the Commissioner shall be cause for rejection of the application unless the applicant supplies the needed information within thirty (30) days of the written request by the Commissioner for such information.

(c) No variance shall be approved unless the applicant presents adequate proof to the Commissioner's satisfaction that:

(i) Noise levels occurring during the period of the variance will not constitute a danger to the public health; and

(ii) Compliance with the Regulations would impose an arbitrary or unreasonable hardship upon the applicant without equal or greater benefits to the public.

(d) In making a determination on granting a variance, the Commissioner shall consider:

(i) The character and degree of injury to, or interference with, the health and welfare or the reasonable use of property which is caused or threatened to be caused.

(ii) The social and economic value of the activity for which the variance is sought.

(iii) The ability of the applicant to apply best practical noise control measures, as defined in these Regulations.

(e) Following receipt and review of an application for a variance, the Commissioner shall fix a date, time and location for a hearing on such application.

(f) The Commissioner shall cause the applicant to publish at his/her own expense all notices of hearings and other notices required by law, including, but not limited to, notification of all abutters of record.

(g) Within sixty (60) days of the receipt of the record of the hearings on a variance application, the Commissioner shall issue his/her determination regarding such application. All such decisions shall briefly set forth the reasons for the decision.

(h) The Commissioner may, at his/her discretion, limit the duration of any variance granted under these Regulations. Any person holding a variance and needing an extension of time may apply for a new variance under the provisions of these Regulations. Any such application shall include a certification of compliance with any condition imposed under the previous variance.

(i) The Commissioner may attach to any variance any reasonable conditions he/she deems necessary and desirable, including, but not limited to:

(i) Requirements for the best practical noise control measures to be taken by the owner or operator of the source to minimize noise during the period of the variance.

(ii) Requirements for periodic reports submitted by the applicant relating to noise, to compliance with any other conditions under which the variance was granted or to any other information the Commissioner deems necessary.

(j) The filing of an application for a variance shall operate as a stay of prosecution, except that such stay may be terminated by the Commissioner upon application of any party if the Commissioner finds that protection of the public health so requires.

(k) In any case where a person seeking a variance contends that compliance with any provision of these Regulations is not practical or possible because of the cost involved either in installing noise control equipment or changing or curtailing the operation in any manner, he/she shall make available to the Commissioner such financial records as the Commissioner may require.

(l) A variance may include a compliance schedule and requirements for periodic reporting of increments of achievement of compliance.

(Effective June 15, 1978)

Sec. 22a-69-7.2. Transference

No person who owns, operates or maintains a stationary noise source shall transfer a variance from one site to another site.

(Effective June 15, 1978)

Sec. 22a-69-7.3. Responsibility to comply with applicable regulations

Approval of a variance shall not relieve any person of the responsibility to comply with any other applicable Regulations or other provisions of federal, state or local laws, ordinances or regulations.

(Effective June 15, 1978)

Sec. 22a-69-7.4. Violations and enforcement

(a) No person shall violate or cause the violation of any of these Regulations.

(b) Each day on which a violation occurs or continues after the time for correction of the violation given in the order has elapsed or after thirty (30) days from the date of service of the order, whichever is later, shall be considered a separate violation of these Regulations.

(c) Qualified personnel of the Office of Noise Control shall, with or without complaints, conduct investigations and ascertain whether these Regulations have been complied with. Whenever such personnel determines that any of these Regulations have been violated or there has been a failure to comply therewith, they shall make and serve upon the person(s) responsible for the violation a written order specifying the nature of the violation or failure and affording a reasonable time for its correction or remedy. Prior to the issuance of such order, such personnel shall make a reasonable effort in light of the circumstances to correct a violation or achieve compliance by means of conference, conciliation and persuasion as required by statute. Unless the person(s) against whom an order has been served files a written answer thereto with the Commissioner within thirty (30) days after the date of service of the order and requests a hearing thereon, such order shall become final and effective in accordance with the Connecticut Administrative Procedures Act and the rules, practices, and procedures of the Department of Environmental Protection.

(Effective June 15, 1978)

I
MACHINE ENCLOSURE LISTING CATALOG

ASSEMBLY MACHINES

Bodine Assembly Machine
Dixon Vibration Machine
Vibration Assembly Machine

Bodine #42 Tapper
Gillman Rod Machine

BAR-CUTTING MACHINES

Abrasive Cut-Off Machine
Bar Cut-Off Saw
DeWalt Cut-Off Saw
Extrusion Cut-Off Saw
Fenn #125 Cut-Off Machine
Fenn #250 Cut-Off Machine
Hardigg Friction Saw
Kasto Band Saw
Lewis Cut-Off Machine
Loma Slab Saw #SL-80
Loma Billet Saw #T-40
Peddinghaus Bar Cut-Off Machine
Taylor/Winfield Cut-Off Machine

Aluminum Cut-Off Saw
Billet Saw
Double End Cut-Off Saw
Fenn #62 Cut-Off Machine
Fenn #200 Cut-Off Machine
Guillotine Saw #SG150
Hydraulic Bar Cutting Machine
Kling Friction Saw
Loma Cut-Off Saw #SL-72
Loma Billet Saw #T-32
Oliver Cut-Off Saw
Stone Cut-Off Saw
Wire Cut-Off Machine

BLOWERS

Action Air 3800 CFM Blower
Fuller 600 H.P. Blower

Aerzen #GM6-15-11 Blower
Hick, Hargreaves #4082 Blower

Courtesy of Neiss, Rockville, Conn.

Hick, Hargreaves #4062 Blower
Ling #V810 Exciter Cooling Blower
NAPCO #VCF-140 Blower
Siemens-Allis 150 H.P. Blower
Sutorbilt Duraflow #7012 Blower
ROOTSPAK #24 URAI Blower
ROOTSPAK #33 URAI Blower
ROOTSPAK #36 URAI Blower
ROOTSPAK #53 URAI Blower
ROOTSPAK #65 URAI Blower
ROOTSPAK #59 URAI Blower
ROOTSPAK #68 URAK Blower
ROOTSPAK #711 URAI Blower
ROOTS #418-RCS-J Blower
ROOTS #616 RCS-J Blower
ROOTS #817 RCS-J Blower
ROOTS #1009 RAS-J Blower
ROOTS #1009 RGS-JH Blower
ROOTS #1021 RAS-JV Blower
ROOTS #12-20 RAS-J Blower
ROOTS #1422 RAS-J Blower
ROOTS #1620 RAS-J Blower

Lamson Blower
M-D Pneumatics #7017 Blower
Siemens-Allis 60 H.P. Blower
Sutorbile Duraflow #7015 Blower
ROOTSPAK #22 URAI Blower
ROOTSPAK #32 URAI Blower
ROOTSPAK #42 URAI Blower
ROOTSPAK #45 URAI Blower
ROOTSPAK #47 URAI Blower
ROOTSPAK #56 URAI Blower
ROOTSPAK #76 URAI Blower
ROOTSPAK #615 URAI Blower
ROOTS #409-RCS-JP Blower
ROOTS #404-RGS-JH Blower
ROOTS #624-RCS-JP Blower
ROOTS #821-RCS-J Blower
ROOTS #1009 RGS-J Blower
ROOTS #1016 DVJ Blower
ROOTS #12-16 RAS-J Blower
ROOTS #1452 Direct Connection
   Blower
ROOTS #1617 RAS-J Blower
ROOTS #2002 RAS-J Blower

BRAIDING MACHINES

Cook #1 Buncher
New England Butt Braider
Wardwell Braider

Hose Braiding Machine
New England Butt Twiner
Wire Braiding Machine

BUFFING MACHINES

Automatic Buffing Machine

Jack Lathe Buffer

COLD HEADERS

Behr 3/16" Cold Header
Blanchard 1/8" Pulsator
Greenbat #0 Cold Header

Behr 1/4" Cold Header
Bliss #4 Case Header
Greenbat 3/16" Cold Header

Greenbat 1/4" Cold Header
Hartford 3/16" Cold Header
Hartford 5/16" Ball Header
National 3/16" Cold Header
National 1/4" Hi-Speed Header
National 3/8" Cold Header
National 1/2" Cold Header
National 5/8" Cold Header
National 13/16" Cold Header
National #1250 Cold Header
National S-3-1/2" Cold Header
Sacma #SP-01, 3/16" Header
Sacma #SP-15 Transfer Header
Sacma #SP-31, 1/2" Header
Waterbury 1/8" Cold Header
Waterbury 3/16" Hi-Pro Header
Waterbury 5/16" Cold Header
Waterbury #1 Cold Header
Waterbury #3 Universal Header
Waterbury #4 Cold Header
Waterbury #5 Parts Former
Waterbury #6 Cold Header
Waterbury #750 Nut Former
Waterbury Ajax Wire Straightener
#0 Toggle Header

Hartford 1/8" Cold Header
Hartford 5/16" Cold Header
National 1/8" Cold Header
National 1/4" Cold Header
National 5/16" Cold Header
National 7/16" Cold Header
National 9/16" Cold Header
National 3/4" Cold Header
National #1000 Cold Header
National 1-1/2" Cold Header
National #141 Slug Header
Sacma #SP-11, 1/4" Header
Sacma #SP-21, 5/16" Header
Wafios 1/4" Cold Header
Waterbury 3/16" Cold Header
Waterbury 1/4" Cold Header
Waterbury 3/8" Cold Header
Waterbury #2 Nut Former
Waterbury #3 Parts Former
Waterbury #4 Transfer Header
Waterbury #5 Nut Former
Waterbury #N4 Nut Former
Waterbury #150 Slug Header
Waterbury #400 Slug Header

COMPRESSORS

Allis Chalmers Compress
Elliot #13 Compressor
Gardner Denver #CDL Compressor
Gardner Denver #MLA Compressor
Joy Centrifugal Compressor
Vilter #8 Compressor

Centac Compressor
Elliot #15 Compressor
Gardner Denver #PDR Compressor
Ingersoll Rand #30 Compressor
Sull-Air Compressor
Worthington Compressor

FORGING MACHINERY

Chambersburg Impactor
Chambersburg #4 Hammer
Greenlee Rota-Forge Machine

Chambersburg Die Forger
Forge Press Hammer
Mesta HE-55 Forge Machine

GRANULATING MACHINES

Allsteel Pelletizer
Blo-Hog Granulator
Cumberland Dicer
Cumberland Pelletizer
Nelmor Granulator

Amacoil Granulator
Connair Pelletizer
Cumberland Granulator
Foremost Granulators
Polymer Granulator

GRINDING MACHINES

Blade Tip Grinder
Cincinnati Magnesium Chipper
Ultrasonic Grinder

Cincinnati Centerless Grinder
Hertlein Flute Grinder

HYDRAULIC PUMPS

Arco Hydraulic Pump
Vacuum Pump

Bell & Gossett Hydraulic Pump

MISCELLANEOUS MACHINES

Acro Bar Feed Loaders
Ball Mill Machine
Blow Molding Machine
Cam Lobe Mill
Carton Cutter/Creaser
Churchill Tracer Lathe
Medical Razor Machine
Engine Test Room
ExCello Boring Machine
Fiberizing Machine
Folder/Sorter Machine
Nilson Fourslide Machine
Hammermill
Millhouse Machine
Packaging Machine
Steam Turbin
Sweed Chopper
Tube Cutter
Tube Forming Machine

Acme Gridley Chucker
Bag Making Machine
Branson Sonic Welder
Cardboard Shredder
Chip Crusher Machine
Deburring Machine
Electron Beam Welder
Envelope Making Machine
Farrel Mixer Unidrive
Film Processing Equipment
Gear Reducer
Hair Brush Stapling Machine
Injection Molding Machine
Nibblers
Pin Making Machines
Stud Welder
Test Stand Enclosure
Tube Finner Machine
Ultrasonic Cutters

Vacuum Pump   Vertislide Machines
Westinghouse Heat Pump   Wheelabrator System
Wiedematic Machines

NAILMAKING MACHINES

Glader Nailmaker   Wafios Nailmaker

PUNCH PRESSES

AIDA 500 Ton Press   AIDA #15 Gap Press
Bihler Press   Blase 100 Press
Bliss #4 OBI Press   Bliss #7 Press
Bliss #21 Press   Bliss #C-22 Press
Bliss #23 OBI Press   Bliss #28 Press
Bliss #C-35 Press   Bliss 40 Ton Press
Bliss 45 Ton Press   Bliss #HP2-45 Press
Bliss #C45 Press   Bliss #C-60 Press
Bliss #HP2-60 Press   Bliss #HP2-75 Press
Bliss #C-110 Press   Bliss #HP2-100 Press
Bliss #HP2-150 Press   Bliss #SC2-150 Press
Bliss #HP2-200 Press   Bliss #SC2-200 Press
Bliss #SC2-250 Press   Bliss #SC2-400 Press
Bliss 600 Ton Press   Bliss #1005 Press
Bliss #1831 OBI Press   Bliss #5012 Press
Bliss #6100 Press   Bliss #6K-225 Coin Press
Bruderer BSTA 25 Press   Bruderer BSTA 30 Press
Bruderer BSTA 60 Press   USI/Clearing 22 Ton Press
USI/Clearing 60 Ton Press   USI/Clearing 75 Ton Press
USI/Clearing 110 Ton Press   USI/Clearing KS-1400 Press
Cleveland 250 Ton Press   Danly 75 Ton Press
Danly 110 Ton Press   Danly H2-150 Press
Danly H2-175 Press   Danly H2-200 Press
Danly H2-250 Press   Danly H2-300 Press
Danly SE2-600 Press   Danly S2-1000 Press
Federal #7 OBI Press   Federal 15 Ton Press
Federal 60 Ton Press   Federal 110 Ton Press
Federal S2-150 Press   Federal S2-200 Press
Havair Press   Henry & Wright S2-100 Press
Henry & Wright S2-150 Press   L & J 60 Ton Press
L & J EM-75 Pres   McKay-Warco SC2-250 Press
McKay-Warco SE4-800 Press   Minster #3 Press

Minster #4 Press	Minster #5 Press
Minster #6 Press	Minster #7 Press
Minster #B-60 Press	Minster #Hb2-30 Press
Minster #P2-30 Press	Minster #P2-45 Press
Minster #P2-60 Press	Minster #P2-75 Press
Minster #P2-100 Press	Minster #P2-125 Press
Minster #P2-150 Press	Minster #P2-200 Press
Minster 60 Ton Pulsar	Minster S2-100 Press
Minster E2-100 Press	Minster E2-200 Press
Minster E2-300 Press	Minster E2-400 Press
Minster E2-600 Press	Minster Lamination Press
Niagara 22 Ton Press	Niagara M-22 Press
Niagara M-45 Press	Niagara M-60 Press
Niagara SA2-100 Press	Niagara SC2-100 Press
Niagara BP2-200	Niagara BP2-300 Press
Oak Lamination Press	Oak 73-LP-300 Lamination Press
Perkins Press	Rovetta #SC2-250 Press
Teledyne Flexopress	Teledyne SA2-200 Flexopress
Teledyne 200 Flexopress	V & O 25 Ton Press
V & O 50 Ton Press	V & O 75 Ton Press
V & O 110 Press	V & O Cam Press
V & O SS2-125 Press	Verson D2-125 Press
Verson 150-B2 Press	Verson 200 Ton Blanking Press
Verson 200 Ton Restrike Press	Verson 200 Ton Transfer Press
Verson SE-200 Press	Verson 300 Ton Press
Verson 400 Ton Press	Verson 500 Ton Press
Verson 800 Ton Press	Verson SC2-800 Press
Verson 1000 Ton Press	Verson TS2-4000 Press
Warco SC2-300 Press	Waterbury #30 DS Press
Waterbury #105 DS Press	Wean 300 Ton Press
Special Battery Press	Special Paper Press

REFINERS

Beloit 24" DD-400	Beloit 20" DD-3000
Beloit 26" DD-3000	Beloit 30" DD-4000
Beloit 34" DD-2000	Beloit 34" DD-3000
Beloit 38" DD-4000	Beloit 46" DD-3000
Beloit 54" DD	Emerson #4 Refiner

SWAGING MACHINES

Abbey-Etna #156 Swager	Abbey-Etna #208 Swager
Abbey-Etna #308 Swager	Abbey-Etna #408 Swager

Abbey-Etna #412 Swager
Fenn #1F Swager
Fenn #2H Hydroformer
Fenn #3½F Swager
Fenn #4½F Swager
Fenn #6F Swager
Fenn #8F Swager
Torrington #433 Swager
Torrington #1033 Swager
Stevans & Bullivant Swager

Abbey-Etna #1515 Swager
Fenn #2F Swager
Fenn #3F Swager
Fenn #4F Swager
Fenn #5F Swager
Fenn #7F Swager
Torrington #12 Swager
Torrington #438 Swager
Torrington #1047 Swager

THREAD-ROLLING MACHINES

Hartford #0-400 Thread Roller
Hartford #5-700 Point Former
Hartford #6-600 Point Former

Hartford #0-500 Thread Roller
Hartford #10-400 Thread Roller
Waterbury 3/8" Thread Roller

TRANSFER PRESSES

Bard Transfer Press
Bliss #62 Duplex
Waterbury #10 Transfer Press
Waterbury #105 Transfer Press
Waterbury #1010 Transfer Press
Waterbury #1210 Transfer Press
Waterbury #2012 Transfer Press
Waterbury #8W8 Horizontal Press

Baird Eyelet Die Set
Waterbury #5 Transfer Press
Waterbury #58 Transfer Press
Waterbury #115 Transfer Press
Waterbury #1012 Transfer Press
Waterbury #1512 Transfer Press
Waterbury #5012 Transfer Press

TUMBLING MACHINES

Sweco Deburring Machine
Ransohoff Batch Tumblers

Sweco Tumbling Barrel

VIBRATORY FEED BOWLS

Moorbin Vibe Bowls
Spiratron Vibe Bowls
Special Battery Bowls

Parts Feeder Vibe Bowls
Syntron Feed Bowl
Vibration Test Machine

WIRE MACHINERY

Cook Annealer Cabinets
Syncro #B-16 Wire Draw
Syncro #C-12 Wire Draw
Syncro #F-13 Rod Machine
Syncro Rod Breakdown Machine
Waterbury #F-7 Rod Machine
Wire Cut-Off Machine
Wire-Twisting Machine

Syncro #AG-16 Wire Draw
Syncro #BG-16 Wire Draw
Syncro #E-11 Rod Machine
Syncro #FX-13 Rod Machine
Vaughn #10DMT Rod Machine
Watson #11 Wire Strander
Wire-Straightening Machine

WOODWORKING MACHINES

Diehl Four-Head Moulder
Newmann Roughing Planer
Newman Planer/Matcher
Richardson Shaper

Buss Planer
Whitney Planer
Oliver Shaper
Greenlee Double End Tenoner

INDEX